POLYMER SCIENCE AND TECHNOLOGY

MODIFIED BIOPOLYMERS

CHALLENGES AND OPPORTUNITIES

Polymer Science and Technology

Additional books in this series can be found on Nova's website under the Series tab.

Additional e-books in this series can be found on Nova's website under the e-book tab.

POLYMER SCIENCE AND TECHNOLOGY

MODIFIED BIOPOLYMERS

CHALLENGES AND OPPORTUNITIES

DEEPAK PATHANIA,
GAURAV SHARMA
AND
AMIT KUMAR
EDITORS

nova
science publishers
New York

Library of Congress Cataloging-in-Publication Data

ISBN: 978-1-53612-116-2

Published by Nova Science Publishers, Inc. † New York

This book is dedicated to Goddess “Shoolini”

Contents

PREFACE

Biopolymers are the naturally obtained polymers that are degradable in nature. They have a well-marked structure and uniformly distributed set of molecular mass and appear as a long chain of curled up string ball under microscope. This type of polymer is differentiated based on their chemical structure. Under particular conditions of temperature, moisture and oxygen amount, their degradation takes place and the resultant products are non- toxic in nature. Biopolymers such as nucleic acids, carbohydrates, DNA, RNA, proteins, polysaccharides and carbohydrates account for a greater part of the human body as well as the ecosphere. The DNA biopolymer is the most important for humans. The entire body structure as well as genetic behavior that pass from parents to children is based on it. Both DNA and RNA are composed of nucleic acids that alternate in definite patterns to encode huge amount of genetic data.

Biopolymers and their derivatives are antithetical, lush, consistent for continuance, they display fascinating properties and are of increasing liability for distinctive applications. Living evidence is efficient to blend an overwhelming alteration of polymers, which can be divided into proteins and poly(amino acids), polysaccharides e.g. cellulose, sugar and xanthan, biological polyoxoesters e.g. poly(hydroxyalkanoic acids), poly(malic acid) and cutin, polythioesters, inorganic polyesters mutually polyphosphate as the only example, polyisoprenoides such as innate rubber and polyphenols a well-known as lignin.

Biopolymers develop in any organism, and in most organisms they contribute to the by far major division of the cellular restrained matter. Biopolymers utilize a wide chain of distinct essential functions for the organisms such as self-defense and allusion of genetic flea in ear, catalysis of reactions, computerized information of carbon, nitrogen, phosphorus and at variance nutrients, campaign and insurance against the clash of contrasting cells or intrinsic factors, sensors of biotic and abiotic factors, communication by any of the environment and distinctive organisms, mediators of adhesion to surfaces of disparate organisms or of non-living experience. In installation, multiple biopolymers are structural components of cells, tissues, and whole organisms.

To fulfill these divergent functions, biopolymers must unmask these diverse properties. They should specifically interact with a large fluctuation of antithetical substances, components and materials, and they must have extraordinarily affinities and strength to them. Some of these properties are utilized directly or indirectly for contrasting applications. The possibility to produce them from renewable sources makes them applicable for a number of applications.

Basic review has already revealed for all practical purposes knowledge on the enzyme systems catalyzing biosynthesis, indignity and diversification of biopolymers as readily as on the properties of biopolymers. Biopolymers can effectively reduce carbon dioxide levels present in the atmosphere and also decrease the release of carbon into the environment. This is due to bio-degradation of these chemical compounds that can release carbon dioxide which can be reabsorbed by crops. Biopolymers also help in reducing the dependency on the non- degradable materials such as petroleum based plastic materials and thus, reduce the air pollution. Sugar based polymers, such as Polyactides have the tendency to naturally degenerate in the human body without creating any toxic effects. Polyactides are commonly used as surgical implants. This has further resulted in an increased interest in biopolymers for distinct applications in trading, medicine, pharmacy, cultivation, electronics and distinct other areas. However, the recent developments show that our knowledge is still scarce.

The genes for the biosynthesis pathways of multiple biopolymers are still not accessible or identified new biopolymers have seldom been described, and only a fraction of the properties of biopolymers such as biological, chemical, physical and material properties have been investigated. Often determined biopolymers are not ready to be drawn in sufficient amounts. Nevertheless, polymer chemists, engineers and material scientists in academia have dug up biopolymers as chemicals and materials for many recent applications, or they consider biopolymers as the models to design new and cheap polymers.

Chapter 1-This chapter provides insight to possible applications of biopolymer-based materials such as cellulose, aragose, starch, carrageenans, silk, shellac, keratin, gelatin, melanine and other polysaccharides (glucose, starch, PLA, chitosan, pectin, agar, etc.), for electronics and photonics , in components for low temperature fuel cells, rechargeable batteries and Grätzel solar cells, for developing and manufacturing of sensors for different environmental pollutants detection etc.

Chapter 2- This chapter deals with biopolymers produced by living organism, their modifications and applications in the field of biomedical, including tissue engineering, gene delivery, drug delivery and for cell culture. Controlled drug release system has been developed to improve the drug release in the body.

Chapter 3- This chapter presents the effect of heavy metal ions toxicity on human health and their influence on environment pollution in recent years. The chapter discusses the various biopolymers based biocomposite extensively used in recent year for the removal of these heavy metal ions.

Chapter 4- This chapter discuses the polymeric materials based hydrogels and their applications in biological and environmental aspects as implants or materials for removal of toxic pollutants. Chapter also presents brief about physicochemical properties of hydrogels.

Chapter 5- This chapter deals with detail study of biochar and its composites giving information about the biochar magnetization, modification, synthesis techniques and properties. Chapter discus the various applications of biochar as adsorbent, detoxifying agent, soil enrichment agent, photocatalyst etc.

Chapter 6-This chapter describes the relevance of biodegradable biopolymers and presents their considerable importance as most promising materials for vast range of applications. The chapter deals with an overview of development of natural polymers such as starch, cellulose, guar gum, chitin, chitosan, pectin etc. their novel processing methods, characteristics and properties.

Chapter 7- This chapter provides an insight to progression of composite material to biocomposite materials. The chapter describes the biocomposite materials as new class of the engineered materials, successfully used for the various applications in the field of environmental remediation and medical science etc.

Chapter 8- This chapter explains the biological traits of nanocomposites and their role as nanofertilizers, nanopesticides, anticancer and antimicrobials agents. The present chapter we reviewed the role of these composites to reduce the burden of fertilizer and pesticides and, for their use in antimicrobial and anticancer activities.

Chapter 9- This chapter covers the modifications in the biopolymers so as to mould them into hybrid biopolymers having desired properties and explains the impotance of biobased- nanocomposites in the field of food packaging applications due to improved gas and water barrier properties, antimicrobial ability, high mechanical strength and thermal stability. This chapter also covers the impact of biobased- nanocomposites on the human health.

Chapter 10- The chapter deals with the natural fibre and their reinforced biodegradable composite materials. The chapter dicuss the various chemical method of natural fibers treatment and their classification.

Chapter 11- The chapter explains the various bio-inspired polymer composites and their applications in biomedical application podium

Chapter 12- The chapter deals with biopolymer modifications using ionic liquids for industerial and environmental applications. The chapter emphasis on various properties applications of natural polymers.

ACKNOWLEDGMENTS

It is the time to memorize the contributions of those who visibly or invisibly contributed in drafting this book. Firstly we would like to pay our gratitude to goddess Shoolini, without her blessings the dream that we envisaged would have never been manifested. We would like to thank every individual author who spent their valuable time in structuring this concept.

We are extremely grateful to Prof. PK Khosla, Vice-Chancellor, Shoolini University, Solan (HP) for providing all the amenities during the course of work. At this moment of accomplishment, we pay our deep gratitude to Prof. Sunil Puri for providing his encouragement and support.

This book has been kept on track and been viewed till compilation, with the support and encouragement of numerous people including our well wishers, our friends & families, colleagues and various institutions. At the end of this work, it is a pleasant task to express our thanks to all those who contributed in many ways to the success of this work and made it an unforgettable experience for us.

Last but not the least, we would like to thank our coffee cups for beating our lethargy.

Editors

In: Modified Biopolymers: Challenges and Opportunities ISBN: 978-1-53612-116-2
Editors: D. Pathania, G. Sharma and A. Kumar

Chapter 1

BIOPOLYMERS IN DEVICES FOR ENVIRONMENTAL MONITORING AND PROTECTION

Marinela I. Panayotova[1,*] ***and Vladko T. Panayotov***[2]
[1]Department of Chemistry, University of Mining and Geology, Sofia, Bulgaria
[2]Department of Mineral Processing, University of Mining and Geology, Sofia, Bulgaria

ABSTRACT

Our everyday life, development and application of RES, as well as environmental monitoring are unimaginable without electronics and photonics. Nowadays the lifetime of electronic devices is becoming very short. In order to cope with the exponentially growing amount of electronic waste, biodegradable organic electronic and photonic materials are needed and already considered by the world science and innovation community. Biopolymer materials are being developed, studied and already tested as substrates and insulators, semiconductors, and conductors. Organic materials are suited to produce electronics that possess functionalities inaccessible to standard crystalline semiconductors. Biopolymers find their application in solid polymer electrolytes for fuel cells, solar cells and energy storage batteries. The sustainable development of the society includes substitution of synthetic materials for bio-based materials in all areas, including the environment monitoring. This chapter presents:

1. Possible application in components for electronics and photonics of different biopolymer-based materials, such as cellulose and other polysaccharides (glucose, starch, PLA, chitosan, pectin, agar, etc.), DNA, as well as other natural biopolymer materials (silk, shellac, keratin, gelatin, melanine).
2. Possible application in components for low temperature fuel cells, rechargeable batteries and Grätzel solar cells of biopolymer-based materials, such as chitosan,

[*] Corresponding author Email: marichim@mgu.bg.

alginate, cellulose and carboxymethyl cellulose, aragose, starch, carrageenans, sago-based plastics.
3. Application for developing and manufacturing of sensors for different environmental pollutants detection and determination of different biopolymer-based materials, such as chitosan, alginate, cellulose and carboxymethyl cellulose, agar, starch, carrageenans, guar-gum, pectin, collagen, DNA and its aptamers.

Keywords: biopolymer, green electronics, bio-based electronics, bio-based SPE, bio-based PV, bio-based conducting membranes, biosensors, bio-based sensors

1. Materials for Electronics and Photonics

In our days, the lifetime of electronics is becoming shorter. The European Consumer Centre has reported average life-spans for smart phones of 20 months due to style obsolescence. The reported average life-spans for first, second and third generation iPods was 18 months before battery failure [1].

Replacing the devices turned out to be more cost effective than repairing or replacing the damaged subcomponent. Short-life, cheaper electronics are needed. The road is already paved by mature organic light emitting technology. Good progress has also been made in organic photovoltaics (OPV), with reported efficiencies of about 10% or higher in 2012. Printed flexible photovoltaics are being commercialized at a smaller scale for rooftop and small appliance applications [2]. However, the waste generated by such devices poses a growing environmental problem. Electronics built of biodegradable organic materials represents one of the smartest solutions. Many biopolymer-based materials (Figure 1) have been shown to possess needed technological properties, while at same time being biodegradable, safe, nontoxic, as well as low-cost and able to be produced from abundant natural resources and/or waste from the food industry. Such materials are being developed and used for all types of electronic components, such as substrates, insulators, semiconductors, and conductors.

1.1. Substrates and Insulators

Different materials based on bio-polymers, have been identified as suitable substrates for the fabrication of organic electronics. In addition, many of these materials demonstrate excellent insulating properties, which in combination with their easy processability make them suitable as gate electrode insulators for organic field effect transistors (OFETs).

Paper made from plant-derived cellulose, is one of the most well-known and used 'substrate' materials. Paper is the cheapest biodegradable substrate material and enables large-area printing of 'use-and-throw' devices [2]. Arrays of OFETs and OFET circuits have been printed on paper, thus enabling production of flexible devices with performance similar to that of the more traditional substrates [3-4]. Paper substrates have been applied in thermochromic displays for disposable consumer products [5] as well as in paper based OPV. Fully roll-to-roll solution printing, applying gravure and flexographic printing techniques, has been used to prepare printed-paper photovoltaics [6]. The devices use an inverted design, with a printed ZnO/Zn back electrode and a conducting polymer transparent top electrode. Another device described in the literature uses thin semitransparent paper as a substrate, with a conducting polymer transparent electrode, organic active layer, and reflective back electrode. All cell components were fabricated via low-temperature chemical vapor deposition [7]. Both paper based solar cells are low-cost and can be folded repeatedly with no degradation of the array's performance.

Caramelized glucose was studied as a substrate for electronics. It was found that despite its sensitivity to moisture, the film forming characteristics of glucose rivaled those of glass in terms of smoothness [8].

Polylactic acid (PLA) is another polysaccharide-based material which has been recently studied as a substrate material. It has been demonstrated [9] that PLA can be successfully used as a substrate for the realization of printed circuit boards where the conductive paths were fabricated by the lamination of copper foils. Mattana et al. [10], reported the fabrication and characterization of spin-coated biodegradable PLA thin films to be used as substrates for the realization of all-solution-processed organic electronic devices. The produced thin films exhibited good mechanical and dielectric properties and were suitable for the fabrication of disposable electronics. To demonstrate practically, the functionality of such PLA thin films, organic electronic devices were realized on top of them, by means of inkjet printing. A photonic curing procedure was presented as a means for sintering the conductive inks without heating up the PLA substrates. Two types of organic transistors were fabricated on top of the PLA: OFETs, where the PLA film was used not only as a substrate but also as the gate dielectric, and all-inkjet-printed organic electrochemical transistors (OECTs). The second type of transistors exhibited one of the highest transconductance reported in the literature (up to 2.75 mS). The study opened the way for the fabrication of disposable, able to be produced at large-scale and low-cost organic electronic devices. According to the authors, the next step towards the improvement of the biocompatibility or biodegradability of these devices would consist of replacing silver-based conductive patterns with biocompatible (for instance, gold-based inks) or biodegradable (for instance, magnesium and its alloys) metals.

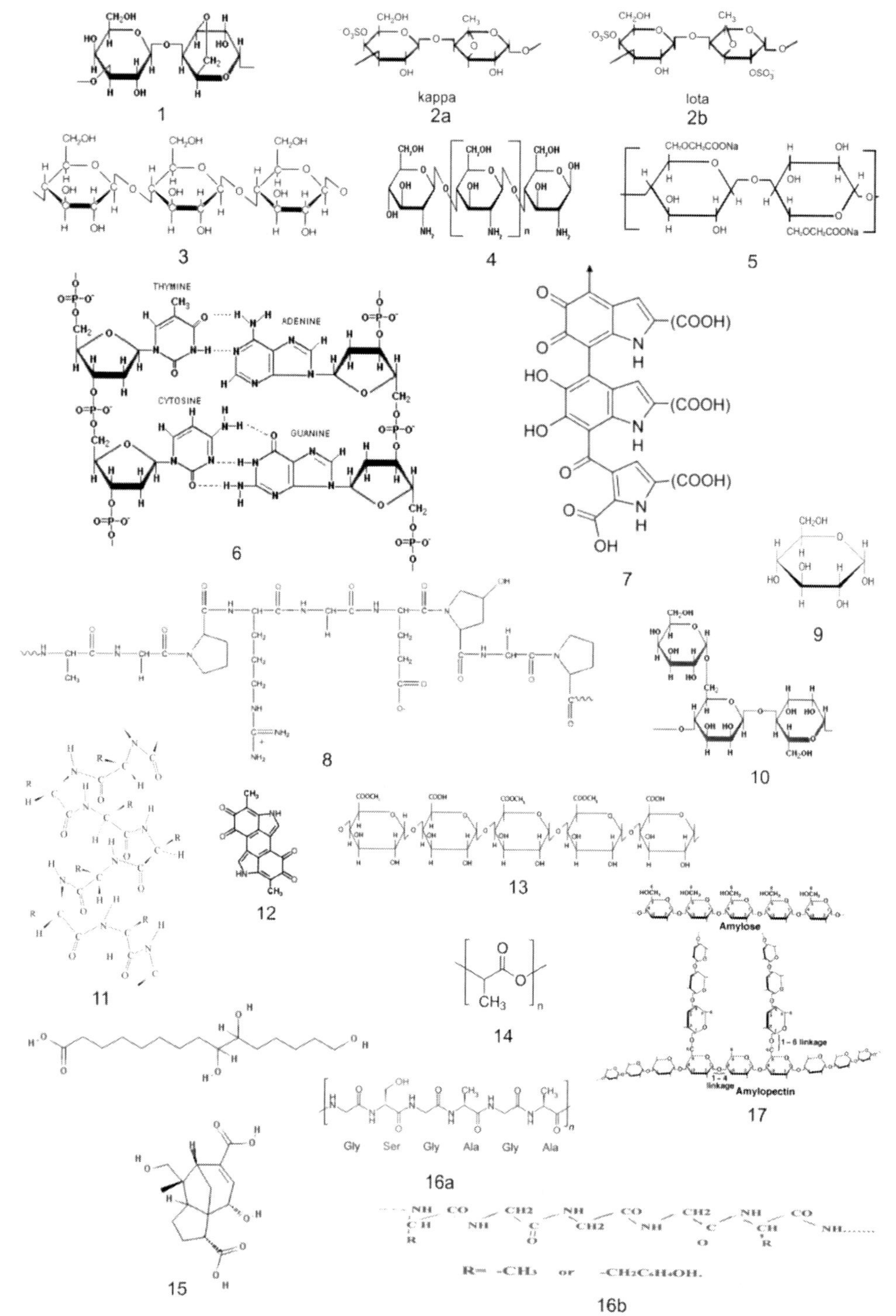

Figure 1. Biopolymers, most often used alone, or in different composite materials in modern electronics, photonics, RES, PVs and environment sensing: 1 - agarose; 2a, 2b - carrageenans, 3 - cellulose, 4 - chitosan, 5 - CMC, 6 - DNA, 7 - eumelanine, 8 - gelatin, 9 - glucose, 10 - guar gum, 11 - keratin, 12 - melanin, 13 - pectin, 14 - PLA, 15 - shellac, 16a - silk-fibroin, 16b - silk sericin, 17 – starch.

Deoxyribonucleic acid (DNA) is a natural, abundant, renewable, and biodegradable biopolymer. It is extracted in large amounts from waste from the food processing industry, and thus, it can be produced on an industrial scale. DNA can be applied in organic electronics and photonics. DNA shows several unique properties, such as the ability to form nanometer-scale molecular film, adjustable length, and self-assembly property [11]. DNA is a negatively charged anionic polyelectrolyte, with sodium ions as counter ions. The electric conductivity of DNA has been intensively studied since the early 1960s. The first theoretical suggestion was that the delocalization of π electrons in nucleobases may lead to an efficient electron transport along the DNA stacks [12]. The DNA conductivity measurements carried out by different research groups showed from relatively good to superconducting behavior [13-15]. A very poor DC conductivity of DNA was also reported [16-18]. The controversy about the DNA conductivity was explained by Genereux et al. [19, 20]. These authors have pointed out that the true mechanism of DNA conductivity is the charge transfer along the π-stacks of the bases. The peripheral part of the DNA molecule with sugars and phosphates behaves essentially as an insulator. Therefore, depending on the manner of the conductivity measurements, i.e., how the measurement apparatus is "connected" to DNA, different conductivity results are obtained. Even more, DNA can be used as a super capacitor, with a variable and controllable capacity [21].

Solution-processed and cross-linked DNA was successfully implemented as a gate dielectric layer for low operating voltage OFETs [22, 23]. The extracted individual nucleobases (guanine, adenine, thymine, and cytosine) have been implemented as gate dielectrics for OFETs. A device was developed where the substrate is caramelized glucose, guanine and adenine form the gate dielectric, and a nontoxic textile dye, indanthrene yellow G is the organic semiconductor [8]. A thin film of adenine, applied in combination with electrochemically-grown aluminum oxide dielectric and C60 fullerene, was used to prepare OFETs with a low operating voltage (~0.5 V) and high semiconductor mobility (~5.5 $cm^2/(Vs)$) [8, 24].

Silk is a biodegradable polypeptide polymer, made of two main proteins; fibroin and sericin. Fibroin consists primarily of repeating units of glycine, serine, and alanine that afford inter-chain hydrogen bonding, providing the mechanical robustness of silk fibers [2]. Silicon-based electronics can be fabricated onto silk [25]. Silk can be used as an effective solution-processed gate insulator for OFETs, supporting very high mobilities of ~23 $cm^2/(Vs)$ in pentacene combined with low-voltage operation [26]. Silk protein can be used as a sustainable material in optics, photonics, electronics and optoelectronic applications [27].

Shellac is a polymeric material, which is naturally produced by female lac beetles, and is harvested from trees in India and Thailand. It is a natural polyester copolymer of terpenic and aleuritic acids. It can be processed from polar organic solutions, such as ethanol. Silk and shellac possess excellent surface smoothness (rms < 1 nm, rivaling glass) when deposited as thin films. Shellac was cast to produce substrate foils (200 – 500 μm thickness), which have been used to make OFETs and complementary type circuits with the natural semiconductor indigo [28].

Fully 'green' OFET devices, made of natural materials have been reported. The substrate is shellac foil, produced by drop casting from an ethanol solution. An electrochemically-grown Al_2O_3 layer passivated with the natural oligoethylene tetratetracontane serves as the gate dielectric. The semiconductor material is indigo. These devices showed excellent stability and balanced electron and hole mobilities of 0.01 $cm^2/(Vs)$ [2].

A low cost composite material was developed, made from hollow keratin fibers (from avian sources) and chemically modified soybean oil. The material showed a low dielectric constant and has the potential to replace the dielectrics in microchips and circuit boards. The coefficient of thermal expansion of the composite was low enough for electronic applications and similar to the value of silicon materials or polyimides used in printed circuit boards [29].

The above-described natural substrates - dielectrics show that nature offers a significant number of materials that could be integrated into various biodegradable organic electronic devices.

1.2. Semi-Conductors

Compared to the exploration and use of biopolymer materials for insulators and substrates, the investigations into biopolymers based semiconductors remain scarce.

Melanins are heterogeneous biopolymers largely present in humans. Among the different types of human melanins the eumelanin (a brown–black pigment) is the most common. Among the physical properties of eumelanin, the electrical conductivity is one of the most interesting in the perspective of technological application. Conductivity values ranging from 10^{-5} to 10^{-13} S/cm, depending on the hydration state of the biopolymer, have been reported [30]. The conductivity depends on temperature. The AC dielectric and photoconductivity studies have shown that charge hopping is the main electric charge transport mechanism in eumelanin. [31]. The electrical conductivity properties of this biopolymer are similar to those of amorphous semiconductor solids. So, it can be considered as an organic semiconductor, which is largely available, biocompatible, cheaper and easier to process, compared to inorganic semiconductors, such as silicon, germanium, GaAs, etc. It can be considered a promising material for sensors and photovoltaic devices, due to the broad band spectral absorbance and charge transport properties [32]. Melanin was employed in various sandwich diode type devices [33, 34]. However, future applications need further charge transport characterization in large temperature range and as a function of different treatments to increase the conductivity values.

Nature offers π-conjugated molecules that can be used as semiconductors. For example, the carotenoids, such as β-carotene, are linear π-conjugated molecules that act as holes transporting semiconductors [2]. Devices using β-carotene and natural gate insulators such as glucose, fabricated on biodegradable plastic substrates, have been developed as demonstrations of truly 'natural' OFETs [8, 35]. However, with mobilities in the range of $1x10^{-4}$ $cm^2/(Vs)$ those devices are not efficient. Attempts were made to utilize solution-processed β-carotene as a semiconductor in solar cells [36].

Transistors with solution-processed chitosan proton conductors were reported. They were fabricated on paper substrates and are promising for portable paper electronics [37].

Studies on the electrical conduction of the DNA molecules reveal that they may act as semiconductors and can be used in rectifying devices [38-41]. Fink and Schonenberger [38] have reported that the transport of electric current by DNA is as efficient as that of a good semiconductor. This observation was later confirmed by other authors. Güllü et al. [40, 41] have shown that the DNA molecule is a p-type semiconductor with a wide band gap of 4.12 eV with a resistivity of 1.6×10^{10} Ω cm.

Biopolymer molecules can be used to control the electrical characteristics of conventional Al/n-InP metal–semiconductor contacts. Devices have been realized, where well-defined thin layers from DNA semiconducting biopolymer were grown by a simple cast method on n-type InP (100) inorganic substrates [11]. The inherently low Schottky barrier height of InP (0.40–0.45 eV) which makes it difficult to fabricate field-effect transistors (FETs) directly on InP, due to the serious leakage current problem through the gate electrode, was overcome, since the DNA increases an effective barrier height to 0.87 eV. An ideal Cu/DNA/n-InP biopolymer-inorganic Schottky sandwich device was prepared by coating a n-InP semiconductor wafer with a biopolymer DNA [42]. The Cu/DNA/n-InP contact showed good rectifying behavior. The ideality factor value of 1.08 and the barrier height (Φ_b) value of 0.70 eV for the Cu/DNA/n-InP device were determined from the forward bias current-voltage characteristics. The obtained Φ_b value for the Cu/DNA/n-InP contact was significantly larger than the value of 0.48 eV of conventional Cu/n-InP Schottky diodes. Modification of the interfacial potential barrier of Cu/n-InP diode was achieved using a thin interlayer of DNA biopolymer. In both presented cases, the effective barrier height increase was attributed to the fact that DNA biopolymer interlayer influences the space charge region of n-InP device.

The above-described examples show that cheap and nontoxic materials of natural origin can compete with the best synthetic organic semiconductors. However, the research into their implementation in real-life devices still remains very limited.

1.3. Conductors

Organic conducting materials combine mechanical robustness with flexibility. They are nontoxic and biodegradable and possess the ability to behave as ionic and electronic conductors, which is not the case with traditional metallic conductors [2].

In the previous pages, melanin, and especially eumelanin, has been classified as a semiconductor, based on findings of different authors. However, as mentioned, its conductivity is heavily dependent on the hydration of the material. It has been found that melanine can be used in thin-film form as a conductor [43]. In this case proton conductivity is the mechanism responsible for charge transport in melanin [44].

Except for fuel cell applications, proton-conducting materials have recently been recognized for their great potential in biocompatible and biodegradable electronics. A proton-conducting polysaccharide thin-film field-effect transistor has been demonstrated [45]. The flow of protonic current is turned on or off by an electrostatic potential applied to a gate electrode. The device is based on maleic-chitosan nanofibres. The protons move along the hydrated maleic–chitosan hydrogen-bond network with a mobility of $\sim 4.9\times10^{-3}$ $cm^2/(V\ s)$. The study has introduced a new class of biocompatible solid-state devices, which can control and monitor the flow of protonic current.

Solid polymer electrolytes (SPEs), which are solid state ionic conductors, have attracted attention because of their advantages, such as high energy density, electrochemical stability, and easy processing. SPEs can be applied in electrical devices such as lithium batteries, electrochemical capacitors and fuel cells [46].

As the ionic conductivities of most of the ionically conducting biopolymers are relatively low, different approaches, such as grafting, crosslinking and plasticizing were proposed to improve them. The plasticization process seems to be very successful in providing samples with high ionic conductivities (in the range of 10^{-4} - 10^{-2} S/cm) and improving also the contact between the electrodes and the electrolytes. The plasticization process using glycerol and water can be applied to obtain natural polymer-based gel electrolytes from chitosan [47], starch [48], pectin [49], agar [50], and gelatin [51]. The main positive feature of this approach consists of the ease of their processing into transparent membranes with good mechanical properties as well as good adhesion to glass and metal surfaces.

SPEs obtained from starch provoked different studies because they are based on abundant, renewable, low price, biodegradable and biocompatible material [52, 53]. Native starch can be processed to thermoplastic starch (TPS) by melting extrusion or casting [54]. TPS, doped with alkali metals, is a promising material for the development of novel SPEs. The conductance of glycerol plasticized amylopectin-rich starch doped with $LiClO_4$ is around 10^{-5} S/cm [52]. High environment humidity significantly improves the conductance of metal halides doped TPS-based SPEs [53].

Conductive TPS has been prepared by melting extrusion using N, N-dimethylacetamide (DMAc) and DMAc/LiCl mixtures with different LiCl contents as a plasticizer [55]. DMAc was found to be a less effective plasticizer for starch. However, DMAc containing high LiCl content showed a strong hydrogen bond. It was found that the high concentration of LiCl could improve the thermal stability of DMAc. On the other hand, the thermal depolymerization of starch became more pronounced with increasing LiCl content. The increased LiCl content enhanced the water absorption of TPS, thus improving the conductance of TPS. The conductance of TPS, containing 18 wt % of LiCl (based on DMAc, the highest studied by authors) could achieve $10^{-0.5}$ S/cm. The authors' conclusion was that DMAc/LiCl plasticized starch would be a promising alternative for the development of new SPEs for a wide variety of potential applications, such as electronic shielding, biosensors, environmentally sensitive membranes.

Multifunctional conductive biopolymer composites have been fabricated by combining different amounts of single walled carbon nanotubes (SWCNTs) and silver nanoparticles (Ag) with a biodegradable polymer matrix - poly(ecaprolactone) (PCL). A solvent cast process was used [56]. Binary PCL/Ag, PCL/SWCNTs and ternary PCL/Ag/SWCNTs composites were prepared and studied. The PCL/Ag composites exhibited poor electrical properties, while in PCL/Ag/SWCNTs ternary films higher values of conductivity were measured compared to both binary composites. The research indicated that Ag particles facilitate the formation of conductive pathways in the presence of SWCNTs, as they act as conductive bridges among nanotube bundles and facilitate the electron transfer. The addition of a small percentage of SWCNTs promoted significantly the electrical properties of PCL/Ag nanohybrid films.

The applied research with entirely 'natural' conductors remains limited, while the field of synthetic conducting polymers is relatively mature and the studies are directed also to modifications of synthetic biocompatible conducting polymers with natural biopolymers. A number of conducting polymer complexes, incorporating biomolecular dopants, have already been prepared, in a variety of ways.

A facile approach was described through which properties (such as electroactivity hydrophilicity, shear modulus, film interfacial roughness) of inherently conducting polymer polypyrrole (PPy) may be varied significantly by varying the concentration of biological dopant dextran sulphate (DS) in the polymerisation electrolyte. This way the polymer properties can be tuned in order to enhance their functionality while preserving their fundamental chemistry [57].

One of the well-known synthetic electron-conducting polymers in organic electronics - the poly(3,4-ethylenedioxythiophene) (PEDOT) was used to produce a conducting polymer complex, namely PEDOT: DNA – a material with a conductivity of 1.0 S/cm. According to authors, the phosphate groups of the DNA are responsible for doping [58].

The polyelectrolyte properties of DNA are well pronounced when dissolved in water because of the presence of mobile bulk ions, some of them forming condensed and diffuse counter ions [59]. As DNA is water soluble, it is possible to prepare samples of thin film formed by solution casting technique. DNA can be plasticized with glycerol allowing obtaining of a processable material and fabrication of membranes with good ionic conductivities [60].

DNA complexes, plasticized with glycerol and adequately doped can be processed into self standing conducting membranes with a high electrical conductivity [61]. To increase the electrical conductivities of the membranes they were additionally doped with lithium perchlorate ($LiClO^{-}_4$), Prussian Blue ($K[Fe^{III}Fe^{II}(CN)_6]$) (PB) and PEDOT: PSS (poly(3,4-ethylenedioxythiophene): poly(styrenesulfonate)) [62]. It was found that for the pure DNA–glycerol membranes adding more of the plasticizer to DNA results in an increase of the ionic conductivity. However, there are limits for glycerol concentration to obtain solid membranes, which have to be respected. The ionic conductivities at room temperature were 3.9 $x10^{-6}$ S/cm and 3 $x10^{-5}$ S/cm for the low glycerol and high glycerol content. Membranes doping with $LiClO_4$, PEDOT: PSS and PB increases substantially the membrane conductivities, by more than two orders of magnitude with respect to DNA–glycerol samples. The measured ionic conductivity at room temperature, depending on dopant used and its concentration, is in the range of 3.5×10^{-4}–10^{-5} S/cm and increases linearly as a function of temperature, reaching 10^{-3} S/cm at 358 K for the most conducting sample.

A water-dispersible conducting polymer (PEDOT: DS) from polythiophene family has been synthesized in a single reaction in high yield [63]. It represents a complex between PEDOT and the sulfonated polysaccharide polyanion dextran sulfate. Drop-cast films of aqueous suspensions of the material display a native conductivity of up to 7 ± 1 S/cm, increasing to 20 ± 2 S/cm, after treatment with ethylene glycol and thermal annealing. The material is not cytotoxic. Water dispersions of PEDOT: DS are successfully processed and deposited onto glass by drop casting, spray coating, inkjet printing and extrusion printing. Laser ablation of dried films allows the creation of patterns with excellent definition. Such processing techniques offer great promise for the production of micro-scale devices. Taken together, these results suggest that PEDOT: DS has exceptional potential as an electromaterial for different applications, including the biointerface.

1.4. Photonic Materials

In past years there has been a growing interest to use the biopolymers as potential materials for opto-electronic devices.

Pure DNA has a limited potential for applications in photonics. Deoxyribonucleic acid denaturates at around 90°C, changing its helical structure from double stranded to single stranded [64], thus limiting the temperature range of its applicability. This biopolymer is soluble in water only, a solvent which does not belong to the preferred ones in photonic devices fabrication technologies. The weak π electron conjugation, only in phenyl rings, provides limited hyperpolarizabilities to this compound. Therefore, the only possible practical use of this biopolymer in a pure state in photonics is as an optically inactive material. DNA can be processed from water solutions to produce films with excellent optical transparency from 400 nm through the NIR region. However, DNA possessing a double strand helical structure, with minor and major groves, provides a large free volume for doping molecules as well as a good protection against photo thermal degradation. In addition, DNA exhibits a higher optical damage threshold [63].

Since the DNA molecule exhibits hydrophilic properties, a common approach to render it soluble in organic solvents is to form DNA-surfactant biopolymers. Ogata and his collaborators [65-68] have shown that DNA reacts with the surfactant cetyltrimethylammonium chloride (CTMA) forming a thermally stable, up to ca. 230°C complex, insoluble in water, but soluble in other solvents, particularly in alcohols, such as ethanol, butanol and isopropanol. The complex can be processed into good optical quality thin films [69] with propagation losses of 0.2 dB/cm at the telecommunication window [70]. DNA–CTMA complexes can be doped with photosensitive molecules to provide desired electric or linear and nonlinear optical properties. Consequently, DNA–surfactant complexes can be used to replace the presently used synthetic polymers as host material for different kinds of applications [71, 72].

DNA–CTMA, due to its tunable conductivity by controlling the molecular mass of DNA, is a choice material for buffer layers in electro-optic modulators. Heckman and her co-authors have prepared and studied thin films of DNA–CTMA–Disperse Red 1 chromophore and demonstrated the films work in a bio electro-optic modulator [73].

The DNA–CTMA complexes were also used in other kind of applications, such as field effect transistors [74], as an electric-field poling layer [75], or matrix for luminophores. Hagen and co-authors [76] presented results of improvement of the electroluminescent properties of biopolymer based organic light emitting diodes (BioLEDs) by using DNA–CTMA complex as an electron blocking layer. Kobayashi and co-workers have demonstrated an effective (a fast turn-on response) DNA-based BioLED containing Tris (bipyridine) ruthenium (II) ion - $Ru(bpy)_3^{2+}$ - as active molecules [77].

The adequately functionalized DNA can be also applied in electrochromic cells for displays and for application in "smart windows" [78]. As already mentioned, ionically conductive polymer membranes based on plasticized DNA with glycerol and containing conducting polymers, such as PEDOT:PSS, POEA (poly (o-ethoxyaniline)):HCl or PB were prepared. The DNA based electrolytes are obtained as solid, free standing films, which greatly facilitates technologies for their practical applications. In order to verify their practical use, small area electrochromic displays with glass/ITO/WO_3/DNA-based electrolyte/CeO_2–TiO_2/ITO/glass configuration were assembled and tested [61]. The samples exhibited a good thermal stability up to 170°C and a semicrystalline structure. The measured transparency was 86% for the pure DNA-based samples and 50% for the DNA samples containing either PEDOT:PSS, POEA:HCl or PB. The membranes, applied in small electrochromic devices, show up to 15% transmissivity change when switching from colored to bleached states. The inserted/extracted charges were 0.7–3.0 mC/cm^2 after 15 s of applied potential of −2.5 V for membranes with a thickness of 72–80 μm. All obtained results indicate that DNA-based gel-polymer electrolytes are promising materials to be applied in electrochromic devices.

The DNA–CTMA complex is also an interesting matrix for lasing applications [79]. Amplified spontaneous emission (ASE) of DNA-based materials was observed by several research groups [80-83]. Mysliwiec and co-workers [84] have observed ASE from MR isomer of spiropyran embedded in DNA–CTMA matrix – an effect which is not observed when using synthetic polymers as the matrix, such as, polymethyl methacrylate. A two-photon lasing was also demonstrated in a DNA–CTMA: chromophore complex [85]. Interesting effects were also observed in holography experiments with thin films of DNA–CTMA, doped with photoactive molecules, such as Disperse Red 1 and Disperse Orange 3 [86].

Nanocomposites consisting of silver nanoparticles (Ag NPs) embedded in DNA biopolymers was fabricated by photo-triggered synthesis, using aromatic surfactants [87]. Three different surfactants were employed, namely benzyl-trimethyl-ammonium chloride, phenyl-trimethyl-ammonium chloride, and vinyl-benzyl-trimethyl-ammonium chloride. Thus, prepared DNA biopolymers can be dissolved in organic solvents and the material processing is compatible with the conventional manufacturing platform of organic devices. The characterization results of the obtained nanocomposites suggest that these DNA biopolymers are promising as optical materials and can find broad applications in organic thin-film devices. The facile technique to fabricate photoinduced DNA–NPs nanocomposite paves the way for their implementation in plasmonic nanostructures, holographic applications, and optical storage devices.

Starch by itself or as a designed dye-doped biopolymeric matrix has been proposed for random lasing generation. Gelatinized biopolymer, doped with Rhodamine 6G, was used as a light amplification medium as well as source of positive feedback which originates from high surface roughness fluctuations. A significant increase in photostability of laser emission, compared to other biopolymeric based systems, was demonstrated when applying the starch as a matrix for the laser dye. Water solubility and ease of processing are the other advantages of this material. The lasing threshold of the studied system is comparable with that of other dye-doped polymeric and biopolymeric matrices. Starch shows a higher stimulated emission stability than DNA based systems [88]. All above-described advantages render starch into fully functional material in photonics applications.

Protein-based biopolymers and hydrogels have been increasingly utilized to obtain miniaturized optical waveguide devices for fields like eco/bio-compatible optical micro/nano waveguides [89] and biosensing [90]. Protein-based biopolymers were more often used to fabricate single-wave guide optical micro/nano-devices with relatively simple geometries and less structure details. The functionalities (particularly novel sensing or actuating functions) of these devices were limited by the inflexible and less-detailed device design and micro/nano-fabrication [89, 90]. Micro/nano-wave guide devices (including Y-junction-based optical micro-splitters/couplers) are needed in many situations for not only traditional integrated photonics [91, 92], but also for novel utilizations including lab-on-a-chip integrated sensors [93, 94]. The femtosecond laser direct writing approach was applied to realize non contact and maskless fabrication of protein-hydrogel-based Y-junction optical micro/nano-splitters (PYOMs) from commercial bovine serum albumin aqueous ink [90, 95, 96]. Thus fabricated PYOMs show distinct features (e.g., facile and flexible customization of light-splitting ratio, environment-stimulus-actuated light-splitting adjustments). These PYOMs hold great promise for frontier areas, such as optical sensing, integrated optofluidics and biophotonics, and novel tunable micro/nano-photonics actuated by environmental signals.

2. Materials with Envisaged Use Both in Electronics and Photonics

3-D spheres from ZnO and radial structures were synthesized on glass or quartz substrates through a rapid, low-temperature (100°C), biopolymer-assisted (i.e., by sodium alginate) self-assembly route without any template [97]. The ZnO 3-D architectures were assembled from 1-D nanorods. The obtained products were with good crystallinity and showed an intensive and sharp UV cathode luminescence emission at 389 nm for ZnO radial structures which renders them an ideal candidate for fabricating ultraviolet light emitting diodes and lasers.

Multifunctional and mechanically robust coatings were produced by the layer by layer assembly of single-walled carbon nanotubes (SWNT) dispersed in DNA and lysozyme (LSZ) on an insulating glass substrate [98]. The coatings were characterized using scanning electrochemical microscopy (SECM) and impedance spectroscopy. SECM analysis demonstrated that subsequent layers were not electrically insulated from each other and that a direct correlation exists between the SECM feedback response and the number of layers. The results show that the combination of natural biopolymers and single-walled carbon nanotubes enable multifunctional conductive coatings where the charge transport can be tailored based on the number of layers and the composition and polarity of the outermost layer. This research indicates that the combination of SWNT with biopolymers is promising for creating new, sustainable biointerfaces with controllable electrical and mechanical parameters for applications in electrochemical processes, electronics and sensing.

Green composites are defined as biopolymers (bio-derived polymers) reinforced with natural fibers. Various fibers can be used, thus contributing to different properties of the composites. The group of biodegradable green composites is defined by appropriate standards [99, 100]. Natural fibers are derived from a variety of renewable sources and include both thermosetting and thermoplastic polymers. However, if these thermosets are formed by polymerization with synthetic monomers – as many are – they are no longer 100% green materials [101].

Bio-thermoplastics, commonly used in green composites include PLA, poly-hydroxy-butyrate (PHB), polysaccharides of plant origin (cellulose and alginate), and materials from animals (chitin) and of microbe origin (hyaluronate). Consumer electronics and toys are potential applications for green composites due to their renewable, low embodied energy, biodegradability and low durability (high rates of obsolescence along with their limited resistance at exposure to moisture during use) [102]. Green composites are fast entering consumers' electronics. The electronics company NEC has been working with green composites since 2004 when they used a PLA/kenaf composite in dummy cards in personal computers [103].

In 2006 NEC Corporation together with UNITIKA Ltd. developed an "eco"' phone with this material and in 2007 began using some green composites in the housing of their personal computers [104]. In June 2012 a report by Juniper Research stated that 392 million "green" handsets (those with at least 50% recycled content and made without certain hazardous chemicals) will be shipped by 2017 [105].

Polysaccharide-based nanocomposites are one of the most studies classes of green composites over the past decade. Most polysaccharides are natural materials produced as energy storage or structural biopolymers by microbes, plants, and animals. They are recovered easily and at relatively low cost. The diverse structures and properties of polysaccharides offer molecular and biological advantages for their use in the preparation of nanomaterials and nanocomposites. Polysaccharides show a number of advantages over nucleic acids and proteins for applications in materials science, since they are generally more stable than nucleic acids and proteins and are usually not irreversibly denatured on heating [106]. Polysaccharides offer a green alternative to synthetic polymers in the preparation of soft nanomaterials. They have also been used in composites with hard nanomaterials, such as metal nanoparticles and carbon-based nanomaterials. Nanocomposites take on the properties of the materials present as well as their size scale. Challenges in preparing nanocomposites include: controlling their fabrication/synthesis; ensuring the compatibility of their different material components; and obtaining desirable/unique properties. Different ways can be used to prepare or synthesize nanocomposites, such as electrospinning, film casting, dip coating, physically mixing, layer-by-layer assembly, ionotropic gelation, colloidal assembly, co-precipitation, in situ preparation, and covalent coupling [107].

It has been shown that basic components, such as the electrode, separator, and electrolyte, can all be integrated into single contiguous nanocomposite units that can serve as building blocks for a variety of thin mechanically flexible energy storage devices [108]. Nanoporous cellulose paper embedded with aligned carbon nanotube electrode and electrolyte constitutes the basic unit. These units can be used to build various flexible supercapacitor, battery, hybrid, and dual-storage battery-in-supercapacitor devices. The thin freestanding nanocomposite paper devices offer mechanical flexibility during operation. The supercapacitors operate with electrolytes including aqueous solvents, room temperature ionic liquids, and bioelectrolytes, and over wide temperature ranges. In comparison with conventional capacitors, the supercapacitors boast higher power density and better cyclability [109-111].

Core sheath multi-walled carbon nanotube (MWCNT)-cellulose fibers have been prepared by coaxial electrospinning [112]. The MWCNT-cellulose fibers formed a cable structure with a conductive core and insulating sheath and fiber mats and demonstrated excellent conductivity because of a conductive pathway of bundled MWCNTs. These nanofibres could be used as a double-layer supercapacitor device because of the thin insulating cellulose sheath. Very high capacitance was measured, caused by the inverse relationship between capacitance and distance between CNT electrodes. The fibers could also be used as a thin and flexible separator for an actuator.

SWCNTs/cellulose nanocomposites, prepared using N-methylmorpholine-N-oxide monohydrate, afford toughened, electro-conducting, transparent SWNT/cellulose films [113]. These transparent films may be used in applications such as transparent electrodes.

In conclusion, it can be stated that biopolymer-based materials are suited to produce all types of electronic components and be used in photonic devices. These materials can not only be sustainable and biodegradable, but can also have functionalities inaccessible to standard materials used up to our time in mass produced electronic and photonic devices.

3. Biopolymer-Based Materials for Preparing Components for RES and Batteries

The rising worldwide consumption of fossil fuels causes a serious depletion of the natural resources and an increasing of environmental pollution. In order to solve these problems, exploitation of renewable energy systems (RES) is needed. Among different RES, the development of fuel cells and photovoltaic cells can profit from use of biopolymer-based materials. The storage methods of electrical energy are also crucially important in the utilization of renewable energy. In the intense research efforts aimed to enhance the efficiency of energy storage systems, biopolymers find their application mainly in rechargeable batteries and capacitors.

3.1. Membranes for Fuel Cells

A fuel cell (FC) is a device that produces electricity and heat by electrochemical reactions. The fuel is not burned and then converted to electrical power and therefore the FC is a more efficient and cleaner energy producing technology compared to classical ones. Reactants are continuously fed into the cell to sustain the chemical reaction and the products are continuously taken out. Therefore, the fuel cell produces power indefinitely (for the rated lifetime of the cell). The reaction that takes place inside a fuel cell depends on the type of fuel used. A FC consists of an anode (the electrode at which oxidation, i.e., a loss of electrons takes place), a cathode (the electrode at which reduction, i.e., a gain of electrons occurs) and an electrolyte (a solid or liquid substance that conducts charged ions from one electrode to the other in a FC). The electrolyte usually defines the type of fuel cell. The electrochemical reactions occur at the interfaces. Electrons released due to oxidation are drawn from the anode to the cathode through an external circuit, producing DC electricity. There they participate in a reduction reaction (usually of oxygen). The main waste product is water. The anode and cathode often (especially in FCs operating at low temperatures) contain catalysts that facilitate the proceeding of electrochemical reactions.

A fuel cell that uses direct oxidation of methanol (CH_3OH) in gaseous or liquid form, is called a direct methanol fuel cell. Among different types of fuel cells, direct methanol fuel cells (DMFCs), owing to their capability of providing high energy density at low temperatures (60 to 130°C), simplified system designs and convenient fuel storage have been proposed as a means of energy supply for future portable electronic devices and replacement for lithium-ion batteries [109].

The membrane electrode assembly (MEA) of DMFCs includes anode, electrolyte, cathode, and catalysts. The electrolyte, currently used, is the proton exchange membrane (PEM). The following reactions are proceeding at cell work:

At anode: $CH_3OH + H_2O \rightarrow CO_2 + 6H^+ + 6e^-$

At cathode: $3/2\ O_2 + 6H^+ + 6e^- \rightarrow 3H_2O$

Overall: $CH_3OH + 3/2\ O_2 \rightarrow CO_2 + 2H_2O$.

The PEM, i.e., the FC electrolyte, is usually considered the most important and often performance-limiting component [114]. An ideal PEM for DMFCs should meet a number of requirements, such as high proton conductivity, long-term chemical and hydrolytic durability under heated and humidified conditions, and low methanol crossover [115]. Perfluorinated ionomers such as Nafion®, with fluoroalkyl ether side chains and sulphonic acid end groups on Teflon™-like backbones are the most commonly used PEMs. They satisfy the requirements of hydrogen fueled FCs exhibiting high proton conductivity below 100°C and high chemical stability, but possess relatively high methanol permeability [116]. The permeation of non-reacted methanol across PEMs hinders oxygen reduction at cathode and significantly reduces the efficiency of the DMFC [117].

Different studies have been carried out to develop alternative materials for Nafion® membranes [118, 119]. Another approach to reduce methanol permeability is to coat Nafion® membranes with methanol barrier layers [120], with layer-by-layer self-assembly thin polyelectrolyte films [121], to prepare different composites, for example by incorporating inorganic particles into Nafion® matrices [122], or to prepare different polymer blends, where by systematic variation of the composition, specific characteristics can be obtained for desired applications.

Because of its high hydrophilicity, facile chemical modification, good chemical and thermal resistance, convenient film forming properties and low price [123] chitosan is the "favorite" biopolymer of researchers to be modified or for preparing blends, composites and coatings – all activities aimed at decreasing the PEM methanol permeability. Below are presented some examples.

Polyelectrolytes based on chitosan were synthesized and assessed for their applicability in DMFCs [123]. A binary cross-linking agent (sulfosuccinic acid / glutaraldehyde) was used for the structural modification of chitosan and membranes comprised various amounts of sulfosuccinic acid. Based on the membrane selectivity evaluation and activation energy measurements of proton conduction, the optimum composition of cross-linking agent was determined. The developed material showed a relatively high proton conductivity of 0.0452 S/cm and a low methanol permeability of $9.6\times10^{-7}cm^2/sec$. Its selectivity value was 47,100 in comparison with the corresponding value of 40,500 for Nafion®117. The fabricated membranes showed acceptable oxidative and hydrolytic stability. A single DMFC performance test revealed a power density of 17 mW/cm^2 at 30°C and 41 mW/cm^2 at 60°C in a 2 M methanol feed. The prepared proton-conducting bioresource membranes have a promising potential in the field of green power generation as a low cost material.

Polyion complex membranes (PICs) for direct methanol fuel cell were made by blending 84% of chitosan and sodium alginate biopolymers and using solution casting and solvent evaporation technique [124]. Chitosan, a deacetylated derivative of chitin, possess reactive amino and hydroxyl groups and is widely accepted as a material with good membrane forming properties, chemical resistance and high intrinsic ionic conductivity in a suitable swollen state of the membrane [125]. Alginate (one of the polysaccharides extracted from seaweed) has shown excellent performance as a membrane material for the dehydration of ethanol mixture [126]. Water solubility and mechanical weakness of the alginic acid membrane were its drawbacks. By alginate blending with chitosan, the drawbacks of the former were overcome and good characteristics of both were combined. Blending of the two polymers resulted in the spontaneous formation of polyion complex due to the occurrence of ionic cross-linking which not only reduced the tendency of swelling but also improved specific properties such as structural strength and thermal and mechanical stability of the material. Obtained membranes were found to be suitable for DMFC applications because of the low methanol permeability (4.6×10^{-8} cm^2/s at 50 vol.% methanol concentration), excellent mechanical and thermal properties and relatively high proton conductivity (0.042 S/cm). According to authors [124], the membranes offer no significant advantages over Nafion®117 with respect to proton conductivity but show lower permeability to methanol when compared to Nafion®117. The described properties, complemented with the cost effectiveness and simple fabrication technique involved in the synthesis of such PICs, make their applicability in DMFC attractive.

Biopolymer composites (chitosan - oxide compounds, such as MgO, CaO, SiO_2, Al_2O_3) were prepared and studied. The introducing of hygroscopic oxides was aimed at ensuring higher membrane humidity. Further, a catalyst layer / membrane ensemble was made, followed by a membrane electrode assembly (MEA), and finally – a demonstration fuel cell - with the aim to study the properties of the obtained composites [127]. It was found that the presence of a hygroscopic additive binds a larger amount of water in the membrane, thus increasing the membrane water content at a given RH and ensuring conductivity enhancement. Measured conductivities were in the range between 10^{-3} and 10^{-2} S/cm. The presence of oxides in the membrane reduced pores and permeability of gas through the membrane. However, the retention of water at temperatures above 80 °C was not possible and this produced a decrease of fuel cell performance at higher temperature. In addition, an increased interfacial resistance between the electrode and the chitosan composite membrane was observed which resulted in poorer fuel cell performance, compared to Nafion®117. The gas diffusion layers, catalyst layers and interfaces accounted for some of the decrease in conductivity from membranes to MEAs. It was found that although a greater conductivity was achieved for the composite membrane, the improper method of MEA making (although a traditional one) led to unsatisfactory fuel cell discharge characteristics. The study has demonstrated that not only intrinsic membrane properties are important, but the compatibility of the membrane and the electrode is a key issue when MEA studies are undertaken.

A high-performance triple-layer proton exchange membrane comprising two structurally modified chitosan layers coated onto Nafion®105 sides was prepared [128]. A mixture of glutaraldehyde (GA) and sulfosuccinic acid (SSA) water solutions were used as a cross-linking agent. The optimum composition of cross-linking agents turned out to be 12 and 2 wt% for SSA and GA, respectively. SEM and EDX studies of the cross-sectional area of fabricated multi-layer membrane showed a high affinity of opposite charged polyelectrolyte layers and tight adherence between layers. Proton conductivity and methanol permeability measurements showed improved transport properties (remarkably reduced methanol crossover and significantly higher membrane selectivity parameter) for the multi-layer membrane compared to Nafion®117 with approximately the same thickness. The performance of fabricated membranes under working conditions was investigated using a laboratory made DMFC single cell. The cell was constructed of four 316 stainless steel sheets (for end plates and flow fields), two carbon papers (for gas diffusion layers) and the membrane electrode assembly. The electrodes were prepared via a catalyst painting technique where Pt-black and Pt/Ru black were used as catalysts for anode and cathode, respectively. Silicon rubber sheets were used to seal the internal sections. Direct methanol fuel cell tests revealed higher open circuit voltage, power density output, and overall fuel cell efficiency for the triple-layer membrane compared to Nafion®117, especially when concentrated methanol solutions were used. A power output of 68.10 mW/cm^2 at 5 M methanol feed was supplied using a multi-layer membrane, which was found to be about 72% more than that of Nafion®117. Owing to the ability to provide high power output, significantly reduced methanol crossover, ease of preparation and low cost, the proposed triple-layer membrane could be considered as a promising polyelectrolyte for high-performance DMFC applications.

In conclusion, it may be stated that biopolymers could be a solution to the task to improve the electrical and mechanical properties of proton exchange membranes for low temperature FCs, and especially - for DMFCs, at a low cost of production.

In addition to membranes, biopolymers can be useful in the production of hydrogen - one of the basic fuels for the FCs. Cellulose nanocomposite films have been proposed for photocatalytic production of H_2 from a water solution under visible light irradiation [129]. Experimental results indicated that the proposed nanocomposite films were characterized by high visible-light photoactivity for H_2 production, enduring photostability, and convenient regeneration. They are proposed as a pathway for the fabrication of portable visible-light-driven photocatalyst for the H_2 production via "green" processes.

3.2. Solid Polymer Electrolyte System for Rechargeable Batteries

Lithium-ion batteries (LIBs) are used in the most portable electronic devices, because of their good characteristics, such as no memory effect, high energy densities and a slow self discharge when the device is not in use. However, the lithium-ion battery drawback is the use of liquid electrolytes with organic solvents, which poses a flammability and self-ignition hazard under high temperatures [130].

The solid polymer electrolyte (SPE) has become a reliable candidate for an alternative electrolyte for energy storage devices. The advantages of batteries, using SPE, are no leakage of electrolyte, non-combustible reaction products at the electrode surface [131], less corrosion and handling problems, a higher power to weight ratio and longer shelf life [132]. Poly (ethylene) oxide [133], poly (acrylonitrile) [134] poly (vinyl chloride)/poly (methyl methacrylate) [135] and poly (vinylidene) fluoride - hexafluoropropylene, in combination with (poly (3-{2-[2-(2-hydroxyethoxy) ethoxy] ethoxy}methyl-3′-methyloxetane) [136] are applied as matrices in SPEs.

The use of biodegradable materials such as SPEs has been studied with the aim to address environmental problems. Chitosan is considered a promising polymer host for solid electrolytes [137-139].

The main drawback with chitosan use is its relatively low ionic conductivity (generally below 10^{-3} S/cm). Various modifications of chitosan have been introduced to overcome the problem, such as adding plasticisers [140] and inorganic fillers [141]. However, these approaches fail to optimize enough the properties of the material, so that it will be effectively used in SPEs. The introduction of plastic crystal solids in the polymer showed a potential to increase significantly its conductivity [142]. Succinonitrile (SN) is one of the non-ionic plastic crystals with good solvent properties and with high polarity which is able to dissolve the lithium salts and provide high ionic conductivity. High conductivity was obtained by increasing the concentration of lithium bis (trifluoromethylsulfonyl) imide (LiTFSI), when it was dissolved in SN [143]. Therefore, solid polymer electrolytes consisting of chitosan, LiTFSI (as a salt) and SN (as a plastic crystal) are prepared by using the solution casting method [144]. Electrochemical impedance spectroscopy is used to examine the ionic conductivity of the films at the temperature range of 303–383 K. Chitosan–LiTFSI containing 50 wt% succinonitrile exhibits high ionic conductivity at a room temperature ($0.4x10^{-3}$) S/cm, a high lithium ion transference number (0.598) and wide electrochemical window (up to 4.7 V). The solid polymer electrolyte lithium battery with $LiFePO_4$ as a cathode displays a stable discharge capacity of 160 mAh/g for up to 50 cycles at a current density of 17 mA/g. The results demonstrate the potential for the chitosan-based solid polymer electrolyte system to be used as an electrolyte in lithium rechargeable batteries.

Lithium-ion batteries based on proton conducting solid polymer electrolytes are still costly, difficult to handle lithium electrodes, with an inherent safety concern. Hence, the proton rechargeable battery can be considered as a cost-effective alternative and appears a good option for lower energy density battery applications, in spite of a smaller electrochemical stability window compared with LIBs [145, 146]. Research in this direction has been devoted to the preparation of solid polymer electrolytes for rechargeable batteries, using as a host polymer, renewable plants-based biopolymer materials, such as starch, cellulose, sago and soy-based plastics [147-149]. A proton-conducting biopolymer electrolyte (BPE) was developed and prepared via a solution casting method by using natural carboxymethyl cellulose (CMC) as the host polymer and NH_4Br - as the ionic dopant, since ammonium salts are considered good proton donors to the biopolymermatrix [150]. In order to determine the potential of the highest proton conducting BPE as an electrolyte system in battery, it was used in the fabrication of Zn + $ZnSO_4 \cdot 7H_2O$/BPE/MnO_2 rechargeable proton battery. Ion conductance was observed in the fabricated BPE and protons (H^+) were the predominant conducting species. The highest ionic conductivity (1.12×10^{-4} S/cm at ambient temperature) was measured for the sample containing 25 wt % NH_4Br. Linear sweep voltammetry results revealed that the BPE is electrochemically stable up to 1.42 ± 0.01 V and suitable for proton battery application. The constructed proton battery gave an average initial cell potential of 1.52 V at ambient temperature and after a small drop, the potential remained constant with an average open circuit potential of 1.36 V. The battery showed good rechargeability. At testing in 10 cycles, it did not show any loss of cell potential. These results imply that CMC–NH_4Br proton-conducting biopolymer electrolyte systems may find practical application in the fabrication of rechargeable proton batteries.

Wide-scale implementation of renewable energy will demand a growth of inexpensive, safe and efficient energy storage systems [151]. Consequently, the extension of batteries to large-scale storage will become a very important task [152].

Sodium-based secondary batteries (NaBs) are considered to be among these task solutions. Sodium based energy storage systems may never fit in a car, and cannot reach LIBs in terms of energy density. However, when gigantic energy storage systems are needed for the electricity grid and/or other non-portable applications, where size does not matter, sodium-based systems can be an effective and intelligent choice [153]. Low-cost storage may enhance the grid utilization without adding more wires, and energy may be stored and released slowly to keep power flowing without interruption. Sodium exhibits a suitable redox potential (–2.71 V) for batteries preparation. It is abundant in the Earth's crust and cheap. The storage capacity of NaBs is not affected by temperature fluctuations, thus this technology might suit the most large scale storage plants in the desert. The research on NaBs' components is not mature, particularly for what concerns the electrolyte, where standard organic liquid electrolytes are mainly used, such as $NaClO_4$ or $NaPF_6$ salts dissolved in propylene carbonate (PC) [154]. Unfortunately, PC is flammable. The application of solid polymer electrolytes presents several advantages over organic carbonate based liquid counterparts: i) internal short circuits, secondary reaction products and their aftermath, such as organic solvent decomposition and related fire hazards may be avoided, ii) the final battery architecture may be designed according to application, iii) simplified production and packing procedures may be adopted, which in turn decrease the production cost [155]. However, the development of solid polymer electrolytes for NaBs is at the very early stage. In this case, the SPE must exhibit sufficient ionic conductivity, good mechanical integrity, thermal stability, a wide electrochemical stability window, appreciable Na^+-ion transference number and good compatibility with the electrodes [156]. Biopolymer materials have already found their place in the preparation of SPEs for NaBs. The incorporation of a pyranose ring based ionic backbone sodium carboxymethyl cellulose (Na-CMC) into a PEO-based polymer matrix containing a suitable sodium salt has been proposed for all solid Na-ion secondary cells that can operate at moderate temperatures [157]. The Na-CMC (an ionic ether derivative of cellulose, which is prepared from alkaline cellulose and chloroacetic acid by etherification) has been selected, as it can improve the mechanical integrity of the final film without affecting the ionic mobility and the electrode/electrolyte interfacial characteristics. Moreover, it shows good solubility in water, low cost, biodegradability, and lack of toxicity. The same Na-CMC has been also used as binder for the active electrode material particles, which enables the overall process, including the electrodes and electrolyte preparation to be carried out through very simple, cheap and environmentally-friendly water based procedures. The properties of the hybrid electrolyte were investigated by means of galvanostatic charge/discharge cycling in lab-scale cell with TiO_2-based or $NaFePO_4$-based working electrodes. The overall electrochemical performance was demonstrated to be stable. In spite of the need for improvements and optimization in terms of specific energy output and long-term performance, the proposed new materials could facilitate the work in the field of large energy storage devices (and especially truly solid Na-ion polymer cells conceived for moderate temperature application) towards safer, biosourced and reliable options.

A cellulose based nanocomposite has been prepared by the casting/solvent evaporation method, where graphene oxide (GO) nanoplatelets (modified with hexamethylene diisocyanate grafting agent) were dispersed in the cellulose matrix [158]. According to the developers, the cellulose modified GO nanocomposite has turned out to be environmentally stable and excellent (in terms of dielectric and ferroelectric properties) next generation material for flexible energy storage and electronic devices.

3.3. Materials for Solar Cells Application

Among many natural resources of sustainable energy, solar energy is one of the most promising sources that can replace fossil fuels and fulfill the rising global demand for energy. Solar energy can also effectively take part in the photocatalysis (to produce H_2 gas via the photolysis of water molecules) or the photoelectrochemical reaction like in solar cells. Biopolymers find their place in developing materials for solar cells.

According to their developers, previously described ionically conductive membranes based on plasticized DNA with glycerol and containing conducting polymers such as PEDOT: PSS, POEA or electrochromic dye PB, may find application in Grätzel type cells for solar energy conversion [60]. Dye sensitized solar cells (DSSCs) or Grätzel cells (invented by O'Regan and Grätzel in 1991) are known as third generation photovoltaics. They are very different from the previous semiconductor devices as they do not rely on a traditional p–n heterojunction to separate photogenerated charge carriers. Because of the low-cost materials and easy fabrication, this technology is expected to take a significant share in the fast growing photovoltaic areas. The maximum stable DSSC reported by Grätzel's group uses liquid electrolytes which have many disadvantages like corrosion, evaporation, volatility, etc.

Gel electrolytes are proposed as alternatives to liquid electrolytes in DSSC applications [159]. Gel polymer electrolytes (GPE) are solid ion conductors formed by dissolving salts in polymers having high molecular weight. They can be prepared in semisolid or solid form. GPEs possess liquid like conductivity which is necessary for electrochemical devices [160]. GPE has already proved their suitability in DSSC [161].

GPE, based on polysaccharides, are well suited for DSSC as electrolyte because from one side they show liquid like conductivity, while from other side they assist in wetting the electrodes [162-168].

The liquid of a dye-sensitized solar cell composed of a nanoporous titanium dioxide (TiO_2) film and a bipyridine–Ru complex dye (N3 dye) was solidified by utilizing agarose [162]. The electric resistance of the solid material, containing the solution of redox electrolytes (I^-/I_3^-), was of the same order of magnitude as the liquid material, and the charge transfer resistance on the electrode surface was even smaller for the solid than for the liquid. The solid type cell exhibited light-to-electricity conversion efficiency, same as the conventional liquid type cell. For both the solid and liquid type cells, the thinner electrolyte layer gave a higher conversion efficiency.

GPEs prepared from agarose and 1-alkyl-3-methyl-imidazolium salts were investigated in order to improve the durability of DSSCs [163]. The gels showed sufficient mechanical strength even at a very small amount of agarose added (1.0-1.5 wt%). The photon to electron conversion efficiency of the DSCs containing ionic gel electrolyte was 2.93% under simulated sunlight (air mass 1.5) with a light intensity of 100 mW/cm^2.

All-plastic DSSCs using TiO_2 paste (SP-210), indium tin oxide-coated polyethylene naphthalate (PEN/ITO) film and a Pt-coated PEN/ITO film counter electrode, were fabricated with 1.0 wt% κ-carrageenan solid electrolyte containing the redox electrolytes solution [164]. The performance of the prepared cell was compared with the characteristics of the liquid type, showing that the values of the open circuit voltage and fill factor were the same, but short circuit current and conversion efficiency were slightly larger for the liquid type electrolyte cell than the solid type.

The effect of lithium iodide concentration (LiI: 0–85 wt%) on the electrochemical behavior of agarose-based polymer electrolytes for DSSC was investigated [165]. It was found that the charge transfer resistance and charge recombination resistance were reduced when the LiI concentration was increased, while the electron lifetime of DSSC was shortened. The addition of LiI enhances the back reaction in DSSC.

Agarose based gel electrolyte doped with KI has been developed by the solution cast method for DSSC application [166]. Complex impedance spectroscopic analysis of the prepared material has shown that doping with KI provided the additional charge carries (cations/anions) which enhanced overall conductivity and the conductivity maximum was obtained near a 60:40 (agarose: KI) composition. IR and XRD studies confirmed the composite nature of the gel electrolyte. DSSC of area 0.25 cm^2 has been developed to study the gel electrolyte applicability. The performance of DSSC was evaluated by the current–voltage curve measurements. DSSC using maximum ionic conducting electrolyte showed an efficiency of 0.54 at 100 mW/cm^2 light intensity. The developed biopolymer-based gel electrolyte seems to be an environmentally friendly promising material for solar cells fabrication.

A bio-polymer electrolyte system has been prepared by mixing NH_4I with potato starch [167]. The prepared material is an ionic conductor with an ionic transference number of approximately 0.95 and ambient electrical conductivity of approximately 2.4×10^{-4} S/cm. According to the authors, the prepared material is humidity immune up to 77% relative humidity and hence - encouraging for possible use in humidity immune devices.

Hydroxypropyl cellulose (HPC), sodium iodide (NaI), 1-methyl-3-propylimidazolium iodide (MPII) as ionic liquid, ethylene carbonate (EC) and propylene carbonate (PC) were used for the preparation of a non-volatile gel polymer electrolyte system (HPC:EC:PC:NaI:MPII) for DSSC applications [168]. The highest ionic conductivity of 7.37×10^{-3} S/cm is achieved after introducing 100% of MPII with respect to the weight of HPC. Dye-sensitized solar cells were fabricated using the prepared gel polymer electrolytes. A photo-electrode layer was prepared with two TiO_2 layers on the Fluorine-doped tin oxide substrate (FTO). A counter electrode was prepared using a commercial platinum solution to coat FTO substrate and then sintered at 450°C for 30 minutes. The prepared GPEs were sandwiched between two electrodes and analyzed under a Sun simulator. The cell prepared with the gel polymer electrolyte with 100 wt.% of MPII ionic liquid showed the best performance and energy conversion efficiency of 5.79%, with short-circuit current density, open-circuit voltage and a fill factor of 13.73 mA/cm^2, 610 mV and 69.1%, respectively.

4. Biopolymer-Based Materials for Environment Monitoring Sensors

Complying with the aim of the sustainable development of the society, the search for biopolymers for developing electrochemical sensors and biosensors aims to replace synthetic materials. The development of materials using naturally-derived biomolecules and biopolymers, whose versatility, biodegradability, and non-toxicity can offer advantages over synthetic materials is an area of intense research. Biosensors are analytical devices made of a biological recognition element that outputs a measurable signal to an interfaced transducer and a transducer which can convert a biological or biochemical signal or response into a quantifiable electrical signal (which is in turn calibrated to a certain scale). Biorecognition typically relies on enzymes, whole cells, antibodies, or nucleic acids, whereas signal transduction exploits electrochemical (amperometric, chronoamperometric, potentiometric, field-effect transistors, conductometric, capacitative), optical (absorbance, reflectance, luminescence, chemiluminescence, bioluminescence, fluorescence, refractive index, light scattering), piezoelectric (mass sensitive quartz crystal microbalance), magnetic, or thermal (thermistor, pyroelectric) interfaces [169, 170].

Biopolymer-based sensors are electrochemical or photonic devices, which use a biopolymer as a matrix for sensing materials, or the biopolymer is layered on the surface or incorporated in synthetic polymers or inorganic materials. Chitosan is the most often studied and applied biopolymer, followed by other polysaccharides, DNA and silk.

4.1. Chitosan

Chitosan is used as films, or as a capping agent to modify different electrodes, in nanocomposites and nanohibrides employed for electrodes' preparation, or it is doped for sensing applications.

Graphite-epoxy resin composite (GrEC) electrodes were modified with chitosan (Chit) films and characterized [171]. Several film modifications were made by using different crosslinking agents: glutaraldehyde (GA), glyoxal (GO), epichlorohydrin (ECH) and 1-ethyl-3-(3-dimethylaminopropyl) carbodiimide (EDC) together with N-hydroxy succinimide (NHS). Immobilization of functionalized carbon nanotubes into chitosan matrices (Chit–CNT) using the same crosslinking agents was also studied. Results showed that the electrode with the best performance (GrEC/Chit–CNT/EDC–NHS) was suitable as a sensor for dipyrone and hydroquinone.

Chitosan-capped gold nanoparticles were prepared in presence of glutamate ions and proposed for sensing ions of heavy metals [172]. The polycationic nature of chitosan enabled attachment of the polymer to the negatively charged gold nanoparticle surfaces through electrostatic interactions. The use of chitosan provided a sufficient steric hindrance, thus ensuring stability of the colloid and also functionalized the nanoparticles for use as sensors. The chelating properties of chitosan and the sensitivity of the optical properties of gold nanoparticles to agglomeration have been employed to detect low concentrations of heavy metal ions (Zn^{2+} and Cu^{2+}) in water. It was proved that the comparison of the optical absorption spectra of the colloidal suspension before and after exposure to metal ions is a good indicator of the concentration of the heavy metal ions.

The preparation of ionophore doped chitosan biopolymer solution using chitosan, glutaraldehyde and p-tert-butylcalix[4]arene-tetrakis(N,N-dimethylacetamide) in acetic acid under mild condition has been described [173]. After the reaction, the ionophore doped chitosan biopolymer thin film was prepared using a spin coating technique. It was found that the absorption of the thin film is high, with an optical bandgap of 3.177 eV. An optical spectroscopy (based on the principle of surface plasmon resonance) was designed in the laboratory to test the capability of the ionophore doped chitosan thin film for sensing metal ions. Studies showed that this thin film can detect metal ions, such as Pb^{2+}, Cu^{2+}, Hg^{2+}, Zn^{2+} and Mn^{2+} in ion solutions at a concentration of 1 mg/L. This ionophore doped chitosan thin film can be considered as a promising optical material for practical application, especially in the field of metal ions sensing.

Chitosan-carbon nanotubes (Chit-CNT) conductive bio-polymer nano-composites transducers have been prepared using the spray layer-by-layer technique and their chemo-electrical properties have been investigated [174]. The prepared transducers demonstrated a high sensitivity to polar vapours such as water and methanol (difficult to obtain with synthetic polymers being mostly non-polar) but also to non-polar ones like toluene. This good selectivity on the whole range of vapour content makes Chit-CNT a good candidate for implementation in "e-nose" devices.

Chitosan gold nanocomposite (CGNC) was synthesized in a single step process and properties of the composite were investigated [175]. The structure of the polymer was pH dependent and sol-like to gel-like transition of the polymer was observed at around pH 4.5. The pH dependent transition of the CGNC in the visible absorption spectrum was observed. The resulting CGNC system was explored for its use as a sensor for the colorimetric detection of lead. A pH-sensitive limit of detection (LOD) was observed, with the LOD being 1 mM at low pH and 1 μM at high pH. The CGNC was electrodeposited on a gold electrode and thus prepared electrode was applied for the detection of lead using cyclic voltammetry. Lead could be detected with a LOD of 10 μM (2 mg/L). The fabrication reproducibility for four sensor electrodes showed a relative standard deviation of 2.09% for the determination of 1 mM lead nitrate. The developed sensor system showed high reproducibility, stability and satisfactory selectivity and was proposed to be used for the routine environmental monitoring of lead levels in soil samples.

The intercalation of biopolymers into layered nanomaterials can be a strategic approach to develop advanced materials. The combination of clays and biopolymers produces bio-nanocomposites that can be used as active phase of electrochemical sensors for the detection of ionic species.

Nanocomposites, based on the intercalation (through cationic exchange and hydrogen bonding processes) of chitosan, a cationic natural polymer, in Na^+-montmorillonite, are bidimensional nanostructured materials provided of anionic exchange sites ($-NH_3^+X^-$) [176, 177]. This fact is due to the special arrangement of the biopolymer as a bilayer when the amount of intercalated chitosan is higher than the cationic exchange capacity of the clay. The resulting chitosan–montmorillonite nanocomposites shrink as the polymer / clay ratio increases, leading to a compact, robust and handy three-dimensional network that can be easily processed to construct bulk-modified sensors for the detection and potentiometric determination of anions. These nanocomposites are combined with graphite particles to provide the system with electronic conductivity. Thus, the nanocomposite–graphite system acts simultaneously as electronic collector as well as ionophore for the anionic recognition. The high affinity between the chitosan and the montmorillonite clay substrate is the basis of the high stability of the intercalated biopolymer against desorption or degradation, and, as a consequence, of the long-term stability of the developed sensors. These devices are more advanced than surface-modified electrodes with regard to robustness, long-term stability and easy surface renewal. The prepared sensors were successfully applied in the potentiometric determination of several anions – CH_3COO^-, Cl^-, F^-, NO_3^-, SO_4^{2-}, $Cr_2O_7^{2-}$, and $Fe(CN)_6^{3-}$. They showed a higher selectivity towards monovalent rather than to di- or trivalent anions and the best potentiometric response towards nitrate ions was observed. Therefore, chitosan–montmorillonite nanocomposites can be considered as excellent materials for the development of bulk-modified potentiometric sensors for detecting anions in aqueous samples.

Chitosan–sepiolite nanocomposites can be processed as films although their appearance, texture and mechanical properties strongly depend on the chitosan/sepiolite ratio and the preparation procedure [178]. For electrodes based on both, pure clay or nanocomposites with low chitosan content, the response presents a positive slope indicating the cation exchange ability of the material. The electrode based on the nanocomposite with a higher amount of chitosan shows a response with a negative slope that indicates the anionic exchanger behavior of this nanocomposite. These results point out that chitosan–sepiolite can be used as the active phase of ion-sensors in which the response can be modulated depending on the composition of the nanocomposite.

4.2. Other Polysaccharides

Cellulose, starch, and other polysaccharides have also been the subject of intense research to assess their behavior as electro-active polymers that can be potentially used for biopolymer based sensors in environment pollution monitoring.

Cellulose, the most abundant renewable biopolymer, is widely applied for synthesizing nanocomposites due to its excellent mechanical properties, microporous structure, chemical stability, biological compatibility, and remarkable hydrophilic properties.

Magnetic biopolymer nanocomposites were prepared by the reaction of Fe_3O_4 nanoparticles, containing carboxymethyl cellulose, with polyvinyl alcohol [179]. The conductivity of the magnetic nanocomposites was increased by adding conducting materials, such as multiwall carbon tubes (MWCNTs) and polyaniline (PANI). Thus, Fe_3O_4/Hydrogel/MWCNTs and Fe_3O_4/Hydrogel/PANI hybrid nanocomposites were prepared and studied. In the case of nanocomposites with PANI, no response toward CH_4 gas was observed either at room temperature or higher temperatures - up to 60°C. The Fe_3O_4/Hydrogel/MWCNTs hybrid nanocomposites were found to be highly sensitive and selective toward CH_4 gas, showing well repeatable signal output and can be used to regularly monitor the levels of oxygen, CH_4, carbon dioxide, CO, etc. within the working environment.

The crystalline particles in cellulose fibers derived by acid hydrolysis possess rigid shape and anisometry and are called cellulose nanocrystals (CNCs). The CNCs produced by using concentrated sulfuric acid possess high crystallinity, stability and mechanical properties with good dispersion [180]. Due to their hydrophilic properties, the CNCs promote nucleation and growth of inorganic pigments in an aqueous medium. A nanostructured cellulose nanocrystal (CNC) sheet, with iron oxide grown on, was fabricated by the hydrothermal method [181]. The iron oxide particles were allowed to grow on the CNC's surface which ensures better bonding at the interface and finally results in a flexible, resistive and reversible sensor with light weight. The sensing performance test for NO_2 molecules demonstrated that the devices were highly sensitive and fully recoverable at room temperature, which was attributed to the excellent access of nitrogen dioxide molecules to the sensor surface via CNC. The high sensor response, flexibility, room temperature operation, low power consumption, durability and low production cost make these devices suitable for environment monitoring.

The optical chemical sensors have drawn much attention in analytical chemistry because of their advantages, such as easy fabrication, good sensitivity, selectivity and low cost.

A colorimetric sensing probe was developed for cyanide anions determination in aqueous media, based on the functional surface modification of natural cellulose materials by the incorporation of a chromogenic probe into cellulose fibers [182]. The dye-sensitized cellulose materials were able to detect cyanide at about 0.01-0.07 μM in an aqueous solution with high selectivity over other anions.

A chemodosimeter-sensitized starch film was developed by organic compounds (based on quinolinium merocyanine derivative) incorporation into a film, based on starch [183]. Immersion of this functionalized film in an aqueous solution of cyanide induced a color change that can be used for the detection of cyanide. The initial qualitative visual study indicated that the film only changed color (from yellow to purple) in the presence of cyanide. Through the selective nucleophilic addition reaction between cyanide anions and the dye chromophore, the film can selectively detect sub-micromolar quantities of cyanide ions by a distinct color change in the presence of various anions (F^-, Cl^-, Br^-, CH_3COO^-, $H_2PO_4^-$, HSO_4^-, SO_4^{2-}, NO_3^-, HCO_3^-, CO_3^{2-}, SO_3^{2-}, SCN^-, ClO_3^-, ClO_2^-, HS^-) in concentration $C = 10^{-3}$ M. In the absence of cyanides and in the presence of all above-mentioned ions, the film showed a band at $\lambda_{max} = 426$ nm. Only the addition of cyanide anions induced a bathochromic shift by the appearance of a new band at $\lambda_{max} = 527$ nm and the disappearance of the starting band. The lack of any significant response to fluctuations in pH over the range 5-9 suggests that this material may be used for cyanide detection in water at pH 5 < pH 9.

Silver nanoparticles (AgNPs) have gained tremendous popularity in the field of sensors because of their good optical, electronic and chemical properties. One unique property of silver nanoparticles includes their utilization in fluorescence and surface plasmon resonance (SPR) and these are used for colorimetric sensors preparation [184]. Silver nanoparticles (<10 nm in size) were synthesized from silver nitrate using as a template the polysaccharide Cyamopsis tetragonaloba, commonly known as guar gum (GG), derived from plants [185]. GG acts as reductive reagent. Aqueous ammonia sensing studies of the prepared polymer/silver nanoparticles nanocomposite (GG/AgNPs NC) were performed by an optical method based on localized surface plasmon resonance (LSPR) - by monitoring the changes in LSPR position and amplitude with a UV–VIS spectrophotometer in the ammonia concentration range of 1 - 100 mg/L. The response time of 2–3 s and the LOD of ammonia in solution of 1 mg/L were found at room temperature. The developed sensor is promising for both environmental and medical applications.

Silver nanoparticles were synthesized by using locust bean gum (LBG) polysaccharide by mixing optimized weight percent of LBG with a known quantity of silver nitrate ($AgNO_3$) at 55–60°C [186]. The size of synthesized AgNPs was in the range of 18–51 nm depending upon the concentration of LBG and $AgNO_3$. Further, a low cost and portable optical fiber based sensor, using LBG stabilized AgNPs, was developed for monitoring the H_2O_2 concentration as low as 0.01 mM.

Charged polysaccharides were associated with smectites (cation exchangers) or layered double hydroxides (LDHs) (anionic exchangers) [178]. Polysaccharides with negatively charged groups such as alginate, pectin, and ι-carrageenan can be incorporated between the layers of LDHs via a co-precipitation or "co-organized assembly" that allows the formation of LDH in the presence of the polymer. Alginate and pectin intercalate ZnAl-LDH as a monolayer, while ι-carrageenan seems to be incorporated as a bilayer or, most probably, as a double helix within the LDH structure. Due to the procedure used to obtain the LDH–biopolymer nanocomposites, an excess of the polymer presents. Thus, reversing the characteristic anionic exchange capacity of the LDH to a cationic exchange capacity can occur. This property has been the basis for LDH–biopolymer nanocomposites application as active phases of carbon paste electrodes or PVC membrane based potentiometric sensors for the determination of Ca^{2+} in aqueous samples [187].

Thin films were prepared by the Layer-by-Layer (LbL) self-assembly technique onto tin-doped indium oxide (ITO), where natural polysaccharides (agar and carrageenan), extracted from the cell wall of red seaweeds were deposited in layers alternating with polyaniline (PANI) [188]. The layers arrangement in these films has a strong effect upon their final properties. Two architectures of bilayers were analyzed; ITO/PANI/ polysaccharide and ITO/polysaccharide/PANI. For comparative purposes, films were also studied where the polysaccharide was replaced by the PVS, thus constituting systems ITO/PANI/PVS or ITO/PVS/PANI films. The results obtained by UV–VIS showed that the agar/PANI films exhibit highly regular growth, and the film's thickness can be controlled by the number of bilayers adsorbed. Results from cycling voltammetry tests showed that the presence of agar or carrageenan improved the electrochemical stability of the conducting polymer (i.e., of PANI) in an acid medium, in comparison to the case where PVS was used, by working as a type of antioxidant for the polymer. Among the systems studied, the most promising one as regards electrochemical measurements was the ITO/agar/PANI system, which was subsequently employed in the electrochemical detection of chromium (VI). In the measurements, they used LbL films containing only one bilayer of the optimized system. The electrochemical measurements for Cr(VI) determination were carried out in a 1.0 mol/L potassium phosphate buffer at pH 1.5. The achieved detection limit was 2.48 μmol/L.

4.3. Deoxyribonucleic Acid

DNA has been applied in nanotechnology and biochemistry for sensing applications due to its capability of molecular recognition and its affinity for metallic ions [189].

A sensor was fabricated via the immobilization of single standard calf thymus DNA onto a chitosan nanobiocomposite film containing iron oxide nano-particles (Fe_3O_4, 22 nm), deposited onto ITO coated glass surface for pyrethroid, cypermethrin and permethrin detection. This disposable nanobiocomposite bioelectrode was stable for about two months under refrigerated conditions and detected cypermethrin and permethrin rapidly (in 50 s) at concentrations of 0.0025-2 mg/L and 1-300 mg/L respectively [190].

The principle underlying the DNA-based biosensor is the ability of a transducer to monitor a change in the nucleic acid's structure occurring after exposure to a target chemical [170]. In general, the nucleic acid biosensors are nonselective and provide an overall indication of a potentially harmful chemical or chemicals' mixture in the studied environment and (depending on the biosensor construction) an estimate of concentration [191].

Atrazine, hydrazine and aromatic amine compounds in fresh and groundwater were determined by using DNA biosensors. General toxicity pollution in wastewater and industrially contaminated soils has been screened using DNA biosensors. Lead, cadmium, nickel, arsenic, copper, iron, chromium, and other heavy metals have been detected nonselectively through DNA biosensing, incorporating both single- and double-stranded DNA as the sensing element [170].

DNA biosensors were proposed for soil samples screening for genotoxic compounds, assembled by immobilising double stranded calf thymus DNA on screen-printed electrodes [192]. The interactions between DNA and environmental pollutants can cause DNA damage which is turned in an electrochemical signal. Studies were performed using benzene, naphthalene and anthracene derivatives as model targets in extracts of soil samples. The effect of these compounds on the surface-confined DNA was found to be linearly related to their concentration in solution. The applicability of such a biosensor was evaluated by analyzing soil samples from an Italian region with an ecological risk. The developed DNA biosensor for qualitative analysis of soil presented a good correlation with a semiquantitative method for aromatic ring systems determination at fixed wavelength fluorescence. The biosensor can be used as an early warning device in areas with ecological risk at *in situ* measurement.

An electrochemical biosensor based on the chemicals' ability to oxidize DNA guanine residues, was prepared by immobilizing double-stranded DNA on a single use disposable screen-printed electrochemical cell, operated by a handheld battery-powered potentiostat [193]. The sensor was applied to measure the chemical contamination of soil (by using electrochemical scans) in the preliminary prepared extract. The magnitude of electrochemical signals from the oxidation of DNA guanine residues by chemicals-pollutants (i.e., of "guanine peaks") in relation to a reference electrode was linearly related to their concentration in solution (i.e., the higher the concentration of the target chemical, the more the damage imposed on the DNA, and the lower the electrochemical measurement of the oxidation signal).

Nucleic acid can be manipulated to create aptamers (short, target-specific oligonucleotides) by using a process named "systematic evolution of ligands by exponential enrichment" (SELEX) [170]. By iteratively incubating nucleic acid with the desired target, the oligonucleotide sequences (or aptamers) with the greatest affinity for the target can be selected. Predominant development and application of aptamers-based sensors is in the clinical fields, but they inevitably will enter the environment sensing and monitoring area.

Selectivity has been demonstrated by using deoxyribozymes (DNAzymes) or ribozymes (RNAzymes). DNAzyme electrochemical biosensor for lead has been developed which is based on a single-stranded DNAzyme adsorbed onto a gold electrode [194]. A colorimetric DNAzyme-based biosensor for lead has also been developed [195]. According to the authors, the detection of Pb^{2+} could be realized within 20 minutes, with a detection limit of 500 nM with enough good selectivity.

DNA aptamers are excellent sensor moieties for Graphene field-effect transistors (GFETs) due to their strong affinity to graphene, relatively short chain-length, selectivity, and a high degree of analyte variability. DNA aptamers increase the selectivity of GFET sensors for small molecules and proteins and compare them with other, state of the art, detection methods [196]. DNA aptamer-based GFET biosensors are highly sensitive, selective electronic devices with distinct advantages over previous aptasensor platforms. Graphene can be combined with single-strand DNA (ssDNA) to create a chemically diverse family of vapor sensors that are promising for use in a "noselike" vapor sensing system [197].

Kybert et al. reported a highly sensitive chemical vapor sensor, based on GFET functionalized with ssDNA that is able to distinguish among chemically similar analytes, including a series of carboxylic acids, and structural isomers of carboxylic acids and pinene [198]. Target analytes were detected without a loss of sensitivity in a large background of a chemically similar, volatile compounds. High binding affinity of the DNA aptamers and low noise from the graphene substrate allowed the device to reach low detection limits (mg/L). These results motivate further development of the DNA/graphene sensor family for use in an electronic olfaction system.

Improved sensitivity (lower LOD) is reported for a GFET aptasensor that has previously been p-doped by employing an oxygen plasma-treated graphene [199]. The device has demonstrated mercury ion detection into picomolar (10 pM) concentrations in the presence of several similar divalent ionic species (Co^{2+}, Zn^{2+}, Sr^{2+}, Cd^{2+}, Ni^{2+}, Pb^{2+}, Li^{2+}) - all in concentration 10 mM.

Probably, the USA patent of Kaplan et al. [200] is the most complete for the present scientific work, related to the preparation of biopolymer-based sensors for environment monitoring. Different sensors (mainly based on silk, but also on chitosan, collagen, gelatin, agarose, chitin, starch, cellulose, hyaluronic acid, DNA, aptamers, etc.) and the methods for their preparation are included in the patent.

Sensors based on biopolymers and their nanocomposites outdate the traditional chemi-resistive sensors based on semiconducting oxides in lower power consumption, room temperature operability, and high performance. The research in this area, although growing rapidly, is still at its infancy and requires further efforts.

All examples described in this chapter, show that cheap and nontoxic materials of natural origin can compete with the best synthetic organic materials, used in our every-day life in areas as consumer electronics, RES, and environmental monitoring. These materials can not only be sustainable and biodegradable, but are also, unique with their versatility and functionalities, inaccessible to standard materials used up to this time in mass production. Although the research into their implementation in real-life devices still remains very limited, they have shown that truly 'green' devices have potential and are expected to make a positive impact in the future. Fast steps are needed in this direction.

REFERENCES

[1] European Consumer Centre. L'Obsolescence programmée ou les dérives de la société de consommation. <http://www.europe-consommateurs.eu/fileadmin/user_ upload/ eu-onsommateurs/PDFs/publications/etudes_et_rapports/Etude-Obsolescence.pdf>.

[2] Irimia-Vladu, Mihai, Eric. D. Głowacki, Gundula Voss, Siegfried Bauer, and Niyazi S. Sariciftci. 2012. "Green and biodegradable electronics." *Materials Today* 15 (7-8): 340-346.

[3] Bollström, R. A. Määttänen, D. Tobjörk, P. Ihalainen, N. Kaihovirta, R. Österbacka, J. Peltonen and M. Toivakka. 2009. "A multilayer coated fiber-based substrate suitable for printed functionality." *Org Electron*10:1020-1023.

[4] Tobjörk, D., and R. Österbaska. 2011."Paper electronics." *Adv Mater* 23: 1935-1961.

[5] Siegel, Adam C., Scott T. Phillips, Benjamin J. Wiley, and George M. Whitesides. 2009. "Thin, lightweight, foldable thermochromic displays on paper." *Lab Chip* 9: 2775−2781.

[6] Hübler, Arved, Bystrik Trnovec, Tino Zillger, Moazzam Ali, Nora Wetzold, Markus Mingebach, Alexande Wagenpfahl, Carsten Deibel, and Vladimir Dyakonov. 2011. "Printed Paper Photovoltaic Cells." *Adv Energy Mater* 1: 1018–1022. doi:10.1002/aenm.201100394.

[7] Barr, Miles C., Jill A. Rowehl, Richard R. Lunt, Jingjing Xu, Annie Wang, Christopher M. Boyce, C. M., Sung G. Im, Vladimir Bulović, and Karen K. Gleason. 2011. "Direct Monolithic Integration of Organic Photovoltaic Circuits on Unmodified Paper." *Adv Mater* 23: 3500–3505. doi:10.1002/adma.201101263.

[8] Irimia-Vladu, Mihai, Pavel Troshin, Melanie Reisinger, Günther Schwabegger, Mujeeb Ullah, Reinhard Schwödiauer, Alexander Mumyatov, Marius-Aurel Bodea, Jeffrey W. Fergus, Vladimir F. Razumov, Helmut Sitter, Siegfried Bauer, and Serdar N. Sariciftci. 2010. "Environmentally sustainable organic field effect transistors." *Organ Electron* 11:1974-1990.

[9] Géczy, Attila, Márk Kovács, and István Hajdu. 2012. "Conductive layer deposition and peel tests on biodegradable printed circuit boards." *Digest Technol Papers SIITME*: 139–142.

[10] Mattana, Giorgio, Danick Briand, Alexis Marette, Andrés V. Quintero, and Nico F. de Rooij. 2015. "Polylactic acid as a biodegradable material for all-solution-processed organic electronic devices." *Organ Electron* 17: 77–86.

[11] Güllü, Ömer. 2010. "Ultrahigh (100%) barrier modification of n-InP Schottky diode by DNA biopolymer nanofilms." *Microelectron Eng* 87: 648–651.

[12] Eley, *Daniel D.* 1968. "DNA Structure and Conductivity." In *Organic Semiconducting Polymers,* edited by John Edward Katon, 259-267. New York: Marcel Dekker.

[13] Endres, Robert G., Daniel L. Cox, and Rajiv R. P. Singh. 2004. "The quest for high-conductance DNA." *Rev Mod Phys* 76:195-214.

[14] Wang, Qiang, and Torsten Fiebig, 2007. "DNA Photonics – Probing Light-Induced Dynamics in DNA on the Femtosecond Timescale." In: *Charge Migration in DNA: Perspectives from Physics, Chemistry, and Biology,* edited by Tapash Chakraborty, 221–248. Berlin: Springer.

[15] Kasumov, A. Yu., M. Kociak, S. Guéon, B. Reulet, V. T. Volkov, D. V. Klinov, and H. Bouchiat. 2001. "Proximity-induced superconductivity in DNA." *Science* 291: 280–282.

[16] Porath, Danny, Alexey Bezryadin, Simon de Vries, and Cees Dekker. 2000. "Direct measurement of electrical transport through DNA molecules." *Nature* 403 (6770): 635–638.

[17] De Pablo, P. J., F. Moreno-Herrero, J. Colchero, J. Gomez-Herrero, P. Herrero, A. M. Bar, P. N. Ordej, J. M. Soler, and E. Artacho. 2000. "Absence of DC conductivity in DNA." *Phys Rev Lett* 85:4992–4995.

[18] Storm, A. J., John van Noort, Simon de Vries, and C. Dekker. 2001. "Insulating behavior for DNA molecules between nanoelectrodes at the 100 nm length scale." *Appl Phys Lett* 79: 3881-3883.

[19] Genereux, Joseph C., and Jacqueline K. Barton. 2010. "Mechanism for DNA charge transport." *Chem Rev* 110:1642–1662.

[20] Genereux, Joseph C., Amie K. Boal, and Jacqueline K. Barton. 2010. "DNA-mediated charge transport in redox sensing and signaling." *J Am Chem Soc* 132:891–905.

[21] Subramanyam, Guru, Carrie M. Bartsch, James G. Grote, Rajech R. Naik, Lawrence Brott, Morley Stone, and Angela Campbell. 2009. "Effect of external electrical stimuli on DNA based biopolymers." *Nano* 4 (2):1–8.

[22] Singh, Birendra, Niyazi S. Sariciftc, James G. Grotea and Frank K. Hopkins. 2006. "Bio-organic-semiconductor-field-effect-transistor based on deoxyribonucleic acid gate dielectric." *J Appl Phys* 100:024514-1-024514-4.

[23] Yumusak, C., Th. B. Singh, N. S. Sariciftci, and J. G. Grote. 2009. "'Bio-organic field effect transistors based on crosslinked deoxyribonucleic acid (DNA) gate dielectric."*Appl Phys Lett* 95: 263304-1-263304-3.

[24] Schwabegger, G., M. Ullah, M. Irimia-Vladu, M. Baumgartne, Y. Kanbur, R. Ahmed, P. Stadler, S. Bauer, N. S. Sariciftci, and H. Sitter. 2011. "High mobility, low voltage operating C_{60} based n-type organic field effect transistors." *Synthetic Met* 161: 2058-2062.

[25] Kim, Dae-Hyeong, Yun-Soung Kim, Jason Amsden, Bruce Panilaitis, David L. Kaplan, Fiorenzo G. Omenetto, Mitchell R. Zakin, and John A. Rogers. 2009. "Silicon electronics on silk as a path to bioresorbable, implantable devices." *Appl Phys Lett* 95: 133701-1-133701-3].

[26] Wang, Chung-Hwa, Chao-Ying Hsieh, and Jenn-Chang Hwang. 2011. "Flexible organic thin-film transistors with silk fibroin as the gate dielectric." *Adv Mater* 23:1630-34.

[27] Tao, Hu, David L. Kaplan, and Fiorenzo G. Omenetto. 2012. "Silk materials--a road to sustainable high technology." *Adv Mater* 24 (21): 2824-37. doi: 10.1002/adma.201104477. Epub 2012 May 2.

[28] Irimia-Vladu M., E. Glowacki, P. Troshin, G. Schwabegger, L. Leonat, D. Susarova, O. Krystal, M. Ullah, Y. Kanbur, M. Bodea, V. Razumov, H. Sitter, S. Bauer, and S. Sariciftci. 2012. "Indigo - A Natural Pigment for High Performance Ambipolar Organic Field Effect Transistors and Circuits." *Adv Mater,* 24 (3): 375-380.

[29] Hong, Chang K., and Richard P. Wool. 2005. "Development of a bio-based composite material from soybean oil and keratin fibers." *J Appl Polym Sci*, 95 (6):1524–1538.

[30] Jastrzebska, Maria J., Heikki Isotalo, Jari Paloheimo, and Henrik Stubb. 1996. "Electrical conductivity of synthetic DOPA-melanin polymer for different hydration states and temperatures." *J Biomat Sci Polym E* 7: 577-586.

[31] Meredith, Paul, and Tadeusz Sarna. 2006. "The physical and chemical properties of eumelanin." *Pigm Cell Res* 19:572-594.

[32] Ligonzo, T., M. Ambrico, V. Augelli, G. Perna, L. Schiavulli, M. A. Tamma, P. F. Biagi, A. Minafra, and V. Capozzi. 2009. "Electrical and optical properties of natural and synthetic melanin biopolymer." *J Non-Cryst Solids* 355:1221–1226.

[33] Bothma, Jacques P., Johannes de Boor, Ujjual Divakar, Paul E. Schwenn, and Paul Meredith. 2008. "Device-Quality Electrically Conducting Melanin Thin Films." *Adv Mater* 20: 3539 –3542.

[34] Ambrico, Marianna, Paolo F. Ambrico, Antonio Cardone, Teresa Ligonzo, Stefania R. Cicco, Rosa Di Mundo, Vincenzo Augelli, and Gianluca M. Farinola. 2011. "Melanin Layer on Silicon: an Attractive Structure for a Possible Exploitation in Bio-Polymer Based Metal–Insulator–Silicon Devices." *Adv Mater* 23: 3332-3336.

[35] Irimia-Vladu M., P. Troshin, M. Reisinger, L. Shmygleva, Y. Kanbur, G. Schwabegger, M. Bodea, R. Schwödiauer, A. Mumyatov, J. Fergus, V. Razumov, H. Sitter, S. Sariciftci, and S. Bauer. 2010. "Biocompatible and Biodegradable Materials for Organic Field Effect Transistors." *Adv Funct Mater* 20: 4069-4076.

[36] Głowacki, Eric D., Lucia Leonat, Gundula Voss, Marius Bodea, Zeynep Bozkurt, Mihai Irimia-Vladu, Siegfried Bauer, and Niyazi S. Sariciftci. 2011. "Natural and nature-inspired semiconductors for organic electronics." *Proc SPIE* 8118: 81180M-1- 81180M-10.

[37] Dou, Wei, Qing Wan, Jie Jiang, Liqiang Zhu, and Qing Zhang. 2012. "Junctionless dual-gate electrostatic modulation of self-aligned oxide channels by chitosan-based proton conductors." *Materials Sci* (cond-mat.mtrl-sci), arXiv:1205.1309v1

[38] Fink, Hans-Werner, and Christian Schonenberger. 1999. "Electrical conduction through DNA molecules." *Nature* 398:407-410.

[39] Güllü, Ömer, Murat Çankaya, Özlem Barış, and Abdulmecit Türüt. 2008. "DNA-modified indium phosphide Schottky device." *Appl Phys Lett* 92:212106-1-212106-3.

[40] Güllü, Ömer, M. Cankaya, Ö. Barış, M. Biber, H. Özdemir, M. Güllüce, and A. Turut. 2008. DNA-based organic-on-inorganic semiconductor Schottky structures." *Appl Surf Sci* 254:5175-5180.

[41] Güllü, O., M. Murat Çankaya, Özlem Barış, and Abdulmecit Türüt. 2008. "DNA-based organic-on-inorganic devices. Barrier enhancement and temperature issues." *Microelectron Eng* 85:2250-2255.

[42] Güllü, Ömer, and Abdulmecit Türüt. 2015. "Electronic parameters of MIS Schottky diodes with DNA biopolymer interlayer." *Materials Science-Poland* 33 (3), 2015: 593-600].

[43] Bettinger, C. J., J. P. Bruggeman, A. Misra, J.T. Borensteain, and R. Langer. 2009. "Biocompatibility of biodegradable semiconducting melanin films for nerve tissue engineering." *Biomaterials* 30: 3050-3057.

[44] Bernardus Mostert, A., Ben J. Powell, Ian R. Gentle, and Paul Meredith. 2012. "On the origin of electrical conductivity in the bio-electronic material melanin." *Appl Phys Lett* 100: 093701.

[45] Zhong, Chao, Yingxin Deng, Anita F. Roudsari, Adnan Kapetanovic, M. P. Anantram and Marco Rolandi. 2011. "A polysaccharide bioprotonic field-effect transistor." *Nature Comm* 2: article number 476.doi:10.1038/ncomms1489.

[46] Stephan, A. Manuel, and K. S. Nahm. 2006. "Review on composite polymer electrolytes for lithium batteries." *Polymer* 47:5952–5964.

[47] Pawlicka, Agnieszka, Marins Danczuk, Wladystaw Wieczorek, and Ewa Zygadlo-Monikowska. 2008. "Influence of plasticizer type on the properties of polymer electrolytes based on chitosan." *J Phys Chem A* 112: 8888-8895.

[48] Marcondes, R. F. M. S., P. S. D'Agostinia, J. Ferreira, E. M. Girotto, A. Pawlicka, and D. C. Dragunski. 2010. "Amylopectin-rich starch plasticized with glycerol for polymer electrolyte application." *Solid State Ionics* 181:586-591.

[49] Andrade, Juliana R., Ellen Raphael, and Agnieszka Pawlicka. 2009. "Plasticized pectin-based gel electrolytes." *Electrochim Acta* 54:6479-6483.

[50] Raphael, Ellen, César O. Avellaneda, Bruno Manzolli, and Agnieszka Pawlicka. 2010. "Agar-based films for application as polymer electrolytes." *Electrochim Acta* 55:1455-1459.

[51] Vieira, Diogo F., and Agnieszka Pawlicka. 2010. "Optimization of performances of gelatin/$LiBF_4$-based polymer electrolytes by plasticizing effects." *Electrochim Acta* 55:1489-1494.

[52] Lopes, L. V. S., D. C. Dragunski, A. Pawlicka, and J. P. Donoso. 2003. "Nuclear magnetic resonance and conductivity study of starch based polymer electrolytes." *Electrochim Acta* 48: 2021–2027.

[53] Ma, Xiaofei, Jiugao Yu, and Kang He. 2006. "Thermoplastic starch plasticized by glycerol as solid polymer electrolytes." *Macromol Mater Eng* 291:1407–1413.

[54] Avérous, Luc. 2004. "Biodegradable multiphase systems based on plasticized starch: A review." *J Macromol Sci – Polymer Rev* C44:231–274.

[55] Ning, Wang, Zhang Xingxiang, Liu Haihui, and Wang Jianping. 2009. "N, N-dimethylacetamide/ lithium chloride plasticized starch as solid biopolymer electrolytes." *Carbohyd Polym* 77:607–611.

[56] Fortunati, E., F. D'Angelo, S. Martino, A. Orlacchio, J. M. Kenny, and I. Armentano. 2011. "Carbon nanotubes and silver nanoparticles for multifunctional conductive biopolymer composites." *CARBON* 49:2370–2379.

[57] Molino, Paul J., Peter C. Innis, Michael J. Higgins, Robert M.I. Kapsa, and Gordon G. Wallace. 2015. "Influence of biopolymer loading on the physiochemical and electrochemical properties of inherently conducting polymer biomaterials." *Synthetic Met* 200: 40–47.

[58] Ner, Y., M.A. Invernale, J.G. Grote, J.A. Stuart, and G.A. Sotzing. 2010. "Facile chemical synthesis of DNA-doped PEDOT." *Synthetic Met* 160:351–353.

[59] Sun, Mingyun, Srdjan Pejanovic, and Jovan Mijovic. 2005. "Dynamics of deoxyribonucleic acid solutions as studied by dielectric relaxation spectroscopy and dynamic mechanical spectroscopy." *Macromolecules* 38:9854–9864, dx.doi.org/ 10.1021/ma051596j.

[60] Pawlicka, Agnieszka, Franciani Sentanin, Alessandra Firmino, James G. Grote, François Kajzar, and Ileana Rau. 2011. "Ionically conducting DNA-based membranes for electrochromic devices." *Synthetic Met* 161: 2329–2334.

[61] Firmino, A., J. G. Grote, F. Kajzar, J.C. M'Peko, and A. Pawlicka. 2011. "DNA-based ionic conducting membranes." *J Appl Phys* 110 (3): 033704-1-033704-5.

[62] Rau, Ileana, James G. Grote, Francois Kajzar, and Agnieszka Pawlicka. 2012. "DNA – novel nanomaterial for applications in photonics and in electronics." *C R Physique* 13:853–864.

[63] Harman, David G., Robert Gorkin III, Leo Stevens, Brianna Thompson, Klaudia Wagner, Bo Wenga, Johnson H. Y. Chung, Marc in het Panhuis, and Gordon G. Wallace. 2015. "Poly (3,4-ethylenedioxythiophene): dextran sulfate (PEDOT:DS) – A highly processable conductive organic biopolymer." *Acta Biomater* 14:33–42.

[64] Vizard, Douglas L., R. Allen White, and Allen T. Ansevin. 1978. "Comparison of theory to experiment for DNA thermal denaturation." *Nature* 275: 250–251. doi:10.1038/275250a0.

[65] Wang, Lili, Jonichi Yoshida, Naoya Ogata, Sono Sasaki and Tisato Kajiyama. 2001. "Self-assembled supramolecular films derived from marine deoxyribonucleic acid (DNA)–cationic surfactant complexes: Large-scale preparation and optical and thermal properties." *Chem Mater* 13:1273–1281.

[66] Watanuki, Amane, Junichi Yoshida, Soichi Kobayashi, Hiroharu Ikeda, and Naoya Ogata. 2005. "Optical and photochromic properties of spiropyran-doped marine-biopolymer DNA–surfactant complex films." *Proc. SPIE* 5724 *Organic Photonic Materials and Devices* VII:234–241. doi:10.1117/12.590105.

[67] Koyama, Tamami, Yutaka Kawabe, and Naoya Ogata. 2002. "Electroluminescence as a probe for electrical and optical properties of deoxyribonucleic acid." *Proc. SPIE* 4464, Organic Light-Emitting Materials and Devices V: 248–255 doi:10.1117/12.457483.

[68] Yoshida, Junichi, Lili Wang, Soichi Kobayashi, Gong-Jian Zhang, Hiroharu Ikeda, and Naoya Ogata. 2004. "Optical properties of photochromic-compound-doped marine-biopolymer DNA–surfactant complex films for switching applications." *Proc. SPIE* 5351, *Organic Photonic Materials and Devices* VI,: 260-265. doi:10.1117/12.528781.

[69] Heckman, Emily M., Joshua A. Hagen, Perry P. Yaney, James G. Grote, and F. Kenneth Hopkins. 2005. "Processing techniques for deoxyribonucleic acid: Biopolymer for photonics applications." *App. Phys Lett* 87: 211115 -1- 211115-6.

[70] Hagen, Joshua, James G. Grote, Naoya Ogata, John S. Zetts, Robert L. Nelson, D. Diggs, F. Hopkins, P. Yaney, and L. Dalton. 2004 "DNA photonics." *Proc. SPIE* 5351 *Organic Photonic Materials and Devices* VI: 77–86. doi:10.1117/12.533171.

[71] Grote, James 2008. "Biopolymer materials show promise for electronics and photonics applications." *SPIE Newsroom*, http://dx.doi.org/10.1117/2.1200 805.1082.

[72] Yaney, Perry P., Emily M. Heckman, Antonio Davis, Joshua Hagen, Carrie M. Bartsch, Guru Subramanyam, James Grote, and F. Kenneth Hopkins. 2006. "Characterization of NLO polymer materials for optical waveguide structures." Proc. SPIE 6117, Organic Photonic Materials and Devices VIII, 61170W:0W1–0W14 doi:10.1117/12.646398.

[73] Heckman, Emily M., James G. Grote, F. Kenneth Hopkins, and Perry P. Yaney. 2006. "Performance of an electro-optic waveguide modulator fabricated using a deoxyribonucleic-acid-based biopolymer." *Appl Phys Lett* 89: 1811161-181116-181116-3. http:// dx.doi.org/10.1063/1.2378400.

[74] Singh, Thokchom Birendra, Niyazi Serdar Sariciftci, and James G. Grote. 2009. "Bio-Organic Optoelectronic Devices Using DNA." *Adv Polym Sci* 223: 189-212.

[75] Yaney, Perry P., Emily M. Heckman, and James G. Grote. 2007. "Resistivity and electric-field poling behaviors of DNA-based polymers compared to selected non-DNA polymers." *Proc. SPIE* 6646, *Nanobiotronics:* 664601-664605.

[76] Hagen, A., W. Li, A. J. Steckl, and J. G. Grote. 2006. "Enhanced emission efficiency in organic light-emitting diodes using deoxyribonucleic acid complex as an electron blocking layer." *Appl Phys Lett* 88:171109-1-171109-3.

[77] Kobayashi, Norihisa, Sei Umemura, Kazuhiko Kusabuka, Takayuki Nakahira, and Hisamitsu Takashi, 2001. "An organic red-emitting diode with a water-soluble DNA-polyaniline complex containing Ru $(bpy)_3^{2+}$." *J Mater Chem* 11:1766-1768.

[78] Firmino, Alessandra, James G. Grote, François Kajzar, Ileana Rau, and Agnieszka Pawlicka. 2011. "Application of DNA in electrochromic cells with switchable transmission." *Nonl. Opt. Quant. Opt. Concepts in Modern Optics* 42:181–202.

[79] Mysliwiec, Jaroslaw, Lech Sznitko, Anna M. Sobolewska, Stanislaw Bartkiewicz, and Andrzej Miniewicz. 2010. "Lasing effect in a hybrid dye-doped biopolymer and photochromic polymer system." *Appl Phys Lett* 96:141106-1- 141106-3.

[80] Kawabe, Y., L. Wang, S. Horinouchi, and N. Ogata. 2000. "Amplified spontaneous emission from fluorescent dye-doped DNA–surfactant films." *Adv Mater* 12:1281–1283.

[81] Leonetti, Marco, Riccardo Sapienza, Marta Ibisate, Claudio Conti, and Cefe López. 2009. "Optical gain in DNA–DCM, for lasing in photonic materials." *Opt Lett* 34 (24): 3764–3766, dx.doi.org/10.1364/OL.34.003764.

[82] Sznitko, Lech, Jaroslaw Mysliwiec, Pawel Karpinski, Krystyna Palewska, Kacper Parafiniuk, Stanislaw Bartkiewicz, Ileana Rau, Francois Kajzar, and Andrzej Miniewicz. 2011. "Biopolymer based system doped with nonlinear optical dye as a medium for amplified spontaneous emission and lasing." *Appl Phys Lett* 99 (3):031107-1-31107-3.

[83] Kawabe, Yutaka, and Ken-Ichi Sakai. 2011. "DNA based solid-state dye lasers *Nonl Opt Quant Opt Concepts in Modern Optics* 42 (3/4): 273–282.

[84] Mysliwiec, Jaroslaw, Lech Sznitko, Stanislaw Bartkiewicz, Andrzej Miniewicz, Zacaria Essaidi, Francois Kajzar, and Bouchta Sahraoui, 2009. "Amplified spontaneous emission in the spiropyran-biopolymer based system." *Appl Phys Lett* 94: 241106-1-2411063.

[85] He, Guang S., Qingdong Zheng, Paras N. Prasad, James G. Grote, and Frank K. Hopkins, 2006. "Infrared two-photon-excited visible lasing from a DNA–surfactant–chromophore complex" *Opt Lett* 31 (3): 359–361.

[86] Mysliwiec, Jaroslaw, Marta Ziemienczuk, and Andrzej Miniewicz. 2011. "Pulsed laser induced birefringence switching in a biopolymer matrix containing azo-dye molecules" *Opt Mater* 33: 1382–1386.

[87] Hung, Yu-Chueh, Ting-Yu Lin, Wei-Ting Hsu, Yi-Wen Chiu, Yu-Sheng Wang, and Ljiljana Fruk. 2012. "Functional DNA biopolymers and nanocomposite for optoelectronic applications." *Opt Mater* 34:1208–1213.

[88] Cyprych, Konrad, Lech Sznitko, and Jaroslaw Mysliwiec. 2014. "Starch: Application of biopolymer in random lasing." *Organ Electron* 15:2218–2222.

[89] Manocchi, Amy K., Peter Domachuk, Fiorenzo G. Omenetto, and Hyunmin Yi. 2009. "Facile fabrication of gelatin-based biopolymeric optical waveguides." *Biotechnol Bioeng* 103 (4):725–733.

[90] Sun, Yun-Lu, Si-Ming Sun, Pan Wang, Wen-Fei Dong, Lei Zhang, Bin-Bin Xu, Qi-Dai Chen, Li-Min Tong, and Hong-Bo Sun. 2015. "Customization of protein single nanowires for optical biosensing." *Small* 11 (24):2869–2876.

[91] Tong, L. M., J. Y. Lou, R. R. Gattass, S. L. He, X. W. Chen, L. Liu, and E. Mazur. 2005. "Assembly of silica nanowires on silica aerogels for microphotonic devices." *Nano Lett* 5 (2): 259–262.

[92] Noh, Y. O., H. J. Lee, Y. H. Won, and M. C. Oh. 2006. "Polymer waveguide thermo-optics witches with −70 dB optical crosstalk." *Opt Commun* 258:18–22.

[93] Estevez, M. Carmen, Mar Alvarez, and Laura M. Lechuga. 2012. "Integrated optical devices for lab-on-a-chip biosensing applications." *Laser Photonics Rev.* 6 (4):463–487.

[94] Wang, Pan, Yipei Wang, and Limin Tong. 2013. "Functionalized polymer nanofibers: a versatile platform for manipulating light at the nanoscale." *Light: Sci Appl* 2: e102-, doi:10.1038/lsa.2013.58.

[95] Sun, Yun-Lu, W. F. Dong, L. G. Niu, Tong Jiang, Dong-Xu Liu, Lei Zhang, Ying-Shuai Wang, Qi-Dai Chen, Dong-Pyo Kim, and Hong-Bo Sun. 2014. "Protein-based soft micro-optics fabricated by femtosecondlaser direct writing." *Light: Sci. Appl.* 3: e129, doi:10.1038/lsa.2014.10.

[96] Sun, Si-Ming, Yun-Lu Sun, Bo-Yuan Zheng, Pan Wang, Zhi-Shan Hou,Wen-Fei Dong, Lei Zhang, Qi-Dai Chen, Li-Min Tong, and Hong-Bo Sun. 2016. "Protein-based Y-junction optical micro-splitters with environment-stimulus-actuated adjustments." *Sensor Actuat B: Chemical* 232:571–576.

[97] Lupan, O., L. Chow, G. Chaic A. Schulte, S. Park, O. Lopatiuk-Tirpak, L. Chernyak, and H. Heinrich. 2008. "Biopolymer-assisted self-assembly of ZnO nanoarchitectures from nanorods." *Superlattice Microst* 43:292–302.

[98] Pedrosa, Valber A., Tony Gnanaprakasa, Shankar Balasubramanian, Eric V. Olsen, Virginia A. Davis, and Aleksandr L. Simonian. 2009. "Electrochemical properties of interface formed by interlaced layers of DNA- and lysozyme-coated single-walled carbon nanotubes." *Electrochem Commun* 11:1401–1404.

[99] The European Community. 2000. "*The European Standard EN, 13432. Requirements for packaging recoverable through composting and, biodegradation.* Brussels: EC.

[100] The European Community. 2006. *The European Standard EN 14995. Plastics – evaluation of compostability – test scheme and specifications.* Brussels: EC.

[101] Li, Fengkui, and Richard C. Larock. 2005. "Synthesis, properties, and potential applications of novel thermosetting biopolymers from soybean and other natural oils." In: *Natural fibers, biopolymers, and biocomposites,* edited by Amar K. Mohanty, Manjusri Misra, and Lawrence T. Drzal, 727–50. Boca Raton: Taylor & Francis.

[102] Dicker, Michael P. M., Peter F. Duckworth, Anna B. Baker, Guillaume Francois, Mark K. Hazzard, and Paul M. Weaver. 2014. "Review Green composites: A review of material attributes and complementary Applications." *Compos Part A-Appl S* 56:280–289.

[103] Iji, Masatoshi, and Yukihiro Kiuchi. ITR. 2012. "Highly functional PLA composites used for electronic products." *Green Innovation Reseach* Laboratories NEC Corporation http://www.innovationtakesroot.com/~/media/ITR2012/2012/presentations/durables/01_Highly-Functional-PLA-Composites_Iji_pdf.pdf.

[104] http://www.nec.co.jp/press/en/0603/2001.html.

[105] "Green handset shipments to reach almost 400 million by 2017." http://www.cellular-news.com/story/Handsets/54325.php.

[106] Dumitriu, Severian (ed.) 2004. *Polysaccharides: Structural Diversity and Functional Versatility; 2nd ed.*, New York: Marcel Dekker.

[107] Zheng, Yingying, Jonathan Monty, and Robert J. Linhardt. 2015. "Polysaccharide-based nanocomposites and their applications." *Carbohyd Res* 405: 23–32.

[108] Pushparaj, V. L., M. M. Shaijumon, A. Kumar, S Murugesan, L. Ci, R. Vajtai, R. J. Linhardt, O. Nalamasu, and P. M. Ajayan 2007. "Flexible energy storage devices based on nanocomposite paper." *Proc Natl Acad Sci USA* 104(34):13574-13577.

[109] Winter Martin, and Ralph J. Brodd. 2004. "What are batteries, fuel cells, and supercapacitors?" *Chem Rev* 104:4245–69.

[110] Simon, Patrice, and Yury Gogotsi. 2008. "Materials for electrochemical capacitors." *Nature Materials* 7:845–54.

[111] Conte, Mario 2010. "Supercapacitors technical requirements for new applications." *Fuel Cells* 10 (5):806–818.

[112] Miyauchi, Minoru, Jianjun Miao, Trevor J. Simmons, Jong-Won Lee, Thomas V. Doherty, Jonathan S. Dordick, and Robert J. Linhardt. 2010. "Conductive Cable Fibers with Insulating surface Prepared by Coaxial Electrospinning of Multiwalled Nanotubes and Cellulose." *Biomacromolecules,* 11:2440–2445.

[113] Kim, Dong-Hun, Soo-Young Park, Junkyung Kim, and Min Park. 2010. *J Appl Polym Sci* 117 (6):3588–3594.

[114] Haile, Sossina M. 2003. "Fuel cell materials and components." *Acta Mater* 51: 5981-6000.

[115] Kerres, Jochen A. 2001. "Development of ionomer membranes for fuel cells." *J Membrane Sci* 185 (1): 3-27.

[116] Neburchilov, Vladimir, Jonathan Martin, Haijiang Wang, and Jiujun Zhang. 2007. "A review of polymer electrolyte membranes for direct methanol fuel cells." *J Power Sources* 169: 221-38.

[117] Argun, Avni A., J. Nathan Ashcraft, and Paula T. Hammond. 2008. "Highly conductive, methanol resistant polyelectrolyte multilayers." *Adv Mater* 20 (8):1539-1543.

[118] Ahmad, H., S. K. Kamarudin, U. A. Hasran, and W. R. W. Daud. 2010. "Overview of hybrid membranes for direct-methanol fuel-cell applications." *Int J Hydrogen Energ* 35 (5): 2160-2175.

[119] Hasani-Sadrabadi Mohammad M., Erfan Dashtimoghadam, Kaveh Sarikhani Fatemeh S. Majedi, and Ghader Khanbabaei. 2010. "Electrochemical investigation of sulfonated poly (ether ether ketone)/clay nanocomposite membranes for moderate temperature fuel cell applications." *J Power Sources* 195 (9):2450-2456.

[120] Hasani-Sadrabadi, Mohammad M., Erfan Dashtimoghadam, Fatemeh S. Majedi, Shahriar H. Emami, and Homayoun Moaddel. 2011. "A high-performance chitosan-based double layer proton exchange membrane with reduced methanol crossover." *Int J Hydrogen Energ* 36 (10): 6105-6111.

[121] Jiang S. P., Z. Liu, Z.Q. Tian. 2006. "Layer-by-layer self-assembly of composite polyelectrolyte/ Nafion membrane for direct methanol fuel cells." *Adv Mater* 18 (8):1068-1072.

[122] Hasani-Sadrabadi Mohammad M., Erfan Dashtimoghadam, Fatemeh S. Majedi, Kourosh Kabiri, Nassir Mokarram, Mehran Solati-Hashjin, and Homayoun Moaddel. 2010. "Novel high-performance nanohybrid polyelectrolyte membranes based on bio-functionalized montmoriloniote for fuel cells applications." *Chem Commun* 46: 6500–6502.

[123] Dashtimoghadam, Erfan, Mohammad M. Hasani-Sadrabadi, and Homayoun Moaddel. 2010. "Modification of chitosan biopolymer as a novel polyelectrolyte membrane for green power generation." *Polym Advan Technol* 21 (10):726-34.

[124] Smitha, B., S. Sridhar, and A. A. Khan. 2005. "Chitosan–sodium alginate polyion complexes as fuel cell membranes." *European Polymer Journal* 41:1859–1866.

[125] Wan, Ying, Katherine A. M. Creber, Brant Peppley, and V. Tam Bui. 2003. "Ionic conductivity of chitosan membranes." *Polymer* 44 (4): 1057-1065.

[126] Shi Yanqiao, Xinwei Wang, and Guanwen Chen. 1996. "Pervaporation characteristics and solution–diffusion behaviors through sodium alginate dense membrane." *J Appl Polym Sci* 61: 1387-1394.

[127] Ramírez-Salgado, Joel. 2007. "Study of basic biopolymer as proton membrane for fuel cell systems." *Electrochim Acta* 52: 3766–3778.

[128] Hasani-Sadrabadi, Mohammad M., Erfan Dashtimoghadam, Nassir Mokarram, Fatemeh S. Majedi, and Karl I. Jacob. 2012. "Triple-layer proton exchange membranes based on chitosan biopolymer with reduced methanol crossover for high-performance direct methanol fuel cells application." *Polymer* 53: 2643-2651.

[129] Ke, Dingning, Shilin Liu, Ke Dai, Jinping Zhou, Lina Zhang and Tianyou Peng. 2009. "CdS/Regenerated Cellulose Nanocomposite Films for Highly Efficient Photocatalytic H_2 Production under Visible Light Irradiation." *J Phys Chem C*, 113 (36):16021–16026 **DOI:** 10.1021/jp903378q.

[130] Yun, Ye Sun, Seung-Wan Song, Sang-Young Lee, Sang-Hern Kim, and Dong-Won Kim. 2010. "Lithium metal polymer cells assembled with gel polymer electrolytes containing ionic liquid." *Curr Appl Phys* 10: e97–e100.

[131] Idris, Nurul H., Md. Mokhlesur Rahman, Jia-Zhao Wang, and Hua-Kun Liu. 2012. "Microporous gel polymer electrolytes for lithium rechargeable battery application." *J Power Sources* 201: 294–300.

[132] Gauthier, M., A. Belanger, B. Kapfer, G. Vassort, and M. Armand. 1987. "Solid Polymer Electrolyte Lithium Batteries." In: *Polymer Electrolytes Reviews-1,* edited by J. R. MacCallum, and C. A. Vincent, 285-333. London: Elsevier Applied Science.

[133] Pai, Sung J., Young C. Bae, and Yang K. Sun. 2005. "Ionic conductivities of solid polymer electrolyte/salt systems for lithium secondary battery." *Polymer* 46 (9): 3111–3118.

[134] Martinelli, A., M. A. Navarra, A. Matic, S. Panero, P. Jacobsson, L. Börjesson, and B. Scrosati. 2005. "Structure and functionality of PVdF/PAN based, composite proton conducting membranes." *Electrochim Acta* 50 (19): 3992–3997.

[135] Choi, Nam-Soon, and Jung-Ki Park. 2001. "New polymer electrolytes based on PVC/PMMA blend for plastic lithium-ion batteries." *Electrochim Acta* 46:1453–1459.

[136] Wu, Feng, Ting Feng, Ying Bai, Chuan Wu, Lin Ye, and Zengguo Feng. 2009. "Preparation and characterization of solid polymer electrolytes based on PHEMO and PVDF-HFP." *Solid State Ionics* 180 (9-10):677–680.

[137] Yahya, M. Z. A., and A. K. Arof. 2003. "Effect of oleic acid plasticizer on chitosan–lithium acetate solid polymer electrolytes." *Eur Polym J* 39 (5): 897–902.

[138] Ng, L. S., and A. A. Mohamad. 2006. "Protonic battery based on a plasticized chitosan – NH_4NO_3 solid polymer electrolyte." *J Power Sources* 163: 382–385.

[139] Karuppasamy, K., C. Vijil Vani, A. Nichelson, S. Balakumar, and X. Sahaya Shajan. 2013. "Effect of nano-chitosan and succinonitrile on the AC ionic conductivity of plasticized nanocomposite solid polymer electrolytes (PNCSPE)." *AIP Conf. Proc.* 1536: 845–846.

[140] Fan, Li-Zhen, Xiao-Liang Wang, Fei Long, and Xun Wang. 2008. "Enhanced ionic conductivities in composite polymer electrolytes by using succinonitrile as a plasticizer" *Solid State Ionics* 179: 1772–1775.

[141] Nan, Ce-Wen, Lizhen Fan, Yuanhua Lin, and Qiang Cai. 2003. "Enhanced ionic conductivity of polymer electrolytes containing nanocomposite SiO_2 particles." *Phys Rev Lett* 91: 266104.

[142] Zuo, Xiang, Xiao-Min Liu, Feng Cai, Hui Yang, Xiao-Dong Shen, and Gao Liu. 2013. "Enhanced performance of a novel gel polymer electrolyte by dual plasticizers." *J Power Sources* 239: 111–121.

[143] Long, S., D. R. MacFarlane, and M. Forsyth. 2003. "Fast ion conduction in molecular plastic crystals." *Solid State Ionics* 161 (1-2):105–112.

[144] Taib, Nur Umira, and Nurul Hayati Idris. 2014. "Plastic crystal–solid biopolymer electrolytes for rechargeable lithium batteries." *J Membrane Sci* 468: 149–154.

[145] Guitton, J., C. Poinsignon, and J. Y. Sanchez. 1992. "All solid-state protonic batteries" In: *Proton Conductors: Solids, Membranes and Gels - Materials and Devices*, edited by Philippe Colomban, 539-550. New York: Cambridge University Press.

[146] Samsudin, A. S., J. J. P. Tham, and M. I. N. Isa. 2013. "Investigation of a Biodegradable Polymer Electrolytes Based on Carboxy Methylcellulose and Its Potential Application in Solid-State Batteries." *Adv Mat Res* 802: 99-103.

[147] Ma, Xiaofei, Jiugao Yu, Kang He, and Ning Wang. 2007. "The effects of different plasticizers on the properties of thermoplastic starch as solid polymer electrolytes." *Macromol Mater Eng* 292 (4): 503-510.

[148] Harun, N. I., R. M. Ali, A. M. M. Ali, and M. Z. A. Yahya. 2012. "Dielectric behaviour of cellulose acetate-based polymer electrolytes." *Ionics* 18 (6): 599-606.

[149] Samsudin, A. S., M. I. A. Aziz, M. I. N. Isa. 2012. "Natural Polymer Electrolyte System Based on Sago: Structural and Transport Behavior Characteristics." *Int J Polym Anal Ch* 17 (8): 600-607.

[150] Samsudin, A. S., H. M. Lai, and M. I. N. Isa. 2014. "Biopolymer Materials Based Carboxymethyl Cellulose as a Proton Conducting Biopolymer Electrolyte for Application in Rechargeable Proton Battery." *Electrochim Acta* 129: 1–13.

[151] Yang, Zhenguo, Jianlu Zhang, Michael C. W. Kintner-Meyer, Xiaochuan Lu, Daiwon Choi, John P. Lemmon, and Jun Liu. 2011. "Electrochemical energy storage for green grid." *Chem Rev* 111 (5): 3577-3613.

[152] Yang, Zhenguo, Jun Liu, Suresh Baskaran, Carl H. Imhoff, J. Jamie D. Holladay. 2010. "Enabling renewable energy - and the future grid - with advanced electricity storage." *JOM* 62 (9): 14-23.

[153] Ponrouch, A., D. Monti, A. Boschin, B. Steen, P. Johansson, and M. R. Palacín. 2015. "Nonaqueous electrolytes for sodium-ion batteries." *J Mater Chem A* 3: 22-42. DOI: 10.1039/C4TA04428B.

[154] Rudola, Ashish, Doron Aurbach, and Palani Balaya. 2014. "A new phenomenon in sodium batteries: Voltage step due to solvent interaction." *Electrochem Commun* 46:56-59.

[155] Stephan, A. Manuel. 2006. "Review on gel polymer electrolytes for lithium batteries." *Eur Polym J*42: 21–42.

[156] Ni'Mah, Yatim Lailun, Ming-Yao Cheng, Ju Hsiang Cheng, John Rick, and Bing-Joe Hwang. 2015. "Solid-state polymer nanocomposite electrolyte of TiO_2/PEO/$NaClO_4$ for sodium ion batteries." *J Power Sources* 278: 375-381.

[157] Francesca Colòa, Federico Bella, Jijeesh R. Nair, Matteo Destro, and Claudio Gerbaldi. 2015. "Cellulose-based novel hybrid polymer electrolytes for green and efficient Na-ion batteries." *Electrochim Acta* 174: 185–190.

[158] Kafy, Abdullahil, Kishor K. Sadasivuni, Hyun-Chan Kim, Asma Akther, and Jaehwan Kim. 2015. "Designing flexible energy and memory storage materials using cellulose modified graphene oxide nanocomposites." *Phys Chem Chem Phys* 17: 5923-5931, DOI: 10.1039/C4CP05921B.

[159] Yang, Hongxun, Miaoliang Huang, Jihuai Wu, Zhang Lan, Sancun Hao, and Jianming Lin. 2008. "The polymer gel electrolyte based on poly (methyl methacrylate) and its application in quasi-solid-state dye sensitized solar cells." *Mater Chem Phys* 110 (1):38–42.

[160] Choi, Myeon-Cheon, Youngkyoo Kim, and Chang-Sik Ha. 2008. "Polymers for flexible displays: From material selection to device applications." *Prog Polym Sci* 33: 581–630.

[161] Singh, Pramod K., Kang-Wook Kim, Ki-II Kim, Nam-Gyu Park, and Hee-Woo Rhee. 2008. "Nanocrystalline porous TiO_2 electrode with ionic liquid impregnated solid polymer electrolyte for dye sensitized solar cells." *J Nanosci Nanotechnol* 8: 5271–5274.

[162] Kaneko, Masao, Takayuki Hoshi, Yuuki Kaburagi, and Hirohito Ueno. 2004. "Solid type dye-sensitized solar cell using polysaccharide containing redox electrolyte solution." *Journal of Electroanal Chem* 572 (1) 21–27.

[163] Suzuki, Kazuharu, Makoto Yamaguchi, Mikio Kumagai, Nobuo Tanabe, and Shozo Yanagida. 2006. "Dye sensitized solar cells with ionic gel electrolytes prepared from imidazolium salts and agarose." *Comptes Rendus Chimie* 9 (5-6): 611–616.

[164] Nemoto, Junichi, Mami Sakata, Takayuki Hoshi, Hirohito Ueno, and Masao Kaneko. 2007. "All-plastic dye sensitized solar cell using a polysaccharide film containing excess redox electrolyte solution." *J Electroanal Chem* 599 (1): 23–30.

[165] Weijia, Weijia, Xueyi Guo, and Ying Yan. 2011. "Lithium iodide effect on the electrochemical behavior of agarose based polymer electrolyte for dye-sensitized solar cell." *Electrochim Acta* 56 (21) 7347–7351.

[166] Singh, Rahul, Nitin A. Jadhav, S. Majumder, B. Bhattacharya, and Pramod K. Singh. 2013. "Novel biopolymer gel electrolyte for dye-sensitized solar cell application." *Carbohyd Polym* 91: 682–685.

[167] Kumar, Manindra, Tuhina Tiwari, and Neelam Srivastava. 2012. "Electrical transport behaviour of biopolymer electrolyte system: Potato starch + ammonium iodide." *Carbohyd Polym* 88: 54–60.

[168] Khanmirzaei, Mohammad H., S. Ramesh, and K. Ramesh. 2015. "Hydroxypropyl Cellulose Based Non-Volatile Gel Polymer Electrolytes for Dye-Sensitized Solar Cell Applications using 1-methyl-3-propylimidazolium iodide ionic liquid." *Sci Rep* 5: 18056; doi: 10.1038/srep18056.

[169] Jianrong Chen, Yuqing Miao, He Nongyue, Wu Xiaohua, and Li Sijiao. 2004. "Nanotechnology and biosensors. *Biotechnol Adv* 22 (7): 505-518.

[170] Ripp, Steven, Melanie L. Diclaudio, and Gary S. Sayler. 2010. "Biosensors as environmental monitors" In *Environmental Microbiology*, edited by Ralph Mitchell and Ji-Dong Gu, 2nd edition, 213–233. NJ, USA: John Wiley & Sons.

[171] Pauliukaite, Rasa, Mariana E. Ghica, Orlando Fatibello-Filho, and Christopher M.A. Brett. 2010. "Electrochemical impedance studies of chitosan-modified electrodes for application in electrochemical sensors and biosensors." *Electrochim Acta* 55: 6239–6247.

[172] Sugunan, A., C. Thanachayanont, J. Dutta, and J. G. Hilborn. 2005. "Heavy-metal ion sensors using chitosan-capped gold nanoparticles." *Sci Technol Adv Mater* 6 (3–40): 335-340.

[173] Fen, Yap W., W. Mahmood Mat Yunus, Nor A. Yusof, Nurul S. Ishak, Nur A. Sheh Omar, and Afiq A. Zainudin. 2015. "Preparation, characterization and optical properties of ionophore doped chitosan biopolymer thin film and its potential application for sensing metal ion." *Optik* 126: 4688–4692.

[174] Kumar, Bijandra, Jean-François Feller, Mickaël Castro, and Jianbo Lu. 2010. "Conductive bio-Polymer nano-Composites (CPC): Chitosan-carbon nanotube transducers assembled via spray layer-by-layer for volatile organic compound sensing." *Talanta* 81: 908–915].

[175] Mathew, Manjusha', S. Sureshkumar, N. Sandhyarani. 2012. Synthesis and characterization of gold–chitosan nanocomposite and application of resultant nanocomposite in sensors." *Colloid Surface B* 93: 143–147.

[176] Darder, Margarita, Montserrat Colilla, and Eduardo Ruiz-Hitzky. 2003. "Biopolymer–clay nanocomposites based on chitosan intercalated in montmorillonite." *Chem Mater* 15 (20): 3774– 3780.

[177] Darder, Margarita, Montserrat Colilla, and Eduardo Ruiz-Hitzky. 2005. "Chitosan–clay nanocomposites: application as electrochemical sensors." *Appl Clay Sci* 28: 199– 208.

[178] Aranda, P., M. Darder, R. Fernández-Saavedra, M. López-Blanco, and E. Ruiz-Hitzky. 2006. "Relevance of polymer– and biopolymer–clay nanocomposites in electrochemical and electroanalytical applications." *Thin Solid Films* 495 (1-2): 104 – 112.

[179] Aldalbahi, Ali, Peter Feng, Norah Alhokbany, Tansir Ahamad, and Saad M. Alshehri. 2016. "Synthesis, characterization, and CH_4-sensing properties of conducting and magnetic biopolymer nano-composites." JECE, dx.doi.org/10.1016/j.jece.2016.05.028.

[180] Wei, Haoran, Katia Rodriguez, Scott Renneckar, and Peter J. Vikesland. 2014. "Environmental science and engineering applications of nanocellulose-based nanocomposites." *Environ Sci: Nano* 1: 302-316.

[181] Sadasivuni, Kishor K., Deepalekshmi Ponnamma, Hyun-U Ko, Hyun Chan Kim, Lindong Zhai, and Jaehwan Kim. 2016. "Flexible NO_2 sensors from renewable cellulose nanocrystals/iron oxide composites." *Sensor Actuat B-Chem* 233:633–638.

[182] Isaad Jalal, and Ahmida El Achari. 2011. "Colorimetric sensing of cyanide anions in aqueous media based on functional surface modification of natural cellulose materials." *Tetrahedron* 67:4939-47.

[183] Isaad, Jalal, Ahmida El Achari, and Fouad Malek. 2013. "Bio-polymer starch thin film sensors for low concentration detection of cyanide anions in water." *Dyes Pigments* 97: 134-140.

[184] Schultz, Sheldon, David R. Smith, Jack J. Mock, and David A. Schultz. 2000. "Single-target molecule detection with nonbleaching multicolor optical immunolabels." *Proc Natl Acad Sci USA* 97 (3): 996–1001.

[185] Pandey, Sadanand, Gopal K. Goswami, and Karuna K. Nanda. 2012. "Green synthesis of biopolymer–silver nanoparticle nanocomposite: An optical sensor for ammonia detection." *Int J Biol Macromol* 51: 583– 589.

[186] Tagad, Chandrakant K., Sreekantha R. Dugasani, Rohini Aiyer, Sungha Park, Atul Kulkarni, and Sushma Sabharwal. 2013. "Green synthesis of silver nanoparticles and their application for the development of optical fiber based hydrogen peroxide sensor." *Sensor Actuat B-Chem* 183: 144– 149.

[187] Darder, M., M. López -Blanco, P. Aranda, F. Leroux, and E. Ruiz-Hitzky. 2005. "Bio-Nanocomposites Based on Layered Double Hydroxides." *Chem Mater* 17 (8): 1969-1977.

[188] De Oliveira Farias, Emanuel A., Marianne C. dos Santos, Natália de A. Dionísio, Patrick V. Quelemes, José R. de Souza Almeida Leite, Peter Eaton, Durcilene A. da Silva, and Carla Eiras. 2015. "Layer-by-Layer films based on biopolymers extracted from red sea weeds and polyaniline for applications in electrochemical sensors of chromium VI." *Mater Sci Eng B* 200: 9–21.

[189] Kwon, Young-Wan, Chang-Hoon Lee, Dong-Hoon Choi, and Jung-Il Jin. 2009. "Materials science of DNA."*J Mater Chem 19*: 1353-1380.

[190] Kaushik, Ajeet, Pratima R. Solanki, Anees A. Ansari, Bansi D. Malhotra, and Sharif Ahmad. 2009. "Iron oxide-chitosan hybrid nanobiocomposite based nucleic acid sensor for pyrethroid detection." *Biochem Eng J* 46:132–140.

[191] Ezeonu, Chukwuma S., Richard Tagbo, Ephraim N. Anike, Obinna A. Oje, and Ikechukwu N. E. Onwurah. 2012. "Biotechnological Tools for Environmental Sustainability: Prospects and Challenges for Environments in Nigeria—A Standard Review." *Biotechnol Res Int* 2012: article ID 450802, 26 pp, doi:10.1155/2012/450802.

[192] Bagni, Graziana, Silvia Hernandez, Marco Mascini, Elena Sturchio, Priscilla Boccia, and Simona Marconi 2005. "DNA biosensor for rapid detection of genotoxic compounds in soil samples," *Sensors* 5: 394–410.

[193] Sassolas, Audrey, Béatrice D. Leca-Bouvier, and Loïc J. Blum. 2008. "DNA biosensors and microarrays." *Chem Rev* 108 (1): 109–139.

[194] Xiao, Yi, Aaron A. Rowe, and Kevin W. Plaxco. 2007. "Electrochemical detection of parts-per-billion lead via an electrode-bound DNAzyme assembly." *J Am Chem Soc* 129 (2): 262–263.

[195] Wei, Hui, Bingling Li, Jing Li, Shaojun Dong, and Erkang Wang. 2008. "DNAzyme-based colorimetric sensing of lead (Pb^{2+}) using unmodified gold nanoparticle probes." *Nanotechnology* 19 (9): article ID 095501, 5 pp, doi:10.1088/0957-4484/19/9/095501.

[196] Green, Nathaniel S., and Michael L. Norton. 2015. "Interactions of DNA with graphene and sensing applications of graphene field-effect transistor devices: A review." *Analyt Chim Acta* 853: 127–142.

[197] Lu, Ye, Brett R. Goldsmith, Nicholas J. Kybert, and A.T. Charlie Johnson, 2010. "DNA-decorated grapheme chemical sensor." *Appl Phys Lett* 97: 083107-1 - 083107-3, doi.org/10.1063/1.348312.

[198] Kybert, Nicholas J., Gang Hee Han, Mitchell B. Lerner, Eric N. Dattoli, Ali Esfandiar, and A. T. Charlie Johnson. 2014. "Scalable arrays of chemical vapor sensors based on DNA-decorated graphene." *Nano Res* 7 (1) 95–103. doi:10.1007/s12274-013-0376-9.

[199] An, Ji Hyun, Seon Joo Park, Oh Seok Kwon, Joonwon Bae, and Jyongsik Jang. 2013. "High-performance flexible graphene aptasensor for mercury detection in mussels." *ACS Nano* 7 (12) 10563–10571.

[200] Kaplan, David L., Fiorenzo Omenetto, Brian Lawrence, Mark Cronin-Golomb, and Irene Georgakoudi. 2013. "Biopolymer sensor and method of manufacturing the same" US 8,529,835 B2, Sep. 10, 25 pp.

In: Modified Biopolymers: Challenges and Opportunities ISBN: 978-1-53612-116-2
Editors: D. Pathania, G. Sharma and A. Kumar

Chapter 2

BIOPOLYMERS FOR *IN VIVO* AND *IN VITRO* CONTROLLED DRUG DELIVERY

***Sangeeta Kumari*[1] *and Raj Pal Singh*[2,*]**
[1]Pharmaceutical and Molecular Biotechnology Research Centre, Department of Science, Waterford Institute of Technology, Waterford, Ireland
[2]Advanced Research Centre in Pharmaceutical Sciences and Applied Chemistry, Bharati Vidyapeeth University, Erandawane, Pune, India

ABSTRACT

Biopolymers are polymeric biomolecules. They are produced by living organism. Biopolymers are nontoxic and biocompatible in nature, e.g., chitosan, gelatin, alginate, etc. Due to these properties biopolymers, have received much attention in the field of biomedical. Biopolymers are potential carriers for drug delivery. Modification of biopolymer made it more efficiently applicable in the field of biomedical. Modified biopolymers are used as support materials in the field of tissue engineering, gene delivery, drug delivery and for cell culture. Controlled drug release system has been developed to improve the drug release in the body. This system is designed to enhance the drug therapy. It also protects the drug from premature elimination or physiological degradation. This system carry drug to specific site of action and minimize drug exposure in other parts of body. Modified biopolymers are widely used in the field of *in vitro* and *in vivo* controlled drug release. This system brings the new advancement in drug delivery system which improves the therapy and treatment.

Keywords: biopolymer, biomedicine, drug delivery, tissue engineering

---

* Corresponding Author Address: Advanced Research Centre in Pharmaceutical Sciences & Applied Chemistry, Bharati Vidyapeeth University, Erandawane, Pune – 411038, Maharashtra, India. Tel.: + 91-20-25437237, Fax: + 91-20-25437237. E-mail: rp.singh.ncl@gmail.com.

1. Introduction

Biopolymers are one of the know area of research in recent years. Basically, biopolymers are polymeric biomolecule. At structural level biopolymers have monomeric units, which are linked to one another by covalent bond and form larger structure. The biopolymers mainly occur in nature and obtained from living organism [1]. These biopolymers could be polysaccharide or polypeptide and polynucleotide. Biopolymers have attracted the attention of scientist worldwide due to its biocompatible and non-toxic behavior [2]. Due to its nontoxic and biocompatible nature, they have wide application in, in *in vitro* [3] and *in vivo* studies [4]. This chapter is mainly focused upon biopolymers: chitosan, alginate and gelatin.

Chitin is naturally occurring mucopolysaccharide obtained from the cell wall of fungi and crustaceans shells, etc. [5]. Chitin is similar to cellulose, like cellulose the hydroxyl group at C-2 position is replace with acetamide group [5]. Chemically chitin is long chain polymer of N-acetylglucosamine (Figure 1).

Chitosan is obtained by deacetylation of chitin. It's a biopolymer. It's a linear polysaccharide of β-(1→4)-linked D-glucosamine (deacetylated unit) and N-acetyl-D-glucosamine (acetylated unit) (Figure 2). [6] The biopolymer chitosan is cationic polysaccharide. Due to its polycationic property, it has brought attention of the researcher for its application in biomedical field. Chitosan has various efficient properties such as mucoadhesive, pharmacological, biological properties including, low immunogenicity, non-toxicity, biodegradability and biocompatibility [7].

The other versatile biopolymer is alginate. They are natural polysaccharide polymers extracted from brown seaweed [8]. Structurally alginate is linear polymer. In alginate D-mannuronic acid and L-guluronic acid are arranged in polymer chain in blocks (Figure 3) [8]. Alginate have wide area of application one of which is gel formation. Alginate have shown great promises in the area of gel formation [9]. Alginate show efficient film forming properties.

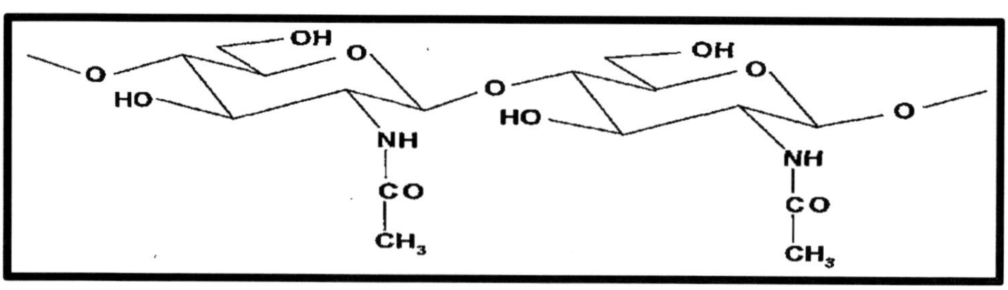

Figure 1. Structure of chitin.

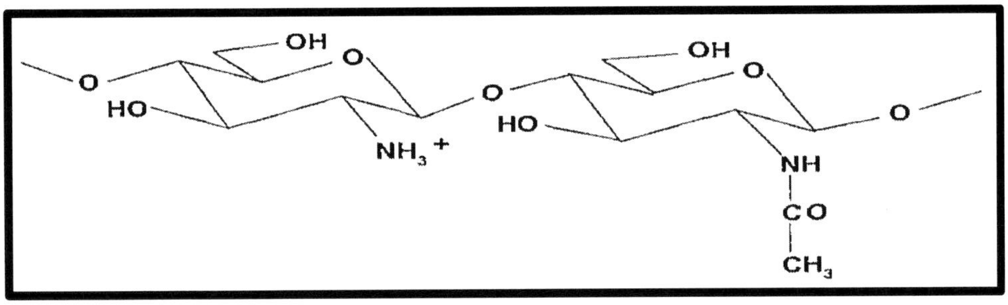

Figure 2. Structure of chitosan.

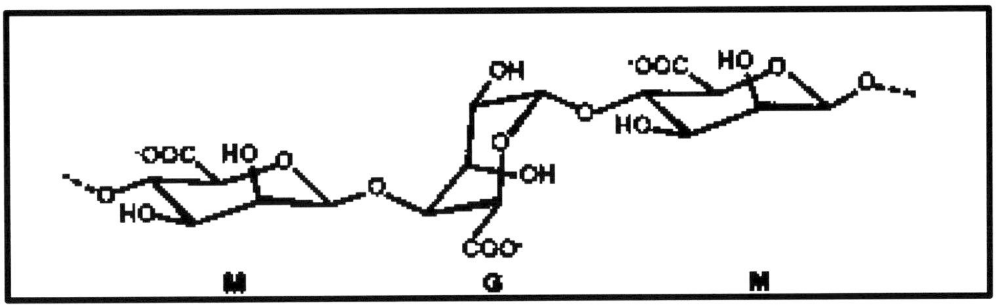

Figure 3. Structure of alginate.

Gelatin is obtained from collagen derived from animal raw material mainly collagen extracted from skin, bone, connective tissue, etc. It has wide application in food, cosmetics, photography, etc. Gelatin has unique cell adhesive structure. It is having high biocompatibility, low immunogenicity, biodegradability [10]. Due to these properties gelatin has grabbed the attention of researcher to apply it in the field of biomedical. These properties made gelatin a suitable candidate for the development of biomaterials for drug delivery and tissue engineering.

1.1. Modification of Biopolymers

Modification of biopolymers increases the performance of biopolymers. Modified biopolymers have much potential then the unmodified biopolymer. It has many advantages and applications in the various field. Therefore recently, the modification of biopolymer is the prime focus for scientific research.

1.2. Modification of Chitosan

The properties of the chitosan can be enhanced with modification. The structure of chitosan shows that it has hydroxy and amino functional groups. The modification of chitosan mainly involves these reactive hydroxyl and amino groups in reaction [11]. Chitosan undergo chemical modification by various reactions such as hydroxy alkylation [12] phosphorylation, [13] enzymatic modification, [14] acylation, [15] sulfation, [16] quaternization, [17] graft polymerisation, [18] thiolation, [19] oligomerisation [20] and alkylation [21] (Figure 4). Modification of chitosan does not change its physiochemical, biochemical properties leads to bring enhanced properties. The properties of chitosan are enhanced by the addition of nanofillers such as nanoparticles. Various chitosan based nanocomposites has been prepared with optical, [22] thermal, [23] electrical [24] and mechanical properties [25]. Modified chitosan nanoparticle based nanocomposites has used for drug delivery applications [26].

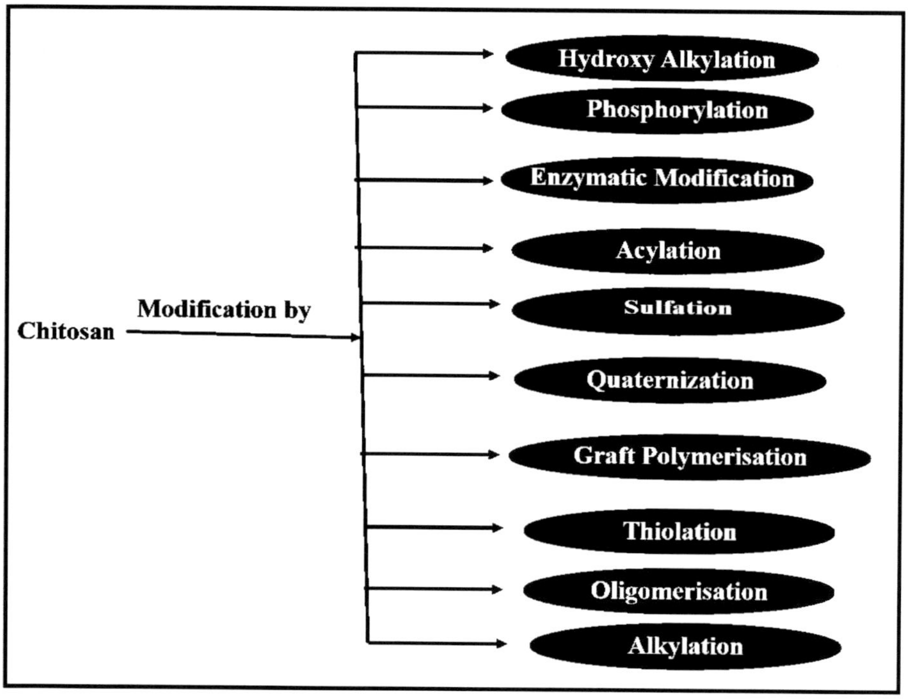

Figure 4. Schematic representation of list of chitosan modification reaction.

1.3. Modification of Alginate

Alginate modification can be performed with various methods. Alginates have hydroxyl and carboxylic group in its molecular structure. These functional groups provide the scope for the modification of alginate. It has two hydroxyl groups at C-2 an C-3 positions and one carboxylic group at C-6 position. Modification of alginate can be achieved by various chemical reaction and process such as: Graft polymerisation, acetylation, sulfation, phosphorylation hydrophobic modification, covalent cross linking and cell signalling molecule (Figure 5) [27].

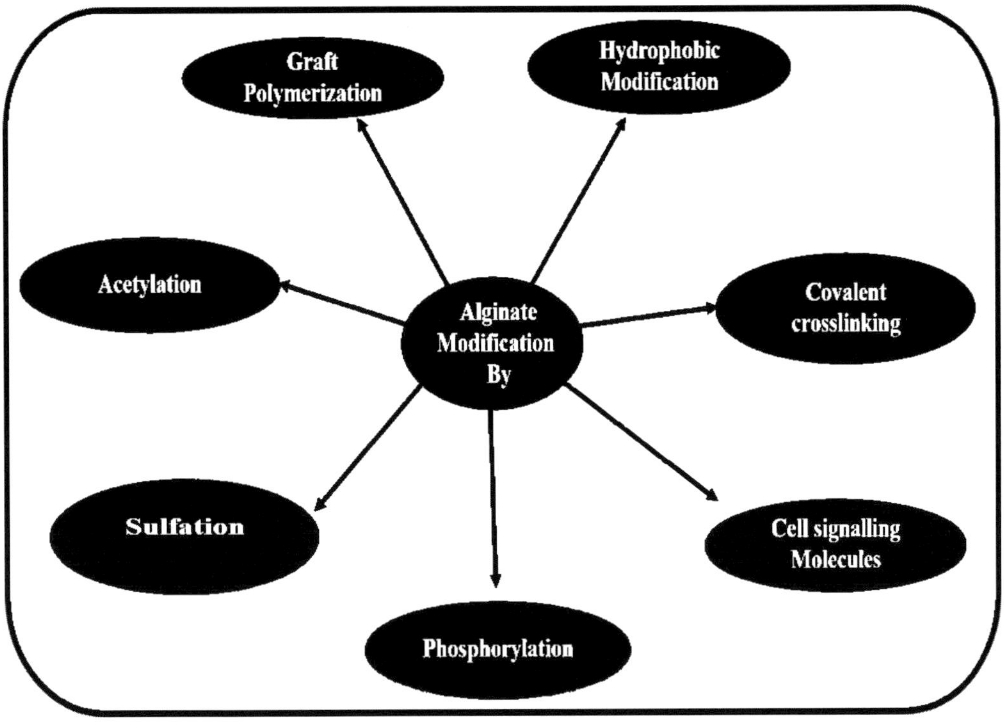

Figure 5. Schematic representation of alginate modification.

Modification of Alginate by various methods is discussed next.

1.3.1. Graft Polymerisation of Alginates

Graft polymerisation is one of important method for the modification of alginates. This method modifies the physical and chemical properties of alginates. Literature have reported various examples of graft polymerisation of alginates. Liu et al. have reported that to study the *in vitro* controlled drug release behaviour of BSA protein from hydrogel they grafted oxidised sodium alginate with poly((2-dimethylamino)ethyl methacrylate) (PDMAEMA) [28]. Aminabhavi et al. reported formation of membranes for pervaporation separation of water tetrahydrofuran mixtures by glutaraldehyde crosslinking with polyacrylamide grafted alginate [29]. Lee et al. synthesised temperature/pH responsive hydrogels upon grafting of poly(N-isopropylacrylamide) (PNIPAAm) on alginate [30].

1.3.2. Acetylation of Alginates

Modification of alginate can be obtained by acetylation. Various examples of acetylation of alginate has been reported in literature. Chamberlain et al. reported modification of alginate with acetylation of the hydroxyl group of alginate [31].

1.3.3. Sulfation of Alginates

Sulfation is another method for the modification of alginates. Yumin et al. by using chlorosulfonic acid in formamide [32] reported the sulfation of sodium alginate. Synthesis of alginate sulfates by using carbodiimide coupling chemistry is reported by Cohen et al. [33].

1.3.4. Phosphorylation of Alginates

Phosphorylation plays an important role in modification of alginate. Recently it has been described that synthesis of derivative of phosphorylated alginate has ability to induce hydroxyapatite (HAP) nucleation and growth [34].

1.3.5. Hydrophobic Modification of Alginates

In alginate, due to the presence of hydroxyl group it's hydrophilic in nature. They are considered to be hydrophilic polysaccharide. Hydrophobic modification of sodium alginate is attempted by Hubert et al. by covalent attachment of short polyether chains to the alginate [35].

1.3.6. Covalent Cross Linking of Alginates

Covalent Cross linking of alginate lead to yields of alginate matrices with improved mechanical properties. Cascone et al. reported that chemical crosslinking of alginates lead to making of beads for affinity and ion exchange chromatography [36].

1.3.7. Modification by Cell Signalling Molecule

Alginates have wide application in the area of bioengineering. Alginates are modified with cell-specific ligands molecules to enhance the chemical interactions of alginate matrices with cells. Skjåk-Bræk et al. reported that for the enhancement of hepatocyte cell, galactose moieties are functionalised to the alginate [37].

1.3.8. Modification of Gelation

Gelatin has wide application in the area of food and cosmatics. Its application is widely explored in the area of biomedical and pharmaceuticals. Functional group in gelatin allows its chemical modification. Chemical modification is obtained by target ligands and coupling with cross linkers [38].

1.4. Application of Biopolymer

Biopolymer has shown extensive application in the field of drug delivery. Applications biopolymer chitosan, Alginate and gelatin is discussed next.

1.5. Application of Chitosan

Chitosan is one of the important natural biopolymer which has been used widely in the field of drug delivery. Due to its biocompatibility and biodegradability it has been efficiently used as a carrier for *in vitro* and *in vivo* drug delivery. Chitosan is polycationic in nature due to cationic charge, it shows feature interaction with mucus layer. Interaction of chitosan to the mucus layer facilitate the interaction of drug with the mucus layer of the various epithelial cells [39, 40]. Chitosan is efficient to widen the tight junction of epithelial cells, it is observed in *in vitro* as well *in vivo* conditions [41, 42]. Chitosan are mainly used as carriers in the form of nanoparticles/macro particles [43-45]. Chitosan as career for drug delivery has been used in cancer therapeutics [46]. Chitosan is safe to use for the *in vivo* application as it shows biodegradability. Due to biodegradability chitosan prevents its accumulation in the body. In body chitosan of high molecular weight is degraded into fragments and later undergo renal clearance [47, 48]. Kidney clears directly, chitosan of certain molecular weight.

1.6. Application of Alginate

Alginates have wide application in drug delivery. Alginic acid, sodium/potassium alginates are extensively explored in the field of biomaterial. These alginates are mainly explored for mucoadhesive biomaterials. They show very efficient biocompatibility, chemical versatility and biodegradability [49]. Alginates have shown efficient application in the area of oral drug delivery. Urbanska et al. reported that chitosan-coated alginate microparticles are used for oral administration in colorectal cancer [50]. Mladenovska and collaborators reported synthesis of coated alginate microparticles loaded with drug for its application in inflammatory bowel disease [51]. Anticancer drug is normally toxic in nature and harmful to healthy cells. So, these drug are normally delivered by using nanoparticles. Li et al. [52] synthesised alginate-chitosan nanoparticles. Encapsulation of drug in these nanoparticles protects prevent the drug from enzymatic degradation and deliver the encapsulated drug to the target site. This lead to controlled release of the drug. Many studies have focused on the preparation of alginate-chitosan nanoparticles containing various anticancer drugs and targeting different types of cancer cells [53].

1.7. Application of Gelatin

Recently gelatin has been used as potential complexation based carriers. Gelatin show potential biodegradability and biocompatibility in physiological environment [54]. Gelatin has very low antigenicity in digestive system and upon degradation it forms harmless metabolic products [55]. It has been reported that the biological performance of gelatin is improved due to the presence of amino acid sequences in the structure [56]. Gelatin has shown wide range of applications such as cancer therapy, [57, 58] angiogenesis therapeutics, [59] tissue engineering [60, 61]. It has been exceptionally adapted as a carrier for drug delivery. Gelatin is modified with poly (ethylene glycol) has extensive application in the area of stealth drug delivery. In this it would be having prolong circulatory time and slower clearance time [62]. Kommareddy and Amiji, has developed PEGylated-thiolatedgelatin for DNA delivery [63]. The importance of thiolation is to stabilize the nanoparticles in bloodstream [63].

CONCLUSION

Biopolymers are biocompatible, biodegradable and nontoxic. They have wide application in the field of biomaterials. They have shown promising results in the area of *in vitro* and *in vivo* drug delivery. The biopolymer discussed in this chapter chitosan, alginate and gelatin are well known area of research and attracted the attention of scientist worldwide. Due their biocompatibility and biodegradability they have wide application in the area of biomedical. Chemical modification of chitosan, alginate and gelatin have shown enhancement in the properties. Various methods have been discussed for the modification of these biopolymer. This chapter concluded that these biopolymers have shown efficient application in *in vitro* and *in vivo* drug delivery. Drug delivery is well known area in biomedical which lead to treatment of many diseases. These biopolymers normally act as carrier for the drugs and deliver the drug *in vitro* and *in vivo*. It has shown that these biopolymers shown promising applications in the field of drug delivery.

REFERENCES

[1] Elnashar. M. M. 2010. *Biopolymers* 632.

[2] Singh A.V. 2011. “Biopolymers in drug delivery: A review.” *Pharmacology online* 1: 666-74.

[3] Kumari, S., and Singh, R. P. (2013). “Glycolic acid-functionalized chitosan-Co3O4-Fe3O4 hybrid magnetic nanoparticles-based nanohybrid scaffolds for drug-delivery and tissue engineering.” *J. Mater. Sci, 48*, 1524-1532.

[4] Croisier, F., & Jérôme, C., 2013. Chitosan-based biomaterials for tissue engineering. *Eur. Pol. J. 49*(4), 780-792.

[5] Kumar, M. N. R., (2000). A review of chitin and chitosan applications. *React. Funct. Polym, 46*(1), 1-27.

[6] Kumari, S., Singh, R. P., Chavan, N. N., & Annamalai, P. K. (2016). Chitosan-based bionanocomposites for biomedical application. *Bioinspir. Biomim*, 1-9.

[7] Kumari, S., & Singh, R. P., (2013). Glycolic acid-functionalized chitosan-Co3O4-Fe3O4 hybrid magnetic nanoparticles-based nanohybrid scaffolds for drug-delivery and tissue engineering. *J. Mater. Sci*, *48*(4), 1524-1532.

[8] Tønnesen, H. H., & Karlsen, J., (2002). Alginate in drug delivery systems. *Drug Dev. Ind. Pharm.*

[9] Lee, K. Y., & Mooney, D. J., (2012). Alginate properties and biomedical applications. *Pro. Poly. Sci. 37*(1), 106-126.

[10] Santoro, M., Tatara, A. M., & Mikos, A. G., (2014). Gelatin carriers for drug and cell delivery in tissue engineering. *J. Cont. re. 190*, 210-218.

[11] Kumari, S., & Singh, R. P., (2012). Glycolic acid-g-chitosan-gold nanoflower nanocomposite scaffolds for drug delivery and tissue engineering. *Int. J. Biol. Macromol. 50*(3), 878-883.

[12] Aranaz, I., Harris, R., & Heras, A., (2010). Chitosan amphiphilic derivatives. Chemistry and applications. *Curr. Org. Chem. 14*(3), 308.

[13] Amaral, I. F., Granja, P. L., & Barbosa, M. A., (2005). Chemical modification of chitosan by phosphorylation: an XPS, FT-IR and SEM study. *J. Biomater. Sci., Polym. Ed. 16*(12), 1575-1593.

[14] Sashiwa, H., & Aiba, S. I., (2004). Chemically modified chitin and chitosan as biomaterials. *Prog. Polym. Sci. 29*(9), 887-908.

[15] Hirano, S., Ohe, Y., & Ono, H., (1976). Selective N-acylation of chitosan. *Carbohydr. Res. 47*(2), 315-320.

[16] Jayakumar, R., Nwe, N., Tokura, S., & Tamura, H., (2007). Sulfated chitin and chitosan as novel biomaterials. *Int. J. boil. Macromol. 40*(3), 175-181.

[17] Tan, H., Ma, R., Lin, C., Liu, Z., & Tang, T., (2013). Quaternized chitosan as an antimicrobial agent: antimicrobial activity, mechanism of action and biomedical applications in orthopedics. *Int. J. mol. Sci. 14*(1), 1854-1869.

[18] Jayakumar, R., Prabaharan, M., Reis, R. L., & Mano, J., (2005). Graft copolymerized chitosan - present status and applications. *Carbohydr. Polym. 62*(2), 142-158.

[19] Chen, M. C., Mi, F. L., Liao, Z. X., & Sung, H. W. (2011). *Chitosan: its applications in drug-eluting devices*. In Chitosan for Biomaterials I (pp. 185-230). Springer Berlin Heidelberg.

[20] Jeon, Y. J., Shahidi, F., & KIM, S. K., (2000). Preparation of chitin and chitosan oligomers and their applications in physiological functional foods. *Food Rev. Int. 16*(2), 159-176.

[21] Bobu, E., Nicu, R., Lupei, M., Ciolacu, F. L., & Desbrières, J. (2011). Synthesis and characterization of n-alkyl chitosan for papermaking applications. *Cellul. Chem. Technol. 45*(9), 619.

[22] Wang, H., Xu, P., Zhong, W., Shen, L., & Du, Q., (2005). Transparent poly (methyl methacrylate)/silica/zirconia nanocomposites with excellent thermal stabilities. *Poly. Deg. Stab. 87*(2), 319-327.

[23] Paul, D. R., & Robeson, L. M., (2008). Polymer nanotechnology: nanocomposites. *Polymer*, *49*(15), 3187-3204.

[24] Mo T. C., H. W. Wang., S. Y. Chen., Y. C. Yeh., (2008). "Synthesis and dielectric properties of polyaniline/titanium dioxide nanocomposites." *Ceram. Int. (*34), 1767-71.

[25] Sarwar, M. I., Zulfiqar, S., & Ahmad, Z., (2008). Polyamide–silica nanocomposites: mechanical, morphological and thermomechanical investigations. *Polym. Int. 57*(2), 292-296.

[26] Kumari, S., & Singh, R. P., (2012). Glycolic acid-g-chitosan-Pt-Fe3O4 nanoparticles nanohybrid scaffold for tissue engineering and drug delivery. *Int. J. Biol. Macromol. 51*(1), 76-82.

[27] Pawar, S. N., (2013). *Chemical Modification of Alginates in Organic Media.*

[28] Gao, C., Liu, M., Chen, S., Jin, S., & Chen, J., (2009). Preparation of oxidized sodium alginate-graft-poly((2-dimethylamino) ethyl methacrylate) gel beads and *in vitro* controlled release behavior of BSA. *Int. J. Pharm. 371*(1), 16-24.

[29] Toti, U. S., & Aminabhavi, T. M., (2004). Synthesis and characterization of polyacrylamide grafted sodium alginate membranes for pervaporation separation of water+ isopropanol mixtures. *J. App. Polym. Sci. 92*(3), 2030-2037.

[30] Kim, J. H., Lee, S. B., Kim, S. J., & Lee, Y. M., (2002). Rapid temperature/pH response of porous alginate-g-poly(N-isopropylacrylamide) hydrogels. *Polymer*, *43* (26), 7549-7558.

[31] Chamberlain, N. H., Cunningham, G. E., & Speakman, J. B., (1946). Alginic acid diacetate. *Nature*, *158*, 553.

[32] Ronghua, H., Yumin, D., & Jianhong, Y., (2003). Preparation and *in vitro* anticoagulant activities of alginate sulfate and its quaterized derivatives. *Carbohydr. Polym. 52*(1), 19-24.

[33] Freeman, I., Kedem, A., & Cohen, S., (2008). The effect of sulfation of alginate hydrogels on the specific binding and controlled release of heparin-binding proteins. *Biomat. 29*(22), 3260-3268.

[34] Coleman, R. J., Lawrie, G., Lambert, L. K., Whittaker, M., Jack, K. S., & Grøndahl, L., (2011). Phosphorylation of alginate: synthesis, characterization, and evaluation of *in vitro* mineralization capacity. *Biomacro. 12*(4), 889-897.

[35] Carre, M. C., Delestre, C., Hubert, P., & Dellacherie, E., (1991). Covalent coupling of a short polyether on sodium alginate: synthesis and characterization of the resulting amphiphilic derivative. *Carbo. Polym. 16*(4), 367-379.

[36] Grasselli, M., Diaz, L. E., & Cascone, O., (1993). Beaded matrices from cross-linked alginate for affinity and ion exchange chromatography of proteins. *Biotechnol. Tech.*, *7*(10), 707-712.

[37] Donati, I., Vetere, A., Gamini, A., Skjåk-Bræk, G., Coslovi, A., Campa, C., & Paoletti, S. (2003). Galactose-substituted alginate: preliminary characterization and study of gelling properties. *Biomacrom. 4*(3), 624-631.

[38] Su, K., & Wang, C. (2015). Recent advances in the use of gelatin in biomedical research. *Biotechnol. Let. 37*(11), 2139-2145.

[39] Deacon, M. P., S. McGurk, C. J. Roberts, P. M., Williams, S. J. B., Tendler, Martyn C., Davies, S. S.. Bob Davis, and S. E. Harding., (2000). "Atomic force microscopy of gastric mucin and chitosan mucoadhesive systems." *Biochem. J.* 348: 557-563.

[40] Chopra, S., Mahdi, S., Kaur, J., Iqbal, Z., Talegaonkar, S., & Ahmad, F. J., (2006). Advances and potential applications of chitosan derivatives as mucoadhesive biomaterials in modern drug delivery. *J. Pharm. Pharmacol. 58*(8), 1021-1032.

[41] Phillips, W., Goins, B., & Bao, A., (2008). *Long-Circulating Liposomes with.* In Multi. Pharm. Nanocar. (pp. 431-456). Springer New York.

[42] Grenha, A., Al-Qadi, S., Seijo, B., & Remuñán-López, C., (2010). The potential of chitosan for pulmonary drug delivery. *J. D. Deliv. Sci. Technol. 20*(1), 33-43.

[43] Amidi, M., Mastrobattista, E., Jiskoot, W., and Hennink, W. E. (2010). "Chitosan-based delivery systems for protein therapeutics and antigens." *Advanc. Drug Deliv. Rev. 62*(1), 59-82.

[44] Park, J. H., Saravanakumar, G., Kim, K., and Kwon, I. C. (2010). "Targeted delivery of low molecular drugs using chitosan and its derivatives." *Advanc. Drug Deliv. Rev. 62*(1), 28-41.

[45] Chiellini, F., Piras, A. M., Errico, C., and Chiellini, E., (2008). "Micro/nanostructured polymeric systems for biomedical and pharmaceutical applications." *Nanomedicine. 3*(3), 367-393.

[46] Ghaz-Jahanian, M.A., Abbaspour-Aghdam, F., Anarjan, N., Berenjian, A. & Jafarizadeh- Malmiri, H., (2015). Application of chitosan-based nanocarriers in tumor-targeted drug delivery. *Mol. Biotech.* 57(3), 201-218.

[47] Braz, L., Rodrigues, S., Fonte, P., Grenha, A., & Sarmento, B., (2011). *Mechanisms of chemical and enzymatic chitosan biodegradability and its application on drug delivery.* Biodegradable Polymers: Processing, Degradation and Applications.

[48] Dash, M., Chiellini, F., Ottenbrite, R. M., and Chiellini, E., (2011). "Chitosan - A versatile semi-synthetic polymer in biomedical applications." *Progress in polymer science. 36*(8), 981-1014.

[49] Sosnik, A. (2014). "Alginate particles as platform for drug delivery by the oral route: state-of-the-art." *ISRN pharmaceutics.*

[50] Urbanska, A. M., Karagiannis, E. D., Guajardo, G., Langer, R. S., and Anderson, D. G., (2012). "Therapeutic effect of orally administered microencapsulated oxaliplatin for colorectal cancer." *Biomats. 33*(18), 4752-4761.

[51] Mladenovska, K., Cruaud, O., Richomme, P., Belamie, E., Raicki, R. S., Venier-Julienne, M. C., and Goracinova, K., (2007). "5-ASA loaded chitosan-Ca-alginate microparticles: Preparation and physicochemical characterization." *Int. J. Pharm. 345*(1), 59-69.

[52] Li, P., Dai, Y. N., Zhang, J. P., Wang, A. Q., and Wei, Q., (2008). "Chitosan-alginate nanoparticles as a novel drug delivery system for nifedipine." *I*nt. J. Biomed. Sci. *4*(3), 221-228.

[53] Das, R. K., Kasoju, N., & Bora, U., (2010). Encapsulation of curcumin in alginate-chitosan-pluronic composite nanoparticles for delivery to cancer cells. *Nanomedicine: Nanotechnol. Biol. Med. 6*(1), 153-160.

[54] Sanjoh, M., Hiki, S., Lee, Y., Oba, M., Miyata, K., Ishii, T., & Kataoka, K. (2010). pDNA/poly(L-lysine) Polyplexes Functionalized with a pH-Sensitive Charge-Conversional Poly(aspartamide) Derivative for Controlled Gene Delivery to Human Umbilical Vein Endothelial Cells. *Macromol. Rapid Commun. 31*(13), 1181-1186.

[55] Elzoghby, A. O., (2013). Gelatin-based nanoparticles as drug and gene delivery systems: reviewing three decades of research. *J. Controlled Release. 172*(3), 1075-1091.

[56] Wang, H., Boerman, O. C., Sariibrahimoglu, K., Li, Y., Jansen, J. A., and Leeuwenburgh, S. C., (2012). Comparison of micro-vs. Nanostructured colloidal gelatin gels for sustained delivery of osteogenic proteins: Bone morphogenetic protein-2 and alkaline phosphatase. *Biomats. 33*(33), 8695-8703.

[57] Madan, J., Pandey, R.S., Jain, U.K., Katare, O. P., Aneja, R., and Katyal, A., (2013). Sterically stabilized gelatin microassemblies of noscapine enhance cytotoxicity, apoptosis and drug delivery in lung cancer cells. *Colloids Surf. B Biointer. 107*, 235-244.

[58] Gaowa, A., Horibe, T., Kohno, M., Sato, K., Harada, H., Hiraoka, M., and Kawakami, K., (2014). "Combination of hybrid peptide with biodegradable gelatin hydrogel for controlled release and enhancement of anti-tumor activity *in vivo*." *J. Controlled Release. 176*, 1-7.

[59] Kimura, Y., and Tabata, Y., (2010). "Controlled release of stromal-cell-derived factor-1 from gelatin hydrogels enhances angiogenesis." *Biomater. Sci., Polym. Ed. 21*(1), 37-51.

[60] Wang, H., Zou, Q., Boerman, O. C., Nijhuis, A. W., Jansen, J. A., Li, Y., and Leeuwenburgh, S.C., (2013). "Combined delivery of BMP-2 and bFGF from nanostructured colloidal gelatin gels and its effect on bone regeneration *in vivo*." *J. Controlled Release. 166*(2), 172-181.

[61] Kim, K., Lam, J., Lu, S., Spicer, P. P., Lueckgen, A., Tabata, Y., and Kasper, F.K., (2013). "Osteochondral tissue regeneration using a bilayered composite hydrogel with modulating dual growth factor release kinetics in a rabbit model." *J. Controlled Release. 168*(2), 166-178.

[62] Knop, K., Hoogenboom, R., Fischer, D., and Schubert, U. S., (2010). "Poly (ethylene glycol) in drug delivery: pros and cons as well as potential alternatives." *Angew. Chem. Int. Ed. 49*(36), 6288-6308.

[63] Kommareddy, S., & Amiji, M. (2007). "Poly(ethylene glycol)-modified thiolated gelatin nanoparticles for glutathione-responsive intracellular DNA delivery. *Nanomedicine 3*(1), 32-42.

In: Modified Biopolymers: Challenges and Opportunities ISBN: 978-1-53612-116-2
Editors: D. Pathania, G. Sharma and A. Kumar

Chapter 3

REMOVAL OF HEAVY METAL IONS BY ADSORPTION THROUGH BIOPOLYMERS

Saruchi[1,*], Vaneet Kumar[2] and P.S. Bedi[2]
[1]Department of Biotechnology, DAV University, Jalandhar, India
[2]Chemistry Lab, Department of Applied Sciences,
CT Group of Institution Jalandhar, India

ABSTRACT

Effect of heavy metal ions toxicity on human health and its influence on environment pollution in recent years has attracted the attention of the researchers worldwide. Heavy metals contamination has increased public concerns because of their highly toxic nature in relatively low concentration, non-biodegradable and capacity of bioaccumulation. The increasing demand for the removal of these heavy metals ions from industrial effluents has stressed the development and the testing of new sorbing materials. Biopolymers based biocomposite have been extensively used in recent year for the removal of these heavy metal ions, because of their interactions with metal ions. Chitin, chitosan, cellulosics, seeweeds etc. based biopolymers are widely studied and reported for metal ion sorption. Chitin is a biopolymer and is the main structural component of crustaceans exoskeleton and fungi cell wall. Acetylglucosamine units are themain repeating unit of chitin. Glucosamine units are the repeating units in case of chitosan, chitosan is commercially synthesized by deacetylation of chitin. Chitosan have excellent properties for the uptake or removal of heavy metal ions because of the presence of both amine and hydroxyl groups in chitosan. Metal cation can be easily sorbed by chelation on hydroxyl and amine groups of chitosan in the neutral medium. Modified chitosan based biopolymer are reported for the removal of heavy metals like Pb, Hg, Cd, Cu, Zn, Ni, Cr etc. Sorption of these heavy metal ions is also affected by the degree of deacetylation and crystallinity of chitosan as well as swelling and diffusion properties of chitosan. Resuability and selectivity of

[*] Correspondence to: suruchinitj15@gmail.com; vanet2106@gmail.com.

biopolymer can be enhanced by modifying the natural biopolymer like chemical grafting and cross-linking. There are lots of natural and modified biopolymer or their composites, which are used for the removal of toxic heavy metals, this chapter is primarily concerned with the removal of heavy metal using biopolymers.

Keywords: heavy metal ions, biopolymer, chitin, chitosan, composites, gum tragacanth

1. Introduction

As the World population increases and civilized there is increase in the development of industrial applications, environmental pollution and it becomes more critical issue. The use of heavy metal started from the ancient time and now the water pollution through heavy metals attain greater attention all over the World due to their increased use in chemical industries. Heavy metal discharged from the different industries affect the quality of water and poor quality water is a great threat to human health. Some of the most common heavy metals which are discharged from different industries like textile, electroplating, paper, tanning and dental operation are mercury, chromium, nickel, copper, lead, cadmium, arsenic etc [1-3]. These toxic heavy metal generally found in high concentrations in landfill leachate and was potential source of water pollution due to their insolubility, non-biodegradability, accumulation and biomagnifications in the food chains. As these heavy metals accumulate in the living tissue throughout the food chain and human being is on top in food chain, so these heavy metals is a great threat to human health [4-6].

Table 1. The most hazardous heavy metals (As per the MCL standards)

Heavy metal	Toxic effects	MCL (mg/L)
Mercury	Nervous, circulatory system, rheumatoid arthritis, kidney, minamata disease	0.00003
Lead	Nervous, circulatory system, kidney and fetal brain damage	0.006
Arsenic	Visceral cancer, skin problem, vascular disease	0.050
Cadmium	Kidney, renal, carcinogenic	0.01
Nickel	Chronic asthma, carcinogenic, nausea, dermatitis	0.20
Chromium	Carcinogenic, diarrhea, nausea, headache	0.05
Zinc	Neurological problem, depression	0.80
Copper	Insomnia, liver effect, Wilson disease	0.25

The removal of toxic heavy metals present in wastewaters from industrial effluents, mine waters and water supplies has received much attention as its high concentration would endanger public health and the environment if discharged without adequate treatment.

Heavy metals are elements such as As (Arsenic), Cu (Copper), Cd (Cadmium), Hg (Mercury), Pb (Lead), Zn (Zinc), Ag (Silver), Cr (III) (Chromium), Fe (Iron), Co (Cobalt), Ni (Nickel), these are associated with toxicity and are natural components of the Earth's crust. They cannot be degraded or destroyed naturally. They enter in our bodies in very small amount via food, drinking water and air. Some of the heavy metals like copper, selenium, zinc etc. are essential in trace amount to maintain the metabolism of the human body [7-9]. However if they are present in higher concentrations they may lead to poisoning. The lists of most hazardous heavy metals are given in Table 1 [10].

2. Effect of Different Heavy Metals on Environment

2.1. Copper

The risks associated with various metals is given in Table 1. (Source: WEB_1 2007). The most widely distributed free metal in nature is copper and more commonly it is present in various ores in different form such as cuprite (Cu_2O), chalcopyrite ($CuS \cdot FeS$), azurite ($Cu(OH)_2 \cdot 2CuCO_3$) and malachite $Cu_2CO_3(OH)_2$. These are also present in deposits form like cupric chloride and cupric arsenide. The copper is mainly used in the production of alloys with zinc, nickel and tin, it is also used as a catalyst in different chemical industry, in electrochemical industry it is used in the form of wires, transformers, generators, heat exchanger and production of piping for water supply. Copper salts are used as fungicides and biocides for controlling slime in human and animal waste. Copper can form different complexes. Some of these organic complexes are essential to life especially haemocyanin [6, 7].

2.2. Cadmium

Cadmium mainly present in ores in combination with zinc, lead and copper. Cadmium compound is generally used as stabilizers in PVC products, several alloys, colour pigment and generally used in re-chargeable nickel-cadmium batteries. Emission of cadmium increased significantly during the 20th century, major reason of this is that cadmium-containing products are rarely re-cycled and are generally discarded together with household waste. The major source of cadmium exposure is cigarette smoking. In case of non-smokers, food is the main source of cadmium exposure. Recent study revealed that adverse health effects of cadmium exposure may occurred at lower exposure level than formerly expected, mainly it effects nearly all the system in the body including bones leading to cardiovascular, reproductive, kidneys, brains , eyes fractures. If an individual inhale fumes or particles of cadmium, it can be life threatening, it causes acute pulmonary effects, death are uncommon, sporadic cases still occur. Cadmium exposure damage the kidney and the first sign is dysfunctioning of tubular and its evidence is the excretion of large amount of low molecular weight protein or enzyme. High cadmium exposure for long term leads to skeletal damage. The itai itai disease was first reported in Japan in 1950, which was caused by using cadmium-contaminated water for irrigation of rice field. In Europe many individuals are exceed these exposure levels and the difference is very small for large groups. Therefore, to reduce the cadmium exposure to the general population is required in order to minimize the risk of adverse health effects [11-14].

2.3. Lead

Galena (PbS) is the main source for lead production, lead is rarely found as a free metal in nature. It is also found as cerrusite ($PbSO_3$) and anglesite ($PbSO_4$). It is commonly used as non-ferrous metals. There are lots of applications of lead due to its resistance behavior to corrosion and malleability and are used in building constructions, corrosive liquid container, making storage batteries which are generally recycled, radiation shielding, cable sheathing, pipework, ammunition, solder, pigments in paints and ceramics, antibacterial substances, catalysers, wood preservatives, anti-knock compounds for addition to petrol, particularly tetraethyl lead, $Pb(C_2H_5)_4$ etc. The major source of lead contamination is the exhaust from vehicle and it is released in the form of lead halides and oxides, but incomplete combustion release alkyl lead compounds. Iron, steel, nickel smelters and copper production are the other source of lead emission in the environment. Long-term-low-level lead exposure in children leads to diminished intellectual capacity. Acute exposure causes proximal renal tubular damage and long term exposure leads to kidney damage. The symptom of lead poisoning are sleeplessness, restlessness, irritability, headache, behavioural disturbances, learning and concentration problems, acute psychosis, confusion and reduced consciousness, memory deterioration, reduced ability to understand etc [15-17].

2.4. Arsenic

Arsenic is mainly present in soil, water, air, metalloid, rock etc. Groundwater which is used for drinking purpose in several countries contains inorganic arsenic, whereas organic arsenic compounds are found in fish, which is the main reason of human exposure to arsenic. Energy production from fossil fuel and smelting of non-ferrous metals are the main industrial processes, which causes arsenic contamination of air, soil and water. Wood preservative and pesticides are the other sources of contamination. Intake of large quantities of arsenic leads to cardio-vascular, central nervous system disturbances, gastrointestinal symptom, haemolysis, hepatomegaly, bone marrow depression, polyneuropathy and eventually death. Arsenic exposure through drinking water show risk of mortality from lungs, skin cancer, pigmentation changes, hyperkeratosis. It also causes chromosomal abnormalities, sister chromatid exchange etc [18-20].

2.5. Mercury

Cinnabar (HgS) the mercury compound was used in pre-historic cave painting due to its red colors, metallic mercury was used as a cosmetic to lighten the skin and in medical mercury amalgam used for filling teeth, diuretics and syphilis. Exposure to inorganic mercury gives rise to lung damage, depression, sleep disturbance, anxiety and restlessness. Acute exposure to organic mercury leads to nervous system damage. High dose of mercury may leads to death. The Methyl mercury poisoning from fish contaminated by mercury discharge near sea leads to minamata disease in Japan in 1950 [21, 22].

3. Different Methods Used for the Removal of Heavy Metals

The separation or removal of heavy metal ions from water is complicated by the number of variables like pH, presence of organic substances, solution composition, low concentration and complex mixture of the heavy metals that must be considered. There are different technique such as chemical precipitation, reverse osmosis, coagulation–flocculation, electro-dialysis flotation, ion exchange and ultra-filtration, which are used for the removal of heavy metal from Industrial Wastewater.

3.1. Chemical Precipitation

This is the most traditional and simple method of heavy metal removal. This method involves the use of different chemicals to change the physical state of dissolved and suspended solids and facilitate their removal by sedimentation. Chemical precipitation is employed for most of the metals. Common precipitants include hydroxide $(OH)^-$ and sulfide. Generally metals are precipitated as metal hydroxides by the addition of lime or caustic to a pH having minimum solubility.

This method is not effective for large solution volumes having very low concentrations of metal ions and it require large amount of chemicals to reduce metals to an acceptable level for discharge. Excessive sludge production also requires further treatment, which increases the sludge disposal cost, poor settling, the aggregation of metal precipitates and slow metal precipitation are the other drawback of this method [23-26].

Figure 1. Different techniques used for the removal of heavy metal from waste water.

3.2. Solvent Extraction

Solvent-extraction method is used for the removal of organometal from waste water. The extractant which is used for the removal organometal is capable of exchanging or forming chelates with the metal ion. Proper mixing with extractant leads to metal ion in organic phase and then phases are allowed to separate, this leads to the stripping of metal ion from the loaded organic phase. The solution having concentrated metal ions can be purified or disposed depending on the requirement. This method has lots of advantage like fast kinetics, high selectivity and capacities for metal ions but also have many disadvantages like finite aqueous solubility of the extractant, solvent and modifiers not only adds to the cost of the process through loss of reagents, but it contaminate the water with toxic organics compounds. This method is not appropriate for diluted metal ion solution due to the requirement of large volume of extractants [27].

3.3. Coagulation–Flocculation

The general principle of coagulation-flocculation process is to destabilize the colloidal particles by adding a coagulant, which results in sedimentation. Coagulation is followed by the flocculation of the unstable particles so as to increase the particle size into bulky floccules. Repulsive forces between the particles can be overcome by the addition of Ferric/alum salts. The main disadvantage of this method is its high operational cost due to chemical consumption and the sludge generated from coagulation–flocculation may hinder its absorption as a global strategy for waste water treatment [28, 29].

3.4. Reverse Osmosis

This technique is employed for the removal of dissolved constituents as well as recovery of precious metal from metal finishing industry. It is effective for treating wastewater containing single or mixture of metal ions. It works when two solutions with different solute concentrations are separated with the help of a semipermeable membrane, this leads to a chemical potential difference across the membrane. Thus, water will diffuse from the lower concentration gradient side to the higher concentration gradient through the membrane. The system with a finite volume, flow will continues until the pressure difference balances the chemical potential difference. As it is balanced by the pressure difference, it is termed as osmotic pressure. If the pressure gradient is in opposite direction and greater than the osmotic pressure, it is imposed across the membrane and it flow from the more higher to lower concentrated gradient and termed reverse osmosis. pH of the solution must be adjusted so as to protect the reverse osmosis membranes. Complete recovery and reuse of the solution will not achieve by reverse osmosis. Pretreatment like pH adjustment, media filtration, equalization and anti-precipitant additions is required prior to the reverse osmosis [30, 31].

3.5. Evaporation

This is the common methods used for the concentration of aqueous solutions. The use of this method for effluent treatment is rare and it is used where the effluent contains a high concentration of inert substances from electroplating operations, especially chromium plating. In this method the liquor is evaporated to a metal concentration, which makes the concentration for direct re-use in the plating bath [32].

3.6. Ultrafiltration

This technique used to reduce the amount of chemicals used for treatment, has lesser space and labor requirements. Disadvantages are that uses more electricity, required pretreatment, and membranes replacement [33, 34].

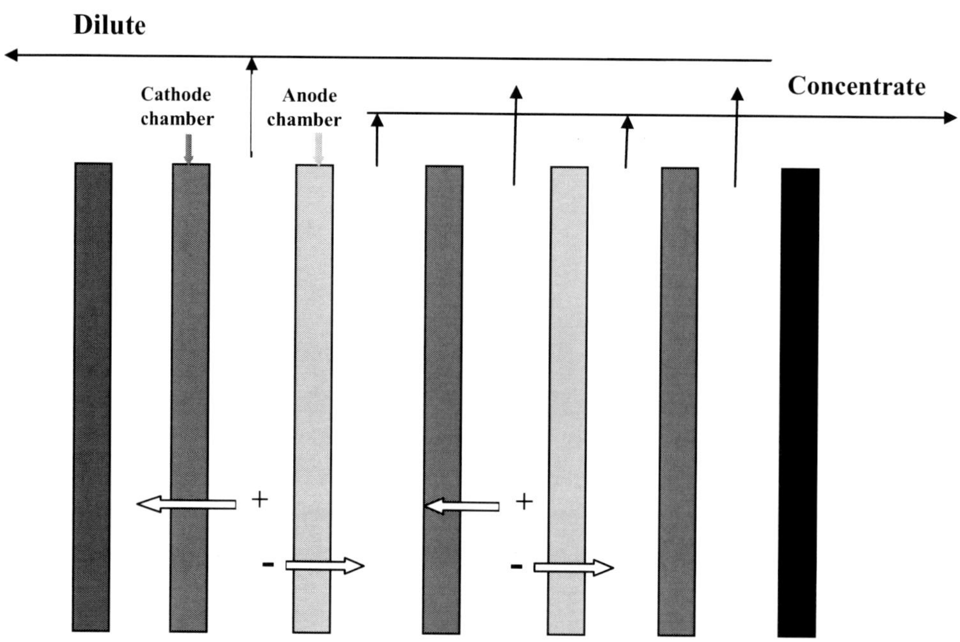

Figure 2. Schematic presentation of electro dialysis process.

3.7. Electrodialysis

In this process, solutions containing ionic components are separated with the help of ion-selective semipermeable membranes. This process takes place either in a continuous or a batch mode. The main drawbacks with the electrodialysis process is that wastewater treatment include chemical precipitation of salts having low solubility on the surface of the membrane is required. The membrane fouling is also a problem with this process. Activated carbon pretreatment is required to reduce the membrane fouling [35]. It has not been widely utilized in full-scale treatment of metal wastes. The schematic diagram of electro dialysis is given in Figure 2.

3.8. Flotation

This process is utilized to separate solids or dispersed liquids from a liquid phase with the help of bubble attachment. As the bubble rises the attached particles are separated from the suspension of heavy metal. Flotation may be vacuum air, dispersed-air, dissolved-air, electro and biological flotation. Out of all these flotation, dissolved-air flotation is the most commonly used method for metal-contaminated wastewater treatment. Foaming is employed to separate the metal impurities in adsorptive bubble separation. The contaminants are separated from wastewater in a foaming phase. It has the potential for industrial application although it is a physical separation process [36].

3.9. Ion Exchange

In the recent years, different types of ion exchanger have been utilized for the heavy metal removal through adsorption process. This is one of the most effective techniques for the removal of heavy matter from waste water. The different ion exchanger has different functional groups, which are capable of ion exchanging or complexing with different heavy metal ions. This process is extensively employed for the effluent polishing after primary treatment, metal finishing bath purification and of recovering precious metals. The main drawback of this process is the slow kinetics as compared to that of solvent extraction and most of the ion exchange resin are very expensive and cannot be discarded once they have been loaded with heavy metals [37-40].

3.10. Adsorption

As discussed above different conventional methods for the removal of heavy metal ions, adsorption is another effective and widely used method for heavy metal removal. This technique has a good potential for treatment of wastewater and industrial residues, because it is cost-effective, efficient and easy for application. There are different adsorption materials like activated carbons, polymeric synthetic resins, clays, metal oxides and some natural biopolymers, which are used for the removal of heavy metal.

Activated carbon is the oldest and most commonly used adsorbents for organic contaminants, but has been rarely used in the treatment of inorganic adsorbates. The activated carbon is used for the removal of $Cu^{2+,}$ Cd^{2+}, Zn^{2+}, Pb^{2+}, Hg^{2+}, Co^{2+}, Mn^{2+}, Ca^{2+}, Ni^{2+}, Cr^{3+} still has some disadvantages for removal of heavy metals. The adsorption capacity and kinetics for many metals like Mn^{2+}, Pb^{2+}, Co^{2+}, Zn^{2+} etc. ions are poor, activated carbon is expensive and regeneration of activated carbon is another challenge. Therefore, the search for more effective adsorbents has attracted the interests of researcher worldwide for removal of heavy metal. The naturally occurring biopolymer is one of the best options for the removal of heavy metal from waste water. So a sub-branch of adsorption i.e., biosorption aims to use of the cheaper biological materials as an adsorbent for the removal of heavy metal ions [41, 42]. There are three general types of adsorption: physical, chemical, and exchange adsorption. Physical adsorption is comparatively nonspecific and is due to the operation of weak forces of attraction or van der Waals'' forces between molecules. The adsorbed molecule is not affixed to a particular site on the solid surface, but is free to move about above the surface. In addition, the adsorbed material may condense and form several superimposed layers on the surface of the adsorbent. Chemica adsorption, on the other hand, is the result of much stronger forces, comparable with those leading to the formation of chemical compounds.

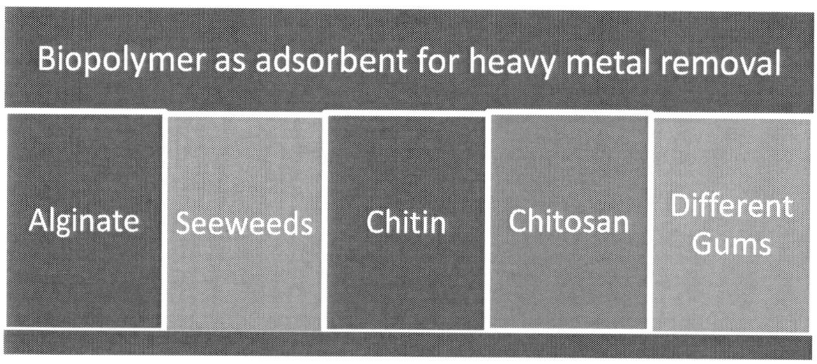

Figure 3. Different biopolymer used for the removal of heavy metal.

3.11. Bioadsorption

Biosorption through naturally synthesized biopolymer attain a great interest for the removal of heavy metals from last two decade. It is non-metabolically mediated process, which removes metals, compounds by materials of living or dead biomass, it is an important means for purifying industrial waste water containing heavy toxic metal. It has many advantages like cheap raw materials, more effective than other conventional technique and easy recovery of the heavy metals as adsorbed heavy metal ions can easily be desorbed. There are lots of naturally occurring biopolymer, which can be used as adsorbate like alginate, seaweeds, chitin, rice hulls, chitosan etc. They all are low in cost and are very effective for the adsorption of heavy metal from waste water. The Biopolymers have been extensively utilized for the removal of heavy metals from the last 30 years because of their interactions with metal ions. These biopolymers are commonly used in there naturally occurring and chemically modified forms and have been effectively used for the removal of heavy metal ions from wastewater. There demand is increasing day by day because of their high ability, selectivity and low-cost as compared to the high cost of synthetic resins [43]. The different biopolymer used for the removal of heavy metal ions is given in Figure 3. The different characteristics of these biopolymer as an adsorbent can be explained as given below:

3.11.1. Seaweeds

These are present in many coastal countries. Brown marine algae used in sorption columns for the removal of cadmium and the maximum adsorption capacity was found to be approximately 67 mg Cd/g seaweed. Other biological materials like fungi, green seaweed and unicellular green algae were also used for the removal of cadmium and mercury. Modified seaweeds were used for the uptake of copper, nickel, zinc, cadmium and lead. These has been found to have relatively high adsorption capacities than other conventional methods. The main drawback with seaweed is its regeneration, as its regeneration is usually not effective [44, 45].

3.11.2. Alginate

Calcium alginate is glycuronans and produced from microorganisms like *Pseudomonas aeruginosa* and mostly it was extracted from brown seaweeds. The maximum binding capacity of sodium alginate for copper is 138.3 mg Cu/g. Alginate are very effective bioadsorbent for heavy metal ions like Cd^{2+}, but it is also very effective for the removal of light metal ions like Ca^{+2} and Mg^{+2} from water [46]. In a later research Nestle and Kimmich (1996) the adsorption mechanism for heavy metal removal through alginate can be regarded as ion exchange. Heavy metal ions, which are present in the solution was absorbed by alginate and the light metals like Ca^{2+}, Mg^{2+}, K^+ and Na^+ or H^+ ions present on alginate will released into the solution [47].

3.11.3. Chitin and Chitosan

Chitin is the second most abundant natural biopolymer next to cellulose. Chitin is found in the shells of crab, shrimp and lobster, in the exoskeleton of marine zooplankton and in the cell wall of yeast, mushroom and other fungi [48]. It is considered as cellulose derivative and have Beta- (1 $\rightarrow$ 4)-2-acetamido-2-deoxy-D- glucopyrnaose structural units. N-deacetylation of chitin leads to the formation of chitosan. The difference between them usually lies in the degree of deacetylation. When the degree of deacetylation is more than 70% then it is chitosan. The biopolymer chitosan has the highest adsorption capacity for heavy metals because of the presence of amino/or hydroxy groups. Due to bifunctional nature, its sorption capacity for heavy metal increases many times. The metal ions removal by chitosan occurred through different mechanisms such as metal cations chelation on free electronic doublet of amine groups, or through ion exchange [49, 50].

There is one lone pair on nitrogen atom of amino group and two lone pairs of electron on oxygen atom of hydroxyl group, which can bind a metal ion or proton forming a complex. As the oxygen atom has high electro negativity than nitrogen atom, thus the nitrogen has a higher affinity to donate the lone pair of electrons for forming complex with metal ion than the oxygen atoms. Therefore, it showed that in chitosan amine groups are responsible for the sorption of heavy metal cations by a chelation and the mechanism is as follows:

$$R - NH_2 + H^+ \longrightarrow R - NH_3^+$$
$$R - NH_2 + M^{n+} \longrightarrow R - NH_2\,M^{n+}$$
$$R - NH_3 + M^{n+} \longrightarrow R - NH_2 + M^{n+} + H^+$$

Chemical modification of the natural biopolymer also affects the heavy metal sorption capacity. Chitosan membrane synthesized through solvent casting method has been reported to form complexes with different metal ions like Cu^{+2}, La^{+3}, Zr^{+4}, Fe^{+3} and Mo^{+6} for sorption of As^{+5} [51, 52].

Chitosan-iron complex showed better sorption efficiency (91± 2%) for As^{+5} as compared to chitosan La^{+3} (80 ±2%) in seawater. The sorption of heavy metal ions by chemical modified chitosan is also influenced by the pH, ionic strength and acetylated units, so all these parameters are optimized to get the best results. Chemically modified chitosan biopolymer have strong affinity for copper as compared to zinc, cadmium, nickel etc., with selectivity coefficient in the range of 10 to 1000. Chitosan biopolymer used for the sorption of copper in the presence of citrate at different metal-ligand ratio and it was found that in acidic condition copper uptake took place by electrostatic attraction between anionic copper-citrate complexes and protonated amine groups of chitosan. There is no sorption was found in very acidic condition (below pH 3) due to the competition between counter ions formed by dissociation of the acid used for pH control and dissociated anionic ligand. The sorption of copper was maximum when the anionic copper-complexes fraction exceeded that of anionic copper-free ligand [53, 54].

Copper adsorption was studied on different adsorbents like chitosan, single-walled carbon nanotubes, multi-walled carbon nanotubes and activated carbon by varying different parameters such as time, adsorbent dose, pH, initial concentration of Cr^{+6} and the competing anions has been reported, which showed that maximum adsorption takes place by activated carbon followed by chitosan at low pH. High efficiency of activated carbon was due to its functionalized group and porous nature, while that of chitosan was due to presence of its protonated amines. A biopolymer blend of chitosan/cellulose beads was synthesized via homogeneous dissolution of chitosan and cellulose in N-Methylmorpholine- N-oxide and these beads were used for the the sorption of metal ion like Ni^{+2}, Cu^{+2} and Fe^{+3} [55, 56].

3.11.4. Chitosan/a-Alumina Composite

The alumina have been widely used as an adsorbent due to its acidebase dissociation behavior, which leads to sorb positive as well as negative ions depending on the medium pH. Chitosan/a-alumina biocomposite synthesized by chemical crosslinking of chitosan and oxalic acid. The sorption capacity of the composite for arsenic was found to be 79.4 mg/g. Boddu, *et al.* (2008), found that pH also influenced the sorption process and maximum adsorption was found at pH 4. The sorption was higher for arsenic (V) as compared to arsenic (III) at pH 4.0 [57, 58].

3.11.5. Manganese Copper Ferrite/Polymer (AA, MA, VA) Composite

Polymers like acrylic acid (AA), methylacrylate (MA) and vinyl acetate (VA) utilize for the modification of the natural polysaccharides through chemical cross-linking. Aluminum doped manganese copper ferrite/AAeMAeVA composite synthesizd by chemical co-precipitation (59). The synthesized composites were effective for the removal of arsenic from aqueous solution and the maximum sorption capacity was found to be 0.053 mg/g. The sorption energy (40.98 kJ/mol) was greater than standard value (16 kJ/mol), which confirms that adsorption process is chemisorption type [59, 60].

3.11.6. Gum Tragacanth Based Biopolymer

Gum tragacanth is the gummy exudation obtained from the stems of several species of *Astragalus,* belonging to the sub-genus *tragacantha.* The plants of this group are low perennial shrubs, remarkable for their leaves having a strong, persistent and spiny petiole. *Gum tragacanth* is an anionic carbohydrate which consists of two major fractions: tragacanthin (water soluble) and bassorin (water swellable). Both the fractions can be easily separated. Bassorin and tragacanthin composition differ particularly in terms of their uronic acid and methoxyl content. Grafting and network formation of *Gum tragacanth* with different vinyl monomers under variable synthetic conditions is an important method to develop a range of polymers with diversified properties and applications. The main building blocks of *Gum tragacanth* are D-xylose, L-arabinose, D-galacturonic acid, D-galacturonic acid methylester, D-galactose and L-fucose [61, 62]. The chemical structure of tragacanthic acid is depicted in Figure 4.

Gum tragacanth-zirconium (IV) tungstoiodophosphate was synthesized through chemical cross-linking using methylmethacrylate as monomer, glutaraldehyde, ascorbic acid-potassium persulfate as cross-linker initiator system. The synthesized biopolymer was characterized through SEM technique, and it is quite clear from the result that grafting modify the backbone (Figure 5). The synthesized biopolymer is very effective for the removal of heavy metals. The study showed the high preference for the adsorption of the heavy metal in the following order: $Pb^{2+} > Cu^{2+} > K^{+}$ i.e., synthesized biopolymer is lead selective. The percentage Pb^{+2} ion removal was increased with increase in the dosage level and maximum Pb^{+2} ion removal was found to be 98.5% with 5g/L of the sample dose [40, 41].

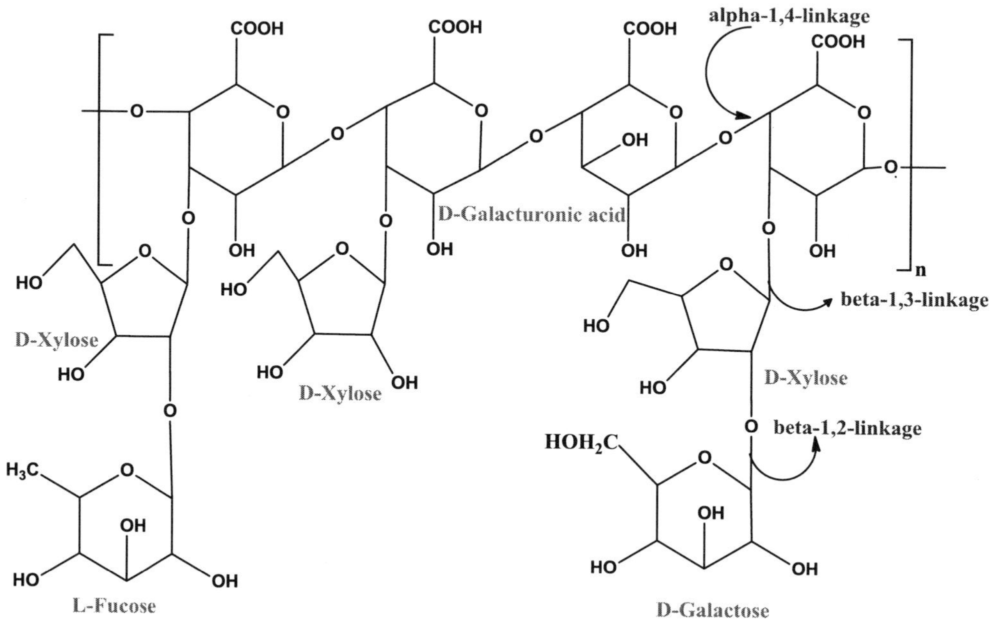

Figure 4. Structure of *tragacanthic acid.*

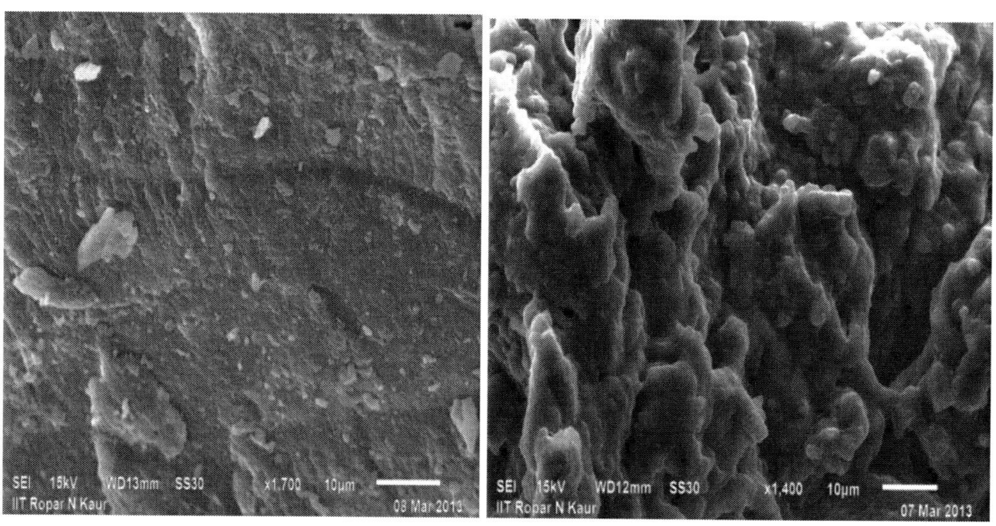

(a) (b)

Figure 5. SEM images of (a) *Gum tragacanth* and (b) Pb^{+2} ion treated modified *Gum tragacanth* [63, 40].

Concluding Remarks and Future Scope

Different forms of phosphoric acid modified Zeolites can also be used for the removal of cationic heavy metal pollutants present in the industrial effluents. Recently a systematic investigation has been carried out to calculate the performance of locally available synthetic zeolites for removal of cesium, strontium and thorium from aqueous solution. Without a doubt, such low cost adsorbents offer a lot of promising benefits for their commercial use in the future point of view. Although zeolites have been mostly exploited for adsorption application, the phosphoric acid forms have not been exploited for removal of heavy metal ion from aqueous solution. They were subjected to a detailed study in batch experiment, for the evaluation of their performance towards elimination of the heavy metal ions. Further, the uptake of hydrolysable was taken for different cationic ions by zeolites. Zeolite encumbered with heavy metal ions on being subjected to hydrothermal treatment, the phosphoric acid modified zeolites were evaluated for further uptake of heavy metal ions.

Environmental pollution due to heavy metal ions is of great concern due to their toxicity, bio-accumulating nature, impact on human health and environment. Various human activities like ore mining, industrial processes etc. are the reason of increased heavy metals deposition in terrestrial and aquatic environment. As the research on biosorption of heavy metals from different techniques is going on from the past decade, it gave a better understanding of the better heavy metal removal technique. Natural biopolymer based products has been recognized as a potential alternative to existing conventional technologies for the removal of heavy metal due to their better sorption properties and low cost. These natural biopolymers can be easily modified chemically, enzymatically etc., as per the requirements.

The obvious advantages of using biopolymer based products in a biosorption process as compared to conventional techniques are their low cost, environment friendly nature, ease in modification, biodegradable nature and their effectiveness in removing heavy metal ions from the waste effluents. Thus, from the foregone discussion it can be concluded that heavy metals removal from waste water using biosorption process can be a solution to clean the environment. This advance technology of biosorption using modified biopolymer is not only cheaper and effective, but also quick and easy to operate.

Other unique modes of operation can be tried for maximum adsorption of heavy metals. Also behaviour of the adsorbent have to be tested with real industrial effluents where different types of heavy metals are present and analyzing the adsorption capacity.

REFERENCES

[1] Bradl, H., (2002). "Heavy Metals in the Environment: Origin, Interaction and Remediation." *London: Academic Press.*

[2] He, Z.L., Yang, X.E. and Stoffella, P.J., (2005). "Trace elements in agroecosystems and impacts on the environment". *J Trace Elem Med Biol.* 19, 125–140.

[3] Brruti, A., Fernandez-Olmo, I. and Irabien, A., (2010). "Evaluation of the contribution of local sources to trace metals levels in urban PM 2.5 and PM 10 in the Cantabria region (Northern Spain)". *J Environ Monit.* 12, 1451–1458

[4] WHO/FAO/IAEA. World Health Organization. Switzerland: Geneva; 1996. "Trace Elements in Human Nutrition and Health".

[5] Stern, B.R., (2010). "Essentiality and toxicity in copper health risk assessment: overview, update and regulatory considerations". *Toxicol Environ Health A.* 73, 114–127.

[6] Harvey, L.J. and McArdle, H.J., (2008). "Biomarkers of copper status: A brief update". *Br J Nutr.* 99, 510–513.

[7] Tchounwou, P., Newsome, C., Williams, J. and Glass, K., (2008). "Copper-induced cytotoxicity and transcriptional activation of stress genes in human liver carcinoma cells". *Metal Ions Biol Med.* 10, 285–290.

[8] Beyersmann, D. and Hartwig, A., (2008). "Carcinogenic metal compounds: recent insight into molecular and cellular mechanisms". *Arch Toxicol.* 82, 493–512.

[9] Yedjou, C.G., Tchounwou, P.B., (2006). "Oxidative stress in human leukemia cells (HL-60), human liver carcinoma cells ($HepG_2$) and human Jerkat-T cells exposed to arsenic trioxide". *Metal Ions Biol Medical.* 14, 298–303.

[10] Barakat, M.A., (2011). "New trends in removing heavy metals from industrial wastewater". *Arabian Journal of Chemistry.* 4, 361-377.

[11] Paschal, D.C., Burt, V., Caudill, S.P., Gunter, E.W., Pirkle, J.L. and Sampson, E.J., (2000). "Exposure of the U.S. population aged 6 years and older to cadmium 1988–1994". *Arch Environ Contam Toxicol.* 38, 377–383.

[12] Agency for Toxic Substances and Disease Registry (ATSDR) Draft Toxicological Profile for Cadmium. Atlanta, GA: 2008.

[13] Satarug, S., Baker, J.R., Urbenjapol, S., Haswell-Elkins, M., Reilly, P.E. and Williams, D.J., (2003). "A global perspective on cadmium pollution and toxicity in non-occupationally exposed population". *Toxicol Lett.* 137, 65–83.

[14] Benguella, B. and Benaissa, H., (2002). "Cadmium removal from aqueous solutions by chitin: Kinetic and equilibrium studies". *Water Res.* 36, 2463– 2474.

[15] Yedjou, G.C. and Tchounwou, P.B., (2008). "N-acetyl-cysteine affords protection against lead-induced cytotoxicity and oxidative stress in human liver carcinoma ($HepG_2$) cells". *Intl J Environ Res Public Health.* 4, 132–137.

[16] Tchounwou, P.B., Yedjou, C.J., Foxx, D., Ishaque, A. and Shen, E., (2004). "Lead-induced cytotoxicity and transcriptional activation of stress genes in human liver carcinoma cells ($HepG_2$)". *Mol Cell Biochem.* 255, 161–170.

[17] Esalah,O.J., Weber, M.E. and Vera, J.H., (2000). "Removal of lead, cadmium and zinc from aqueous solutions by precipitation with sodium di-(n-octyl) phosphinate". *Can. J. Chem. Eng. 78*, 948–954.

[18] Yedjou, G.C. and Tchounwou, P.B., (2007). "*In vitro* cytotoxic and genotoxic effects of arsenic trioxide on human leukemia cells using the MTT and alkaline single cell gel electrophoresis (comet) assays". *Mol Cell Biochem.* 301, 123–130.

[19] Tchounwou, P.b., Centeno, J.A. and Patlolla, A.K., (2004). "Arsenic toxicity, mutagenesis and carcinogenesis -a health risk assessment and management approach". *Mol Cell Biochem.* 255, 47–55.

[20] Centeno, J.A., Tchounwou, P.B., Patlolla, A.K., Mullick, F.G., Murakat, L., Meza, E., Gibb, H., Longfellow, D. and Yedjou, C.G., (2005). "Environmental pathology and health effects of arsenic poisoning: a critical review". *Australia: CSIRO Publishing Corp.*

[21] Sutton, D.J. and Tchounwon, P.B., (2007). "Mercury induces the externalization of phosphatidylserine in human proximal tubule (HK-2) cells". *Intl J Environ Res Public Health.* 4, 138–144.

[22] Sutton, D.J., Tchounwou, P.B., Ninashvili, N. and Shen, E., (2002). "Mercury induces cytotoxicity, and transcriptionally activates stress genes in human liver carcinoma cells". *Intl J Mol Sci.* 3, 965–984.

[23] Huisman, J.L., Schouten, G. and Schultz, C., (2006). "Biologically produced sulphide for purification of process streams, effluent treatment and recovery of metals in the metal and mining industry". *Hydrometallurgy.* 83, 106-113.

[24] Ozverdi, A. and Erdem, M., (2006). "Cu^{+2}, Cd^{+2} and Pb^{+2} adsorption from aqueous solutions by pyrite and synthetic iron sulphide". *J. Hazard. Mater.* 137, 626-632.

[25] Esalah, O.J., Weber,M.E. and Vera, J.H.,(2000). "Removal of lead, cadmium and zinc from aqueous solutions by precipitation with sodium di-(n-octyl) phosphinate". *Can. J. Chem. Eng.* 78, 948–954.

[26] Lertlapwasin, R., Bhawawet, N., Imyim, A. and Fuangswasdi, S., (2010). "Ionic liquid extraction of heavy metal ions by 2-aminothiophenol in 1-butyl-3-methylimidazolium hexafluorophosphate and their association constants". *Sep. Purif. Technol. 72*, 70–76.

[27] Matlock, M.M., Howerton, B.S., Aelstyn, M.A.V., Nordstrom, F.L. and Atwood, D.A., (2002). "Advanced mercury removal from gold leachate solutions prior to gold and silver extraction: a field study from an active gold mine in Peru". *Environ. Sci. Technol.* 36, 1636-1639.

[28] Emamjomeh, M.M. and Sivakumar, M., (2009). "Review of pollutants removed by electrocoagulation and electrocoagulation/flotation processes". *J. Environ. Manag.* 90, 1663–1679.

[29] Ghurye, G., Clifford, D. and Tripp, A., (2004). "Iron coagulation and direct microfiltration to remove arsenic from groundwater". *AWWA*. 96, 143–152.

[30] Greenlee, L.F., Lawler, D.F., Freeman, B.D., Marrot, B., Moulin, P., (2009). "Reverse osmosis desalination: Water sources, technology, and today's challenges". *Water Res.* 43, 2317–2348.

[31] Benit, Y. and Ruiz, M.L., (2002). "Reverse osmosis applied to metal finishing wastewater". *Desalination.* 142, 229–234.

[32] Barakat, M.A., (2011). "New trends in removing heavy metals from industrial wastewater". *Arab. J. Chem.* 4, 361–377.

[33] Yurloval, L., Kryvoruchko, A. and Kornilovich, B., (2002). "Removal of Ni (II) ions from wastewater by micellar-enhanced ultrafiltration". *Desalination. 144*, 255–260.

[34] Boricha, A.G., Murthy, Z.V.P., (2010). "Preparation of N,O-carboxymethyl chitosan/cellulose acetate blend nanofiltration membrane and testing its performance in treating industrial wastewater". *Chem. Eng. J.* 157, 393–400.

[35] Mahmoud, A. and Hoadley, A.F., (2012). "An evaluation of a hybrid ion exchange electrodialysis process in the recovery of heavy metals from simulated dilute industrial wastewater". *Water Res. 46*, 3364–3376.

[36] Belkacem, M., Khodir, M. and Abdelkrim, S., (2008). "Treatment characteristics of textile wastewater and removal of heavy metals using the electroflotation technique". *Desalination.* 228, 245-254.

[37] Berber-Mendoza, M.S., Leyva-Ramos, R., Alonso-Davila, P., Fuentes-Rubio, L. and Guerrero-Coronado, R.M., (2006). "Comparison of isotherms for the ion exchange of Pb (II) from aqueous solution onto homoionic clinoptilolite". *J. Colloid Interface Sci.* 301, 40-45.

[38] Inglezakis, V.J. and Grigoropoulou, H.P., (2003). "Modeling of ion exchange of Pb2þ in fixed beds of clinoptilolite". *Microporous Mesoporous Mater.* 61, 273-282.

[39] Alyuz, B. and Veli, S., (2009). "Kinetics and equilibrium studies for the removal of nickel and zinc from aqueous solutions by ion exchange resins". *J. Hazard. Mater.* 167, 482-488.

[40] Saruchi, and Kumar, V., (2016). "Adsorption kinetics and isotherms for the removal of rhodamine B dye and Pb^{+2} ions from aqueous solutions by a hybrid ion-exchanger". *Arabian Journal of Chemistry.* 5, 241-246.

[41] Saruchi, Kumar, V., Kaith, B.S. and Jindal, R., (2016). "Synthesis of Hybrid Ion Exchanger for Rhodamine B Dye Removal: Equilibrium, Kinetic and Thermodynamic Studies". *Ind. Eng. Chem. Res.* 55, 10492–10499.

[42] Pan, B.J., Pan, B.C., Zhang, M.W., Lv, L., Zhang, Q.X. and Zheng, S.R., (2009). "Development of polymeric and polymer-based hybrid adsorbents for pollutants removal from waters". *Chem. Eng. J.* 151, 19–29.

[43] Aydln, Y.A. and Aksoy, N.D., (2009). "Adsorption of chromium on chitosan: Optimization, kinetics and thermodynamics". *Chem. Eng. J.* 151, 188–194.

[44] Sheng, P.S., Ting, Y.P., Chen, J.P. and Hong, L., (2004). "Sorption of lead, copper, cadmium, zinc and nickel by marine algae biomass: characterization of biosorptive capacity and investigation of mechanisms". *J Colloid Interface Sci.* 275, 131-141.

[45] Siddiquee, S., Rovina, K., Azad, S.A., Naher, L., Suryani, S. and Chaikaew, P., (2015). "Heavy Metal Contaminants Removal from Wastewater Using the Potential Filamentous Fungi Biomass: A Review". *J Microb Biochem Technol.* 7, 384-393.

[46] Liu, Y.H., Xia, F.Q., Du, B., Zhang, Q.J., Wang, P., Wang, D.C. and Xia, Y.Z., (2010). "Removal of copper from aqueous solution by carbon nanotube/calcium alginate composites". *J. Hazard. Mater.* 177, 876-880.

[47] Nestle, N.F.E.I. and Kimmich, R., (1996). "Concentration-Dependent Diffusion Coefficients and Sorption Isotherms. Application to Ion Exchange Processes As an Example". *J Phys. Chem.* 100, 12569-12573.

[48] Benguella, B. and Benaissa, H., (2002). "Cadmium removal from aqueous solutions by chitin: Kinetic and equilibrium studies". *Water Res.* 36, 2463– 2474.

[49] Boddu, V.M., Aburri, K., Talmott, J. and Smith, E., (2003). "Removal of hexavalent chromium from wastewater using a new composite chitosan biosorbent". *Environ. Sci. Technol.* 37, 4449–4456.

[50] George, Z.K., Margaritis, K. and Nikolaos, K.L., (2009). "Copper and chromium (VI) removal by chitosan derivatives: Equilibrium and kinetic studies". *Chem. Eng. J.* 152, 440–448.

[51] Akkaya, R. and Ulusoy, U., (2008). "Adsorptive features of chitosan entrapped in polyacrylamide hydrogel for Pb^{2+}, UO_2^{2+}, and Th^{4+}". *J. Hazard. Mater.* 151, 380–388.

[52] Fajardo, A.R., Lopes, L.C., Rubira, A.F. and Muniz, E.C., (2012). "Development and application of chitosan/poly(vinyl alcohol) films for removal and recovery of Pb(II)". *Chem. Eng. J.* 183, 253–260.

[53] Fan, D., Zhu, X., Xu, M. and Yan, J., (2006). "Adsorption properties of chromium (VI) by chitosan-coated montmorillonite". *J. Biol. Sci.* 6, 941–945.

[54] Fan, L., Luo, C., Lv, Z., Lu, F.and Qiu, H., (2011). "Preparation of magnetic modified chitosan and adsorption of Zn^{2+} from aqueous solutions". *Colloids Surf. B. Biointerfaces* 88, 574–581.

[55] Fan, L., Luo, C., Sun, M., Li, X. and Qiu, H., (2012). "Highly selective adsorption of lead ions by water-dispersible magnetic chitosan/graphene oxide composites". *Colloids Surf. B. Biointerfaces.* 103, 523–529.

[56] Chang, Y.C., Chang, S.W. and Chen, D.H., (2006). "Magnetic chitosan nanoparticles: Studies on chitosan binding and adsorption of Co(II) ions". *React. Funct. Polym.* 66, 335–341.

[57] Baddu, V.M., Abburi, K., Talbott, J.L., Smith, E.D. and Haasch, R., (2008). "Kinetic and thermodynamics aspects od adsorption of arsenic onto granular ferric hydroxide (GFH)". *Water Research.* 42, 3371-3378.

[58] Chang, Y.C. and Chen, D.H., (2005). "Preparation and adsorption properties of monodisperse chitosan-bound Fe_3O_4 magnetic nanoparticles for removal of Cu(II) ions". *J. Colloid. Interface Sci.* 283, 446–451.

[59] Malana, M.A., Qureshi, R.B. and Ashiq, M.N., (2011). "Adsorption studies of arsenic on nano aluminium doped manganese copper ferrite polymer (MA, VA, AA) composite: kinetics and mechanism". *Chemical Engineering Journal.* 172, 721–727

[60] Chen, A., Zeng, G., Chen, G., Hu, X., Yan, M., Guan, S., Shang, C., Lu, L., Zou, Z. and Xie, G., (2012). "Novel thiourea-modified magnetic ion-imprinted chitosan/TiO_2 composite for simultaneous removal of cadmium and 2,4-dichlorophenol". *Chem. Eng. J.* 191, 85–94.

[61] Kaith, B.S., Saruchi, Rajeev, J., and Bhatti, M.S., (2012). "Screening and RSM optimization for synthesis of a Gum tragacanth–acrylic acid based device for in situ controlled cetirizine dihydrochloride release". *Soft Matter.* 8, 2286-2293.

[62] Saruchi, Kaith, B.S., Jindal, R. and Kumar, V., (2015). "Biodegradation of Gum tragacanth acrylic acid based hydrogel and its impact on soil fertility". *Polymer Degradation and Stability.* 115, 24-31.

[63] Saruchi, Kaith, B.S., Jindal, R. and Kapur, G.S., (2014). "Synthesis of Gum tragacanth and acrylic acid based hydrogel: its evaluation, for controlled release of antiulcerative drug pantoprazole sodium, characterization and evaluation of Gum tragacanth and acrylic acid hydrogel for sustained calcium chloride release-enhancement of water holding capacity of soil". *Journal of the Chinese Advanced Materials Society.* 2, 110-117.

In: Modified Biopolymers: Challenges and Opportunities ISBN: 978-1-53612-116-2
Editors: D. Pathania, G. Sharma and A. Kumar

Chapter 4

BIOPOLYMER DRIVED HYDROGELS AND THEIR DIVERSE APPLICATIONS: A REVIEW

***Z. M. Siddiqi*[1,*] *and D. Gupta*[2]**
[1]Jubail University College (JUC),
Jubail Industrial City, KSA, Saudi Arabia
[2]School of Chemistry,
Shoolini University of Biotechnology and Management Sciences,
Solan, India

ABSTRACT

A group of polymeric materials constitute hydrogel products. The hydrophilic structure of these hydrogels makes them capable of holding large amounts of water in their three-dimensional networks. Extensive employment of these products is considered to be of prime importance in a number of industrial and environmental areas of application. Due to their higher water absorption capacity, long service life, and wide varieties of raw chemical resources, natural hydrogels are gradually replaced by synthetic hydrogels. The primary objective of this article is to review the literature concerning the classification of hydrogels on the basis of their physical and chemical characteristics and to analyse the technical feasibility of their utilization. The study also included technologies adopted for hydrogel production together with process design implications, blocks diagrams, and optimized conditions of the preparation process. An innovated category of recent generations of hydrogel materials was also presented in some details.

Keywords: hydrogel, polymers, drug-delivery, classification, industrial applications

---

* Corresponding Author Email: siddiqizm2005@hotmail.com.

1. Introduction

In this brief review the materials of interest are primarily hydrogels. The polymer networks extensively swollen with water are known as hydrogels. Hydrogels are networks of polymer chains which are usually referred to as hydrophilic gels [1]. Over the years, researchers have defined hydrogels in many different ways like, it is a polymeric material that exhibits the ability to swell and retain a significant fraction of water within its structure, without dissolving in water. Due to their exceptional properties in a wide range of applications, hydrogels have been materials of considerable interest [2–4]. They possess a high degree of flexibility like natural tissue due to their large water content. The ability of hydrogels to absorb water arises from hydrophilic functional groups attached to the polymeric backbone, while their resistance to dissolution arises from cross-links between network chains. Natural hydrogels were gradually replaced by synthetic hydrogels which have long service life, high capacity of water absorption, and high gel strength. Purely synthetic components are used to synthesize hydrogels. Hydrogels are found to be stable in sharp and strong fluctuations of temperature conditions [5]. Recently, hydrogels have been defined as two or multicomponent systems consisting of a three-dimensional network of polymer chains and water that fills the space between macromolecules. Depending on the properties of the polymer (polymers) used, as well as on the nature and density of the network joints, such structures in an equilibrium can contain various amounts of water; typically in the swollen state. The mass fraction of water in a hydrogel is much higher than the mass fraction of a polymer. So to achieve high degrees of swelling, it is good to use synthetic polymers that are water-soluble when in non-cross-linked form. The numbers of classical chemical ways are used to synthesize hydrogels. These include one-step procedures like polymerization and parallel cross-linking of multifunctional monomers, as well as multiple step procedures involving synthesis of polymer molecules having reactive groups and their subsequent cross-linking, by reacting polymers with suitable cross-linking agents [6].

1.1. Classification of Hydrogel Products

The hydrogel products can be classified according to a variety of bases, as described below.

1.1.1. Classification Based on Source

On the basis of natural or synthetic origins hydrogels can be classified into two groups [7]. Natural or synthetic polymers are used to prepare hydrogels [8]. Those based on natural polymers possess weak mechanical properties. These disadvantages are corrected on the one hand by their biocompatibility and biodegradability. Due to the fact that polysaccharides come from living organisms; they are biocompatible, nontoxic, and possess major physicochemical properties necessary for controlled release applications [9]. The most studied polysaccharides are alginate, dextran, gellan, xantha, hyaluronic acid, and chitosan derivatives [10-20]. Synthetic hydrogels, on the other hand, do not possess these inherent bioactive properties. Fortunately, synthetic polymers usually have well defined structures that can be modified to yield tailorable degradability and functionality. Table 1 shows natural polymers as well as synthetic monomers that are commonly used in hydrogel fabrication.

1.1.2. Classification According to Polymeric Composition

On the basis of polymeric compositions hydrogels are described as below:

- Homopolymeric hydrogels are those polymer networks derived from a single species of monomer [23]. Depending on the nature of the monomer and polymerization technique homopolymers may have a cross-linked skeletal structure.
- Copolymeric hydrogels are comprised of two or more different monomer species with at least one hydrophilic component. These monomeric species are arranged along the chain of the polymer network in a random, block or alternating configuration [24].
- Multipolymer Interpenetrating polymeric hydrogel (IPN) is made of two independent cross-linked synthetic or natural polymer components, contained in a network form. One component is a cross-linked polymer and another component is a non-cross-linked polymer in semi IPN hydrogel [25].

Table 1. Natural polymers and synthetic monomers used in hydrogel fabrication [21, 22]

Natural Polymer	Synthetic Monomer
Chitosan	Hydroxyethyl methacrylate (HEMA)
Alginate	N-(2-hydroxypropyl) methacrylate (HPMA)
Fibrin	N-vinyl-2-pyrrolidone (NVP)
Collagen	Vinyl acetate (VAc)
Gelatin	Polyethylene glycol diacrylate/dimethacrylate (PEGDA/PEGDMA)
Hyaluronic acid	Polyethylene glycol acrylate/methacrylate (PEGA/PEGMA)
Dextran	Methacrylic acid (MAA)
	Acrylic acid (AA)
	N-isopropyl acrylamide (NIPAAm)

1.1.3. Classification Based on Configuration

The hydrogels that depend on physical structure and chemical composition can be classified as follows:

- Amorphous (non-crystalline).
- Semicrystalline: A complex mixture of amorphous and crystalline phases.
- Crystalline.

1.1.4. Classification Based on Type of Cross-Linking

Based on the chemical or physical nature of the cross-link junctions hydrogels can be divided into two categories.

- Chemically cross-linked networks which have permanent junctions.
- Physical networks which have transient junctions that arise from physical interactions such as ionic interactions, hydrogen bonds, or hydrophobic interactions or polymer chain interactions [26].

1.1.5. Classification Based on Physical Appearance

Hydrogels on the basis of appearance are classified as matrix, film, or microsphere. The different appearance of hydrogels depends on the technique of polymerization involved in the preparation process.

1.1.6. Classification According to Network Electrical Charge

On the basis of presence or absence of electrical charge located on the crosslinked chains, hydrogels may be categorized into four groups as follows:

- Nonionic (neutral).
- Ionic (including anionic or cationic).
- Amphoteric electrolyte (ampholytic) containing both acidic and basic groups.
- Zwitterionic (polybetaines) containing both anionic and cationic groups in each structural repeating unit.

Hydrogels that are formed from natural polymers include proteins such as collagen and gelatine and polysaccharides such as starch, alginate, and agarose. Synthetic polymers that form hydrogels are traditionally prepared using chemical polymerization methods.

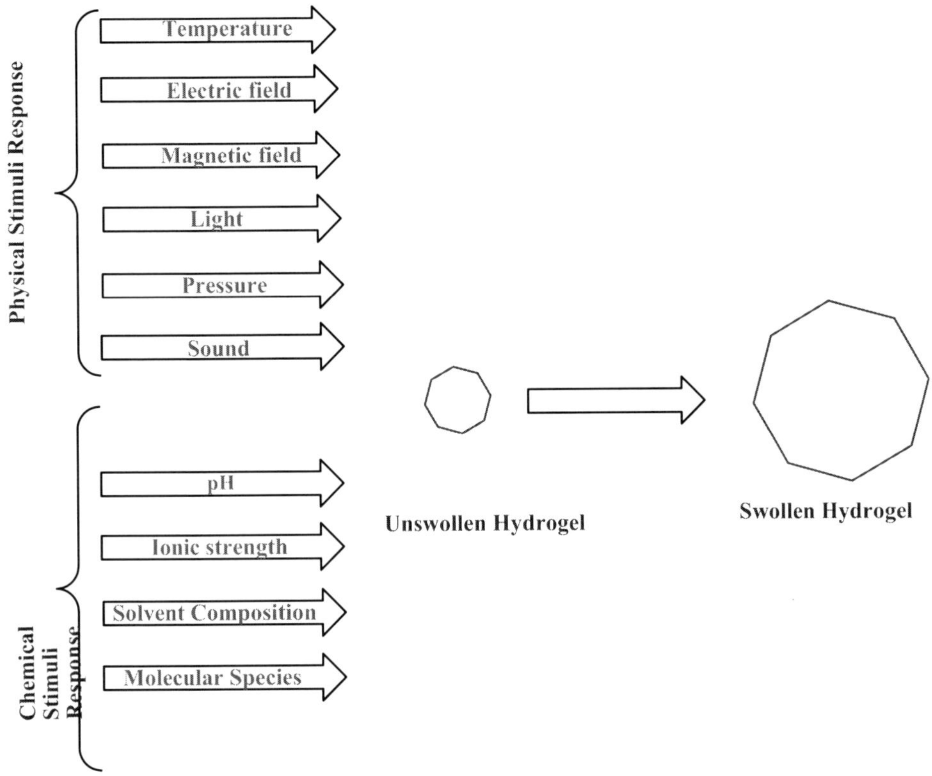

Figure 1. Stimuli response swelling hydrogel.

1.2. Hydrogel Product Sensitive to Environmental Conditions

As mentioned above the three-dimensional crosslinked hydrophilic polymer networks which are capable of swelling or de-swelling reversibly in water and retaining a large volume of liquid in a swollen state are known as hydrogels. In order to shrink or expand with changes in external environmental conditions, hydrogels can be designed with controllable responses. Figure 1 shows that in response to a variety of physical stimuli (temperature, electric or magnetic field, light, pressure, and sound) and chemical stimuli (pH, solvent composition, ionic strength, and molecular species) hydrogels may perform dramatic volume transition. In response to the changes in the external environment the extent of swelling or de-swelling of the hydrogel could be so drastic that the phenomenon is referred to as 'volume collapse' or 'phase transition' [27]. For the past four decades synthetic hydrogels have been a field of extensive research, and today it still remains a very active area of research.

1.3. Utilization of Hydrogel Products

In 1960, Wichterle and Lim established the first synthetic hydrogels which can be utilized in many ways [28]. The hydrogel technologies may be applied to agriculture, hygienic products, drug delivery systems, sealing, coal dewatering, artificial snow, food additives, pharmaceutical and biomedical applications, tissue engineering and regenerative medicines, diagnostics, wound dressing, separation of biomolecules or cells and barrier materials to regulate biological adhesions and Biosensor [29-43]. Hydrogel products for hygienic applications are mainly based on acrylic acid and its salts. For preparationof agricultural hydrogel products, acrylamide is considered as main component [44, 45].

Various publications on the subject of synthetic methods and applications of hydrogels have discussed in detail. For example, Peppas edited a comprehensive review of the chemistry and various synthetic schemes employed for hydrogel preparation [46]. More recently, Khoylou published findings that hydrogels are produced by radiation polymerization and grafting [47]. The preparation and chemical properties of hydrogels employed in agricultural applications have been described by Mi-Ran Park [48]. The potential of hydrogels in sensor utilizations has been reviewed byVijayalakshmi and Kenichi [49]. In general, synthetic polymers or natural polymers are used to synthesize hydrogels. Compared to natural polymers, synthetic polymers are hydrophobic in nature and chemically stronger. Their mechanical strength results in a slow degradation rate. On the other hand, it provides durability and the balance of these opposing properties allows for an optimal design, which should be implemented [50]. In the most succinct sense, hydrophilic polymeric networkcross-linked to produce an elastic structure is called hydrogel. Thus, hydrogel can be produced by any technique which can be used to create a cross-linked polymer. Hydrogels can be produced by copolymerization/cross-linking free-radical polymerizations by hydrophilic monomers reacting with multifunctional cross-linkers. Both natural and synthetic water-soluble linear polymers are cross-linked to form hydrogels in many ways:

- Linking polymer chains via chemical reaction.
- Using ionizing radiation to generate main-chain free radicals which can recombine as cross-link junctions.
- Physical interactions such as entanglements, electrostatics, and crystallite formation.

Various polymerization techniques including bulk, solution, and suspension polymerization can be used to form gels. Monomer, initiator, and cross-linker are the three integral parts of the hydrogen's preparation. The final hydrogen's properties and the heat of polymerization can be controlled by using diluents such as water or other aqueous solutions. To remove impurities left from the preparation process the hydrogel mass needs to be washed [51-54]. Usually hydrogels are prepared from polar monomers. According to the starting material used, hydrogels can be divided as natural polymer hydrogels, synthetic polymer hydrogels, and combinations of the two classes. Graft polymerization, cross-linking polymerization, networks formation of water-soluble polymer, and radiation cross-linking methods are used as preparative methods for hydrogels. There are many types of hydrogels; mostly, they are lightly cross-linked copolymers of acrylate and acrylic acid, and grafted starch-acrylic acid polymers prepared by inverse suspension, emulsion polymerization, and solution polymerization.

1.4. Preparation of Hydrogels

Hydrogels can be prepared by using many different ways, which are described below:

1.4.1. Use of Crosslinkers

Multifunctional co-monomers act as cross linking agents for copolymerization of monomers. The polymerization reaction can be carried out in bulk, solution or suspension which is initiated by a chemical initiator. Cross linking of linear polymers is done by using chemical compounds shown in Figure 2(a) or by irradiation methods Figure 2(b). After the polymerization is completed monomers used here contain an ionizable group that can be ionized or can undergo a substitution reaction. Thus, weakly acidic groups like carboxylic acids or weakly basic groups like substituted amines or a strong acidic and basic group like sulfonic acid and quaternary ammonium compounds may be present in synthesized hydrogels. Glutaraldehyde, calcium chloride and oxidized konjac glucomannan (DAK) are different cross-linkers that can be incorporated. They prevent burst release of the medicaments by imparting sufficient mechanical strength to the polymers [55].

1.4.2. Use of Gelling Agent

In the preparation of hydrogels different gelling agents like glycophosphate-1,2-propanediol, glycerol, trehalose, mannitol etc., have been used. The problems associated with the method are the presence of negative charged moieties and turbidity [56].

1.4.3. Use of Irradiation and Freeze Thawing

The Irradiation method is suitable as well as convenient. But in this method the processing is costly along with the poor mechanical strength of the product. The Freeze thawing method imparts sufficient mechanical strength and stability to the hydrogels. The hydrogels in this method are opaque in appearance with little swelling capacity. However, the microwave irradiation method produces more porous hydrogels than conventional methods [57].

1.4.4. Synthesis of Hydrogel in Industry

The polymerization where formulation of monomer along with initiators and additives has done leads to the formation of the gel. The post treatment is done if needed and the gel is dried, sieved and mixed with other additives. The final product thus formed is packed and dispatched [58].

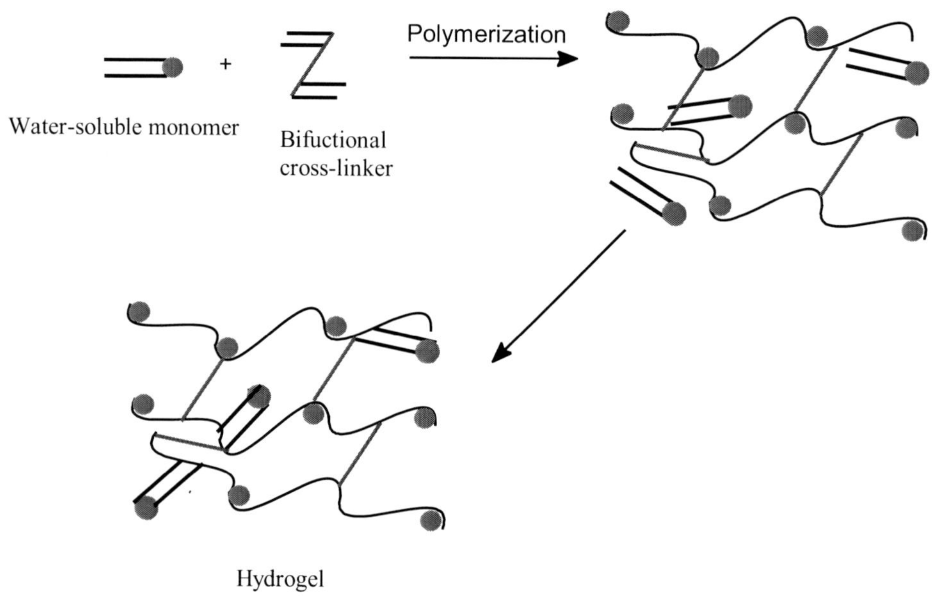

Figure 2(a). Synthesis of Hydrogel by polymerization

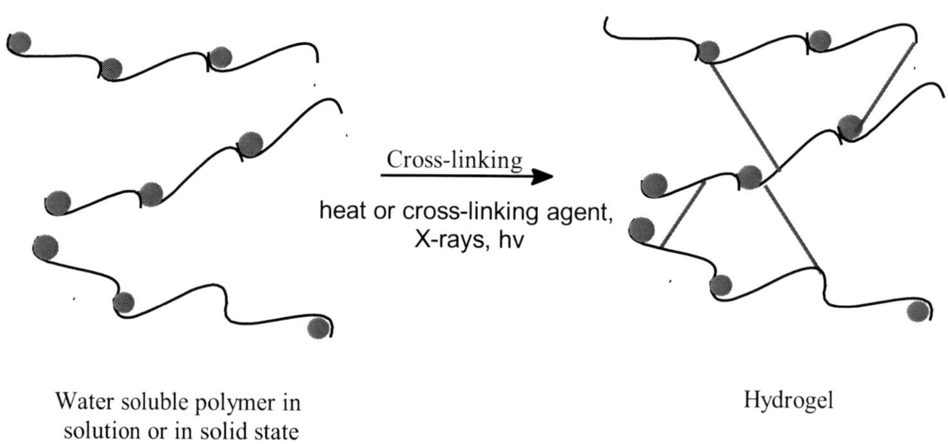

Figure 2(b). Synthesis of Hydrogels by cross-linking of polymers

2. CHARACTERIZATION

The presence of hydrogel in a system can be quantified easily by using various methods by dispersing the polymer in water using a cylindrical vial and visually observe the formation of insoluble material.

2.1. Solubility

2.1.1. Method A

After immersion of hydrogel material in de-ionized water for 16 h [59] or 48 h at room temperature (Nagasawa et al., 2004) its insoluble part is measured in a dried sample and the hydrogel content of the material can be calculated. To ensure the full dispersion of hydrogel material in water, a sample should be prepared at a dilute concentration (typically of ~ 1%). The gel fraction is then measured as follows:

$$Gel\ Fraction\ (Hydrogel\ \%) = W_d/W_i \times 100 \tag{1}$$

where, W_i is the initial weight of dried sample and W_d is the weight of the dried insoluble part of sample after extraction with water.

2.1.2. Method B

A more accurate measure of the hydrogel can bedetermined by measuring the weight retained after vacuum filtration. This method is prescribed by JECFA (Joint Expert Committee on Food Additives) for hydrocolloids which have been modified by changing the solvent from mild alkaline to water [60]. The weight (W_1) of a 70 mm glass-fibre paper is determined by drying in an oven at 105°C for 1 h and subsequently cooled in a desiccator. Depending on the test material, 1-2 wt. % (S) dispersion can be prepared in distilled water followed by overnight hydration at room temperature. Prior to filtration the hydrated dispersion is then centrifuged 5 min at 2500 rpm. At 105°C drying of the filter paper is carried out in an oven, and then cooled to a constant weight (W_2).

The % Insoluble can then be calculated:

$$\%Hydrogel = (W_2 - W_1|S) \times 100 \tag{2}$$

Depending on the test material, different mesh size can also be used, e.g., the use of a 20 mesh steel screen (1041 μm) to determine the gel fraction [61].

2.2. Swelling Measurement

2.2.1. Method A

The swelling of hydrogels has been measured by the Japanese Industrial Standard K8150 method. The roller mixture is used to dry the hydrogel which is immersed in deionised water for 48hours at room temperature. After swelling, the hydrogel is filtered. Stainless steel net of 30 meshes (681 μm) is used for filtration. The swelling is calculated as [62]:

$$Swelling = (W_s - W_d | W_d) \quad (3)$$

where Ws is the weight of hydrogel in swollen state and W_d is the weight of hydrogel indry state. The terms ‘swelling ratio’ (Liu et al., 2005), ‘equilibrium degree of swelling’ (EDS) [63] or ‘degree of swelling’ has been used for more or less similar measurements [64].

2.2.2. Method B

Alternatively, to measure the swelling of hydrogel, in a volumetric vial the dryhydrogel was dispersed into sufficiently high quantity of water for 48 hrs at room temperature. To obtain the layers of water bound material and free unabsorbed water the mixture is then centrifuged. The free water is removed. The swelling can be measured according to Method A above.

2.2.3. Method C

Swelling can also be measured according to the Japanese Industrial Standard (JIS) K7223. At room temperature the dry gel is immersed in deionized water for 14-16 h. After swelling, 100-mesh (149 μm) stainless-steel net is used to filter hydrogel. Swellingis calculated as follows [65]:

$$Swelling = \frac{C}{B} \times 100 \quad (4)$$

where C is the weight of hydrogel obtained after drying and B is the weight of the insoluble portion after extraction with water.

2.3. FTIR

FTIR (Fourier Transform Infrared Spectroscopy) is a useful technique which tells about the chemical structure of a substance. It is based on the principle that the chemical bonds usually can be excited by absorbing infrared light at frequencies that are typical of the types of the chemical bonds. The resulting IR absorption spectrum represents a fingerprint of measured sample. This technique is widely used to investigate the structural arrangement in hydrogel by comparison with starting material [66, 67].

2.4. Scanning Electron Microscopy (SEM)

The information about the sample's surface topography, composition, and other properties such as electrical conductivity can be provided by using SEM. In SEM the magnification can be controlled over a range of up to 6 orders of magnitude from about 10 to 500,000 times. This is a powerful technique. The characteristic 'network' structure in hydrogels can be captured by using this widely used technique [68-71].

2.5. Light Scattering

To determine the molecular distribution and parameters of a polymeric system gel permeation chromatography coupled on line to a multi angle laser light scattering (GPC-MALLS) is a widely used technique. By using this technique in a polymeric system the hydrogels can be quantified [72]. The hydrogels of several hydrocolloids such as gum arabic, gelatin and pullulan have been quantified by using this technique [73-75].

2.6. Other Techniques

By using differential scanning calorimetry (DSC) and nuclear magnetic resonance (NMR) techniques the amount of free and bound water in hydrogels can be characterized and quantified easily. The information about the interchange of water molecules between free and bound states is provided by the proton NMR [76]. The use of DSC is based on the assumption that only the free water may be frozen. The warming the frozen gel represents the melting of the free water and endotherm is measured. Then the value will yield the amount of free water in the hydrogel sample being tested. The difference of the measured total water content of the hydrogel test specimen and the calculated free water content gives the bound water [77]. The formation of the cross-linked network gel structures of hydrogel can also be confirmed by using other techniques like Thermogravimetric analysis [78], X-ray diffraction [79], sol-gel analysis etc. [80-83].

3. Application of Hydrogels

Due to their specific structures and compatibility with different conditions of use hydrogels are used in many fields. The flexibility of hydrogels, which is because of their water content, makes it possible to use them in different conditions ranging from industrial to biological, and the biocompatibility of the materials used to produce them and also their chemical behavior in biological environments, which can be nontoxic, extends their applications to the medical sciences.

Wound Healing: To treat cartilage defects modified polysaccharide found in cartilage is used in the formation of hydrogels. For example, the hydrogel of gelatin and 13 polyvinyl alcohols (PVA) together with blood coagulants are formulated. Silicon hydrogels and polyacrylamides are used to make Soft Contact Lenses. The first commercially available silicon hydrogels adopted two different approaches. In first approach Bausch and Lomb extended its development of silicon monomers with enhanced compatibility of hydrogel forming monomers. The second approach was the development of siloxy monomers containing hydrophilic polyethylene oxide segments and oxygen permeable polysiloxane by Ciba vision.

Industrial Applicability: For industrial effluents like methylene blue dye hydrogels are used as absorbents. Hydrogel beads are used as an adsorbent for dioxins.

Tissue Engineering: Natural hydrogel materials used for tissue engineering include agarose, methylcellulose etc. Macromolecules (phagosomes) are delivered by micronized hydrogels into cytoplasm of antigen-presenting cells. This property is also utilized in cartilage repairing.

Drug Delivery in GI Tract: In the GIT hydrogel delivers drugs to specific sites. Drugs loaded with colon specific hydrogels show tissue specificity and change in the pH. Enzymatic actions cause liberation of drugs. In the presence of micro flora they are designed to be highly swollen or degraded.

Rectal Delivery: For rectal drug delivery hydrogels showing bio-adhesive properties are used. As matrices for drug delivery the xyloglucan gel with a thermal gelling property was explored by Miyazaki et al.

Transdermal Delivery: In the field of wound dressing swollen hydrogels can be used as controlled release devices. For transdermal iontophoresis to obtain enhanced permeation of products viz. hormones and nicotine hydrogel based formulations are being explored [84].

Subcutaneous Delivery: Once the drug has been administered implantable hydrogels are now leading towards the development of biodegradable systems which don't require surgical removal [85, 86].

Novel Hydrogel for Controlled Drug Delivery: For controlled drug delivery HYPAN is the novel hydrogel having useful properties. HYPAN hydrogels are distinguished from others by physical network of crystalline clusters [87].

Hydrogel for Gene Delivery: In the treatment of many genetic and/or acquired diseases and conditions hydrogel versatility has potential application [88].

Cosmetology: When hydrogels implanted into breasts accentuate them for aesthetic reasons. These implants have silicon elastomer shells and are filled with hydroxyl propyl cellulose polysaccharide gel.

Tropical Drug Delivery: Hydrogel formulations are employed to deliver active components like Desonide. It is a synthetic corticosteroid and used as an anti-inflammatory for better patient compliance.

Protein Drug Delivery: Interleukins conventionally administered as injection are now given as hydrogels. They form an in-situ polymeric network and release proteins slowly and show better compliance [89].

Dyes and Heavy metal ions removal: The removal of heavy metal ions from various water resources is of great scientific and practical interest. By using synthetic cross-linked polyacrylate hydrogels the heavy metal toxicity has been removed from aqueous media [90]. The application of these synthetic materials on large scales may not be a practical solution because of the high cost of these materials. Adsorption process is used to remove the pollution caused by heavy metal ions. The process should also possess flexibility in design and operation; offer the advantage of reusing the treated effluent. The metal ion adsorption has been favored by the adsorbents with carboxyl, sulfonic, phosphonic, and nitrogen groups on their surface [91]. The hydrogels were proven to be excellent dye adsorbent materials. Among hydrogel-forming materials, polyelectrolytes have a special significance in heavy metal ions removal. Many applications of polyelectrolytes in this area are due to their ability to bind oppositely charged metal ions to form complexes. Hydrogels are versatile and viable materials that show potential for environmental applications.

Some biopolymer based hydrogels like chitosan, alginate, starch, and cellulose derivatives are used to remove metal ions from aqueous media [92, 93]. Due to the presence of multiple amino (NH_2) and hydroxyl (OH) groups in their structure chitosan-based hydrogels are applicable in the removal of heavy metal ions. This applicability originates from the tendency of metal ions to form chelates with amino groups. The adsorption capacity but also the physicochemical properties of chitosan have been improved by doing chemical modification of these functional groups [94-96]. It is shown in these studies that both adsorption capacity and mechanical resistance of chitosan-based hydrogels will improve after modification of functional groups.

CONCLUSION

This chapter aims to introduce the hydrogels briefly. They are a class of natural or synthetic polymeric materials that have the ability to hold huge amounts of water because of their specific structures and swelling properties. Based on this ability, they found a wide variety of applications, and because of the possibility to modify the polymeric structure to obtain desired functionality, the areas of applications are rapidly expanding. They can be designed in such a way that they can respond to a specific stimulus including pH, temperature, light, etc. at a predefined level and thus be stimuli responsive. Among their amazing characteristics, the biocompatibility and biodegradability make them a powerful candidate to use in biological and environmental applications as implants or materials for removal of toxic pollutants.

References

[1] Ahmed Enas, M., Aggor Fatma, S., Awad Ahmed, M., El-ArefAhmed, T., (2013) An innovative method for preparation of nanometalhydroxide superabsorbent hydrogel. *Carbohydr Polym* 91, 693–8.

[2] Buchholz, F. L., Graham, A. T., (1998) Hydrogel: Preparation, characterization, and applications. *Modern superabsorbent polymer technology. New York: Wiley- VCH* 119, [chapters 1–7].

[3] Brannon-Peppas, L., Harland, R. S., (1991) Absorbent polymertechnology. *J Controlled Release* 17(3), 297–8.

[4] Li, Y., Huang G., Zhang X., Li B., ChenY., Lu T., Tian Jian Lu., Xu F., Magnetic (2013)hydrogels and their potential biomedical applications. *AdvFunct Mater* 23(6), 660–72.

[5] http://vikno.eu/eng/health/health/scientists-develop-synthetichydrogel.html.

[6] Burkert, S., Schmidt, T., Gohs, U., Dorschner, H., Karl-Friedrich, A., (2007) cross-linking of poly(N-vinyl pyrrolidone) films by electron beam irradiation. *Radiat Phys Chem*76(8–9), 1324–8.

[7] Wen, Z., Xing Jin., Yang, Cong., Yuying, Liu., Jun, Fu., (2013) Degradable natural polymer hydrogels for articular cartilage tissue engineering. *J Chem Technol Biotechnol*88(3), 327–39.

[8] Takashi, L., Hatsumi, T., Makoto, M., Takashi, I., Takehiko, G., Shuji, S., (2007) Synthesis of porous poly(N-isopropylacrylamide) gel beads by sedimentation polymerization and their morphology. *JAppl Polym Sci* 104(2), 842.

[9] Davis, K. A., and Anseth K. S., (2002). Controlled release from crosslinked degradable networks. *Critical Reviews in Therapeutic Drug Carrier Systems*19, 385-423.

[10] Tonnesen, H. H., and Karlsen, J., (2002). Alginate in drug delivery systems. *Development and Industrial Pharmacy*28, 621-630.

[11] Liang, H. F., Hang, M. H., Ho, R. M., Chung, C. K., Lin, Y. H., and Chen, C. H., (2004). Novel method using a temperature-sensitive polymer (methycellulose) to thermally gel aqueous Alginate as a pH-senstve hydrogel. *Biomacromolecules*5, 1917-1925.

[12] Hennink, W. E., Franssen, O., Van Dijk-Wolthuis, W. N. E., & Talsman, H. (1997). Dextran hydrogels fort he controlled release of propeties. *Journal of ControlledRelease*48, 107-114.

[13] Stubble, B., Maris, B., Van Den. Mooter, G., De Smendt, S. C., & Demeester, J. (2001). The ınvitra evalution of Azo containing polysaccharide gels for colon delivery. Journal of *Controlled Release*75, 103-114.

[14] Agnihotri, S. A., & Aminabhavi, T. M. (2005). Development of novel interpenetrating network gellan gum-poly(vinyl alcohol) hydrogel microspheres for the controlled release of carvedilol. *Drug Development and IndustrialPharmacy*31, 491–503.

[15] Barbucci, R., Consumi, M., Lamponi, S., and Leone, G., (2003). Polysaccharides based hydrogels for biological applications. *Macromolecular Symposium*204, 37–58.

[16] Ha, D. I., Lee, S. B., Chong, M. S., Lee, Y. M., Kim, S. Y., and Park, Y. H., (2006). Preparation of thermo-responsive and injectable hydrogels based on hyaluronic acid and poly(N-isopropylacrylamide) and their drug release behaviours. *Macromolecular Research*14, 87–93.

[17] Kumari, K., and Kundu, P. P., (2007). Semiinterpenetrating polymer networks of chitosan and L-alanine for monitoring the release of chlorpheniramine maleate. *Journal of Applied Polymer Science*103, 3751–3757.

[18] Guo, B. L., and Gao, Q Y., (2007). Preparation and properties of a pH/temperatureresponsive carboxymethyl chitosan/poly(N-isopropylacrylamide) semi-IPN hydrogel for oral delivery of drugs. *Carbohydrate Research*342, 2416–2422.

[19] Gupta, V. K., Pathania, D., Kothiyal, N. C., and Sharma, G., (2013). Use of Pectin - thorium (IV) tungstomolybdate nanocomposite for photocatalytic degradation of methylene blue. *Carbohydrate Polymer*96, 277-287.

[20] Peppas, N. A., Bures, P., Leobandung, W., and Ichikawa, H., (2000). Hydrogels in pharmaceutical formulations. *European Journal of Pharmaceutical and Biopharmaceutics* 50, 27-46.

[21] Davis, K. A., and Anseth, K. S., (2002). Controlled release from crosslinked degradable networks. *Critical Reviews in Therapeutic Drug Carrier Systems*19, 385-423.

[22] Pathania, D., Gupta, D., Kothiyal, N. C., Eldesoky, G. E., & Naushad, M., (2016). Preparation of a novel chitosan-g-poly (acrylamide)/Zn nanocomposite hydrogel and its applications for controlled drug delivery of ofloxacin. *International journal of biological macromolecules 84*, 340-348.

[23] Salama, A., Abou-Zeid, R. E., El-Sakhawy, M., & El-Gendy, A., (2015). Carboxymethyl cellulose/silica hybrids as templates for calcium phosphate biomimetic mineralization. *International journal of biological macromolecules74*, 155-161.

[24] Yang, L., Chu, J. S., Fix, J. A., (2002) Colon-specific drug delivery: new approaches and in vitro/in vivo evaluation. *Int J Pharm* 235, 1–15.

[25] Maolin, Z., Jun, L., Min, Y., Hongfei, H., (2000) The swelling behaviour of radiation prepared semi-interpenetrating polymer networks composed of polyNIPAAm and hydrophilic polymers. *RadiatPhys Chem* 58, 397–400.

[26] Hacker, M. C., Mikos, A. G., (2011) Synthetic polymers, principles of regenerative medicine. 2nd ed. 587–622.

[27] Shin, J., Braun, P., Lee, W., (2010) Fast response photonic crystal pH sensor based on templated photopolymerized hydrogel inverse opal. *Sens Actuat B: Chem* 150 (1), 183–90.

[28] Wichterle, O., (1960) Hydrophilic gels for biological use. *Nature* 185, 117.

[29] Singh, A., Sharma, P. K., Garg, V. K., Garg, G., (2010)*Hydrogels: a review* 24(2):Article 016.

[30] Amulya, K., Saxena, (2010) synthetic biodegradable hydrogel (PleuraSeal) sealant for sealing of lung tissue after thoracoscopicresection. *J Thoracic Cardiovasc Surg* 139(2), 496–7.

[31] Hamidi, M., Azadi, A., Rafiei, P., (2009) Hydrogelnanoparticles in drug delivery. *Adv Drug Deliv Rev* 60(15), 1638–49.

[32] Sun, X., Zhang, G., Shi, Q., Tang, B., Wu, Z. J., (2002) Preparation andcharacterization of water-swellable natural rubbers. *J ApplPolym Sci* 86, 3212–717.

[33] Chen, X., Martin, B. D., Neubauer, T. K., Linhardt, R. J., Dordick, J. S., Rethwisch, D. G., (1995) Enzymatic and chemoenzymatic approaches tosynthesis of sugar based polymer and hydrogels. *Carbohydr Polym* 28, 15–21.

[34] Kashyap, N., Kumar, N., Kumar, M., (2005) Hydrogels forpharmaceutical and biomedical applications. *Crit Rev TherDrug Carr Syst* 22, 107–49.

[35] Kaihara, S., Matsumura, Shuichi., Fisher John, P., (2008) Synthesisand characterization of cyclic acetal based degradable hydrogels. *Eur J Pharm Biopharm* 68(1), 67–73.

[36] Stamatialis Dimitrios, F., Papenburg Bernke, J., Girone´ s, M., Saiful S., Bettahalli Srivatsa, N. M., Schmitmeier, S., Wessling, M., (2008) Medical applications of membranes: drugdelivery, artificial organs and tissue engineering. *J Membr Sci* 308(1–2), 1–34.

[37] Zhang, L., Li, K., Xiao, W., Zheng, Li., Xiao, Y., Fan, H., (2011) Preparation of collagen–chondroitinsulfate–hyaluronic acid hybrid hydrogel scaffolds and cellcompatibility in vitro. *Carbohydr Polym* 84(1), 118–25.

[38] Saul Justin, M., Williams David, F., (2011) Hydrogels in regenerativemedicine, principles of regenerative medicine. 637–61.

[39] Van der Linden, H. J., Herber, S., Olthuis, W., Bergveld, P., (2003) Patterneddual pH responsive core shell hydrogels with controllableswelling kinetics and volume. *Analyst* 128, 325–31.

[40] Sikareepaisan, P., Ruktanonchai, U., Supaphol, Pitt., (2011)Preparation and characterization of asiaticoside-loaded alginatefilms and their potential for use as effectual wound dressings. *Carbohydr Polym* 83(4), 1457–69.

[41] Wang, F., Li, Z., Khan, M., Tamama, K., Kuppusamy, P., (2010) Injectable, rapid gelling andhighly flexible hydrogel composites as growth factor and cellcarriers. *Acta Biomater* 6(6), 1978–91.

[42] Roy, D., Cambre Jennifer, N., Brent, S., Sumerlin, (2010) futureperspectives and recent advances in stimuli-responsive materials. *Prog Polym Sci* 35(12), 278–301.

[43] Krsko, Peter., McCann Thomas, E., Thach, T. T., LaabsTracy, L., Geller Herbert, M., Libera Matthew, R., (2009) Length-scalemediated adhesion and directed growth of neural cells bysurface-patterned poly(ethylene glycol) hydrogels OriginalResearch Article. *Biomaterials* 30(5), 721–9.

[44] Park, J. H., Kim, D., (2001) Study on foaming water-swellable EPDMrubber. *J Appl Polym Sci* 80, 115–21.

[45] Pourjavadi, A., Harzandi, A. M., Hosseinzadeh, H., (2004) Modifiedcarrageenan 3. Synthesis of a novel polysaccharide-basedsuperabsorbent hydrogel via graft copolymerization of acrylicacid onto kappa-carrageenan in air. *Eur Polym J*40(7), 1363–70.

[46] Peppas, N. A., Mikos, A. G.,(1986) Hydrogels inmedicine and pharmacy – fundamentals, *vol. I. Florida: CRCPress, Inc.* 1–25.

[47] Khoylou, F., Naimian, F., (2009) Radiation synthesis of superabsorbentpolyethylene oxide/tragacanth hydrogel. *Radiat Phys Chem* 78(3), 195–8.

[48] Mi-Ran, P., Chun Chang, J., Sung-Won, A., Min-Hyo, K., Chong-Su, C., Song S.C., (2010) Sustained delivery of humangrowth hormone using a polyelectrolyte complex-loadedthermosensitive polyphosphazene hydrogel. *J ControlledRelease* 147(3), 359–67.

[49] Vijayalakshmi, S., Kenichi, Takahata., (2009) A hydrogel-basedpassive wireless sensor using a flex-circuit inductive transducer. *Sens Actuat A: Phys* 155(1), 58–65.

[50] Tabata, Y., (2009) Biomaterial technology for tissue engineering applications. *J R Soc Interf* 6, 311–24.

[51] Shantha, K. L., Harding, D. R. K., (2002) Synthesis and evaluation ofsucrose-containing polymeric hydrogels for oral drug delivery. *J Appl Polym Sci* 84, 2597.

[52] Raju, K. M., Raju, M. P., (2001) Synthesis of novel superabsorbingcopolymers for agricultural and horticultural applications. *Polym Int* 50, 946–51.

[53] Takeda, H., Taniguchi, Y., (1985) Production process for highly waterabsorbable polymer. *US Patent* 4, 525-527.

[54] Chen, J., Zhao, Y., (2000) Relation between water absorbency andreaction conditions in aqueous solution polymerization ofpolyacrylate superabsorbent polymers. *J Appl Polym Sci* 75, 808–14.

[55] Ta, H. T., Dass, C. R., and Dunstan, D. E., (2008). Injectable chitosan hydrogels for localized cancer therapy. *Controlled Release* 126, 205-216.

[56] Schuetz, Y. B., Gumy, R., and Jordan, O., (2008). A novel thermoresponsive hydrogel of chitosan. *European Journal of Pharmaceutical and Biopharmaceutics* 68, 19-25.

[57] Yang, X., Liu, Q., Chen, X., Feng, Y., and Zhu, Z. (2008). Investigation of PVA/wschitosan hydrogels prepared by combined γ-irradiation and freeze-thawing. *Carbohydrate Polymers*73 (3), 401-408.

[58] Singh, A., Sharma, P. K., Garg, V. K., and Garg, G. (2010). Hydrogels: A review. International. *Journal of Pharmaceutical Sciences Review and Research*4, 97-105.

[59] Katayama, T., Nakauma, M., Todoriki, S., Phillips, G. O. and Tada, M. (2006) Radiation-inducedpolymerization of gum arabic (Acacia sengal) in aqueous solution. *Food Hydrocolloids* 20, 983-989.

[60] Nagasawa, N., Yagi, T., Kume, T. and Yoshii, F. (2004) Radiation crosslinking of carboxymethylstarch. *Carbohydrate Polymers*58, 109-113.

[61] Al-Assaf, S., Dickson, P., Phillips, G. O., Thompson, C. and Torres, J. C. (2009) Compositionscomprising polysaccharide gums. In World Intellectual property Organization, (ed. PCT), *Phillips Hydrocolloid Research Limited (UK), Reckitt Benckiser (UK), United Kingdom.* Vol. WO2009/016362 A2.

[62] Yoshii, F., and Kume, T., (2003) Process for producing crosslinked starch derivatives andcrosslinked starch derivatives produced by the same. *(ed. U. S. Patent), Japan Atomic Energy Research Institute (Tokyo, JP), USA.* 6, 617-448.

[63] Liu, P., Zhai, M., Li, J., Peng, J. and Wu, J. (2002a) Radiation preparation and swelling behavior ofsodium carboxymethyl cellulose hydrogels. Radiation Physics and Chemistry 63, 525-528.

[64] Liu, P., Peng, J., Li, J. and Wu, J. (2005) Radiation crosslinking of CMC-Na at low dose and itsapplication as substitute for hydrogel. *Radiation Physics and Chemistry*72, 635-638.

[65] Valles, E., Durando, D., Katime, I., Mendizabal, E., and Puig, J. E., (2000) Equilibrium swelling andmechanical properties of hydrogels of acrylamide and itaconic acid or its esters. *Polymer Bulletin*44, 109-114.

[66] Mansur, H. S., Orefice, R. L., and Mansur, A. A. P., (2004) Characterization of poly(vinylalcohol)/poly(ethylene glycol) hydrogels and PVA-derived hybrids by small-angle Xrayscattering and FTIR spectroscopy. *Polymer*45, 7193-7202.

[67] Torres, R., Usall, J., Teixido, N., Abadias, M., and Vinas, I. (2003) Liquid formulation of thebiocontrol agent Candida sake by modifying water activity or adding protectants. *Journal of Applied Microbiology*94, 330-339.

[68] Aikawa, K., Matsumoto, K., Uda, H., Tanaka, S., Shimamura, H., Aramaki, Y. and Tsuchiya, S. (1998) Hydrogel formation of the pH response polymer polyvinylacetaldiethylaminoacetate (AEA). *International Journal of Pharmaceutics* 167, 97-104.

[69] Aouada, F. A., de Moura, M. R., Fernandes, P. R. G., Rubira, A. F. and Muniz, E. C., (2005) Opticaland morphological characterization of polyacrylamide hydrogel and liquid crystalsystems. *European Polymer Journal*41, 2134-2141.

[70] El Fray, M., Pilaszkiewicz, A., Swieszkowski, W., and Kurzydlowski, K. J., (2007) Morphologyassessment of chemically modified cryostructured poly(vinyl alcohol) hydrogel. *European Polymer Journal* 43, 2035-2040.

[71] Pourjavadi, A., and Kurdtabar, M., (2007) Collagen-based highly porous hydrogel without anyporogen: Synthesis and characteristics. *European Polymer Journal* 43, 877-889.

[72] Al-Assaf, S., Navaratnam, S., Parsons, B. J., and Phillips, G. O., (2006a) Chain scission of hyaluronan by carbonate and dichloride radical anions: Potential reactive oxidative species in inflammation? *Free Radical Biology and Medicine* 40, 2018-2027.

[73] Al-Assaf, S., Phillips, G. O., and Williams, P. A., (2006b) Controlling the molecular structure of food hydrocolloids. *Food Hydrocolloids* 20, 369-377.

[74] Al-Assaf, S., Phillips, G. O., Aoki, H., and Sasaki, Y., (2007a) Characterization and properties of Acacia senegal (L.) Willd. var. senegal with enhanced properties (Acacia (sen) SUPER GUM(TM)): Part 1--Controlled maturation of Acacia senegal var. senegal to increase viscoelasticity, produce a hydrogel form and convert a poor into a good emulsifier. *Food Hydrocolloids* 21, 319-328.

[75] Al-Assaf, S., Phillips, G. O., Williams, P. A., and Plessis, T. A. D., (2007b) Application of ionizing radiations to produce new polysaccharides and proteins with enhanced functionality. *Nuclear Instruments and Methods in Physics Research B*265, 37-43.

[76] Phillips, G. O., Plessis, T. A. D., SaphwanAl-Assaf, and Williams, P. A., (2003) Biopolymers obtained by solid state irradiation in an unsaturated gaseous atmosphere. *Phillips Hydrocolloid Research limited (ed. U. S. Patent) UK* 6, 610-810.

[77] Hoffman, A. S., (2002) Hydrogels for biomedical applications. *Advanced Drug Delivery Reviews* 43, 3–12.

[78] Lazareva, T. G., and Vashuk, E. V., (1995) Features of rheological and electrophysical properties of compositions based on polyvinyl alcohol and carboxymethylcellulose. *Mechanics ofComposite Materials*31, 524-532.

[79] Singh, B., and Vashishth, M., (2008) Development of novel hydrogels by modification of sterculia gum through radiation cross-linking polymerization for use in drug delivery. *NuclearInstruments and Methods in Physics Research B* 266, 2009-2020.

[80] Torres, R., Usall, J., Teixido, N., Abadias, M., and Vinas, I. (2003) Liquid formulation of the biocontrol agent Candida sake by modifying water activity or adding protectants. *Journal of Applied Microbiology* 94, 330-339.

[81] Mansur, H. S., Orefice, R. L., and Mansur, A. A. P., (2004) Characterization of poly(vinylalcohol)/poly(ethylene glycol) hydrogels and PVA-derived hybrids by small-angle Xray scattering and FTIR spectroscopy. *Polymer* 45, 7193-7202.

[82] Janik, I., Kasprzak, E., Al-Zier, A., and Rosiak, J. M., (2008) Radiation crosslinking and scissionparameters for poly(vinyl methyl ether) in aqueous solution. *Nuclear Instruments &Methods in Physics Research Section B-Beam Interactions with Materials and Atoms* 208, 374-379.

[83] Rosiak, J. M., (1998) Gel / sol analysis ofirradiated polymers. *Radiation Physics and Chemistry* 51, 13-17.

[84] Wach, R. A., Mitomo, H., Nagasawa, N., and Yoshii, F., (2003b) Radiation crosslinking ofmethylcellulose and hydroxyethylcellulose in concentrated aqueous solutions. *Nuclear Instruments and Methods in Physics Research B*211, 533-544.

[85] Xu, G. Y., Chen, A. M., Liu, S. Y., Yuan, S. L., and Wei, X. L., (2002) Effect of C12NBr on theviscoelasticity of gel containing xanthan gum/Cr(III). *Acta Physico-Chimica Sinica* 18, 1043-1047.

[86] Singh, A., Sharma, P. K., Garg, V. K., and Garg, G., (2010). Hydrogels: A review. *International Journal of Pharmaceutical Sciences Review and Research*4, 97-105.

[87] Lee, K. Y., and Mooney, D. J., (2001). Hydrogels for tissue engineering. *Chemical Reviews* 101 (7), 1869-1880.

[88] Van der Linden, H. J., Herber, S., Olthuis, W., and Bergveld, P., (2003). Patterned dual pH responsive core shell hydrogels with controllable swelling kinetics and volume. *Analyst* 128, 325-331.

[89] Azad, A. K., Sermsintham, N., Chandrkrachang, S., and Stevens, W. F., (2004). Chitosan membrane as a wound healing dressing: characterization and clinical applications. *Journal of Biomedical Materials Research Part B Applied Biomaterials* 69(B), 216-222.

[90] Singh, A., Sharma, P. K., Garg, V. K., and Garg, G., (2010). Hydrogels: A review. *International Journal of Pharmaceutical Sciences Review and Research* 4, 97-105.

[91] Guo, K., and Chu, C. C., (2005). Synthesis and characterization of novel biodegradable unsaturated poly (ester amide)/poly (ethylene glycol) diacrylate hydrogels. *Journalof Polymer Science Part A: Polymer Chemistry* 43 (17), 3932-3944.

[92] Bahram, M., Nurallahzadeh, N., Mohseni, N., (2015) pH sensitive hydrogel for Coacervative Cloud Point extraction and Spectrophotometric Determination of Cu(II): Optimization by Central Composite Design. *J. Iran. Chem.. Soc.* 12(10), 1781-1787.

[93] Ramírez, E., Burillo, S. G., Díaz, C. B., Roa, G., Bilyeu, B., (2011) Use of pH-sensitive Polymer Hydrogels in Lead Removal from Aqueous Solution. *J. Hazard. Mater.* 192, 432-439.

[94] Dragan, E. S., (2014) Design and Applications of Interpenetrating Polymer Network Hydrogels. *Chem. Eng. J.* 243, 572-590.

[95] Ramesh, A., Hasegawa, H., Sugimoto, W., Maki, T., Ueda, K., (2008) Adsorption of Gold(III), Platinum(IV) and Palladium(II) onto Glycine Modified Crosslinked Chitosan Resin. *Bioresource Technol.* 99, 3801-3809.

[96] Kandile, N. G., Nasr, A. S., (2009) Environment Friendly Modified Chitosan Hydrogels as a Matrix for Adsorption of Metal Ions, Synthesis and Characterization. *Carbohydr. Polym.* 78: 753-759.

In: Modified Biopolymers: Challenges and Opportunities ISBN: 978-1-53612-116-2
Editors: D. Pathania, G. Sharma and A. Kumar

Chapter 5

WASTE DERIVED BIOCHAR BASED BIO NANOCOMPOSTIES: RECENT PROGRESS IN UTILIZATION AND INNOVATIONS

Preeti Oswal*[1]*, Anamika Rana*[1]*, Renato Cataluna Veses*[2]*,
***Ajay Kumar*[1] *and Amit Kumar*[1,*]**
[1]School of Chemistry, Shoolini University, Solan, India
[2]Instituto de Qu´ımica, Universidade Federal do Rio Grande do Sul (UFRGS), Porto Alegre, Brazil

ABSTRACT

Biochar has been used as a means of remediating contaminated soil, water and waste management. The practical applications of magnetic biochar for contaminant immobilization and removal, however, need further improvements. Hence, further improvements are being introduced into the biochar by coating or impregnation and other modification techniques. In recent times, biochar modifications involving various methods such as acid treatment, base treatment, amination, surfactant modification, impregnation of mineral sorbents, steam activation and magnetic modification have been widely studied. This chapter summarizes and evaluates biochar modification methods, corresponding mechanisms, and their benefits for contaminant management in soil and water by adsorbing organic and inorganic substances. The mechanism of contaminant removal depends on the type of contaminants (i.e., inorganic/organic, anionic/cationic, hydrophilic/hydrophobic, polar/non-polar, etc.), environmental conditions, remediation goals and land use purposes.

Keywords: Biochar, adsorbent, catalysis, modification, soil detoxification, porosity, bio-nanocomposites

[*] Corresponding Author Email: mittuchem83@gmail.com.

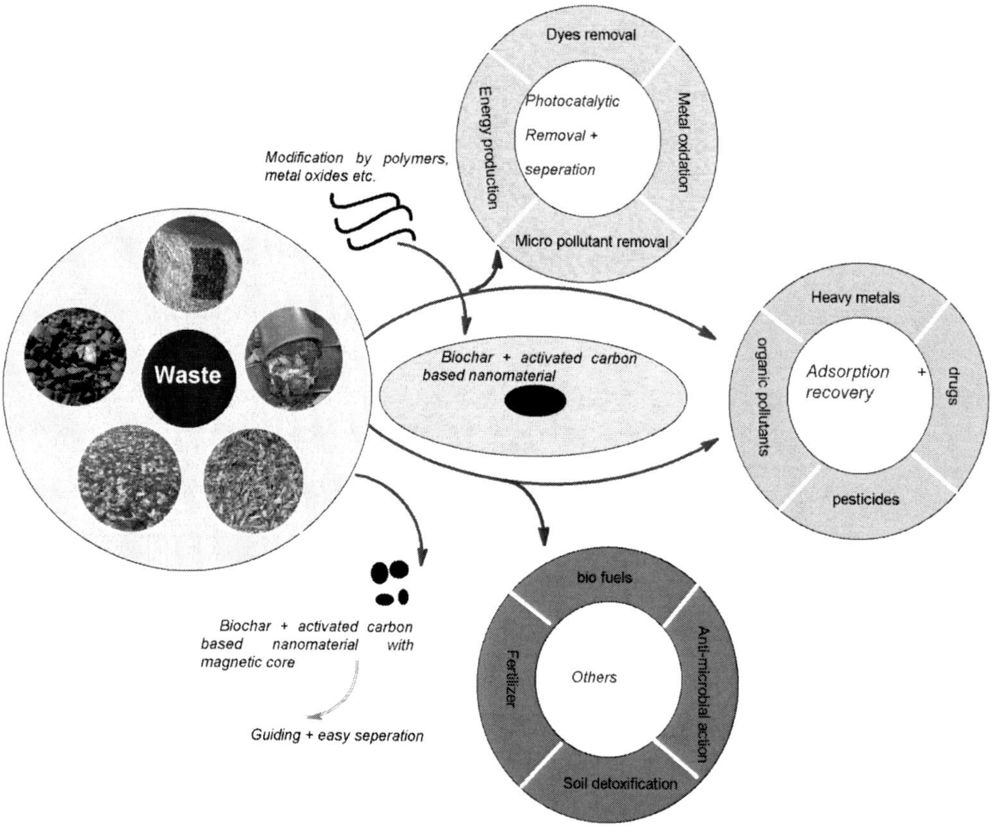

1. INTRODUCTION

1.1. Biochar

Biochar is the carbonaceous solid residue obtained upon heating biomass such as wood, manure, wastes (animals or plants), with little or no oxygen (pyrolysis or "charring") [1]. The International Biochar Initiative (IBI) (http:/cx/www.biochar-international.org/biochar), states "biochar is a solid material obtained from the carbonization of biomass. Char, charcoal, soot, graphitic carbon, ash, coal and black carbon (black C) have all been used to describe the solid residual products" [2]. It has potential as a nutrient recycler, income generator, waste management system, and agent for long-term, safe and economical carbon sequestration and has also been linked to beneficial soil quality improvements [3–7]. Biochar's carbon sequestration benefit results from the fact that carbon from the atmosphere–biosphere pool is transferred to a slower cycling form (i.e., black C) that has the potential to exist for hundreds to thousands of years [8, 9], or possibly longer [10, 11]. Biochar is not a structured homogeneous material; rather it possesses a range of chemical structures and a heterogeneous elemental composition. Many researchers and farmers worldwide are paying attention to its hidden secrets. The increasing curiosity in biochar has initiated multidisciplinary areas for science researches and engineering applications. A large number of studies have emphasized the benefit of using biochar in terms of moderating global warming, enhancing of crop yield and carbon storage [12-17]. The potential to utilize biochar for various applications depends on its properties. For example, biochar with high electrical conductivity, porosity and stability at lower temperatures is preferred as electrode material in microbial fuel cells [18]. Biochar containing relatively high structural bound oxygen groups is preferred in direct carbon fuel cells [19]. Biochar with high porosity and structural bound nitrogen groups is preferred in the development of super capacitors [20]. Furthermore, the high surface area, low ash content of biochar may be preferred as soil amendments.

1.2. Magnetic Biochar

Biochar has rapidly been known as a multifunctional material and has extensively been used for enhancing soil fertility and removal of pollutants from water and soil environments [21]. Because of the surface effective functional groups and higher adsorption potentials, biochar has many advantages for pollutant removal [22, 23]. Biochar made from wood, pyrolysed at 450_C was very effective for the removal of heavy metals like As^{+3},Cd^{+2},Pb^{+2} [24]. However the biochar in powdered form is difficult to detach from aqueous solution. To solve this difficulty, researchers applied innovative adsorbent magnetization technique for non-magnetic adsorbents [25]. The magnetic Fe_3O_4 based adsorbents are highly effective for pollutants removal and the adsorbent can easily be separated from the solution by external magnetic field at the same time [26, 27, 28]. Thus by the introduction of Fe_3O_4 to biochar by the method of chemical co-precipitation is highly effective and it functions as bulk magnetic sorbent with higher adsorptive affinity for organic pollutants [21], organic arsenic [29] and Cr(VI) [30]. Magnetic biochar made from empty fruit branches and functionalized with CNT's possesses high adsorptive affinity for Zn^{+2} removal. Magnetic biochar composites made from paper mill sludge biochar was applied for pentachlorophenol removal from effluent water [31]. For arsenic adsorption Tian et al. synthesized magnetic wheat straw with differentFe_3O_4 contents for arsenic adsorption. Results showed that higher arsenic removal can be achieved with higher Fe_3O_4 content in biochar matrix [32].

2. Production of Biochar

2.1. From Agricultural Wastes

Biochar is a multi-process product whose qualities are dependent on each process and also the material to which the process is applied. Biochar has great potential for managing the waste stream originating from animals or plants; thus, decreasing the associated pollution loading to the environment. The use of waste biomass for biochar production is not only economical but also beneficial. The benefits mainly including energy production and climate change mitigation [33]. Waste biomass that has been extensively used to produce biochar including various types of waste wood [34-36], corn cob and corn stalk [37], wheat residues [38], rice straw [39], and sugar beet tailing [40]. The composition and proportions of different constituents including cellulose, hemicellulose, lignin and small amounts of extractives, vary with the type of biomass [41-43]. These constituents also pyrolyze at different temperatures, with thermal degradation of hemicelluloses and celluloses occurring first (at temperatures of 470-530 K and 510-620 K, respectively) and lignin last (between 550 and 770 K). Due to variations in lignin and recalcitrant carbon content among biochar source materials, the amount of biochar produced from a given amount of biomass varies greatly, with the highest yields typically from lingo cellulosic (wood-derived) biomass. Thus, poor production efficiencies and/or low or seasonal availability of certain feed stocks (i.e., low-lignin biomass) may limit their widespread use. Notably, pyrolyzing the waste biomass, particularly animal manure and sewage sludge, kills any microbes present, thereby reducing the environmental health effects [44]. However, the persistence of toxic heavy metals in biochar developed from sewage sludge and municipal solid waste [45] must be carefully handled before long-term application to soils.

2.2. From Industrial Waste

A large quantity of sludge, dyes and other waste is generated during various manufacturing processes in industries. Treatment, management and disposal of industrial waste is a big challenge everywhere because it contains a large amount of heavy metals and other organic and inorganic pollutants which leads to environmental pollution. Biochars made from industrial waste can be used for multidisciplinary application which alternatively leads to reduce the environmental pollution. A zero valent iron magnetic biochar composite made from the paper mill sludge can be used for the treatment of synthetic and real effluent containing pentachlorophenol (PCP) [46]. A biochar made from Dairy manure efficiently removes Cu^{2+}, Zn^{2+} , Cd^{2+} from waste water generated from various industries [47]. Digested dairy waste biochar removes Pb^{2+} ions from waste water [48].

2.3. From Household Waste

Biochar has a great potential as an adsorbent in water and waste water treatment for the removal of organic and inorganic contaminants as a result of its porous nature, surface area and abundant surface functional groups [49]. Biochar made from household waste like tea waste, cabbage, orange peel had been widely used for the pollutants removal from contaminated soil and water. More than 90% of tea is left after the consumption of tea in our homes. To recycle this tea waste the technologies of pyrolysis and carbonization to form bio-oil and bio-char is the focus of recycling technologies [50]. Biochar prepared from waste tea is a cheap and efficient adsorbent for removal of elemental mercury because of its renewable properties. Biochar made from tea waste also employed for methylene blue, carbofuran, sulphamethazine removal. Biochar made from Korean cabbage waste significantly adsorb crystal violet dye 4.8 times more than activated carbon [51].

3. Modification of Biochar

Biochar can be described as a byproduct of biomass pyrolysis, obtained from fast thermochemical decomposition of lignocellulosic biomass at relatively low temperatures (<973 K). Several methods like acid and alkali modification , oxidation , chemical graft etc. can be employed for biochar modification and these modifications enhances its absorption performance of heavy metals [52-54]. Modified biochars contains many functional moieties like carboxyl, hydroxyl, amino and these functional moieties favours the absorption of pollutants.

3.1. Chemical Modification

There are several methodologies which help us to chemically modify the biochar like acid and base treatment, modification of functional groups, surfactants modification, coating etc. one step modification and two step modification processes are involved. In presence of activating agent the carbonization and activation steps are carried out simultaneously in case of one step modification. Pretreatment of precursor in one step and carbonization in next step is done in two step modification process [55, 56].

Treatment of biochar with strong acids introduces various functional groups like amine, carboxylic groups on carbonized surfaces, thereby enhancing the metal sorption capacity by cation exchange and surface complexion. Strong acids like H_3PO_4, HCL, and H2SO4 on aqueous oxidation increases surface acidities and modify porous structure of biochar [57]. Sorption of pollutants can also be enhanced by using organic acids like oxalic acid through ligand and proton promoted processes [58].

Alkali modification using KOH and NaOH increases O-content and surface basicity [59]. In case of KOH modification, surface area and additional hydroxyl groups are increased. The functional groups present on surface remains unchanged in this modification and the sorption of Cu^{+2}, Cd^{+2} greatly enhanced. Amino modification of biochar also increases Cu^{+2}sorption by strong surface complexion [60], which was highly selective and minimally affected by competing cations.

Modification of biochar via organic solvents like methanol makes the biochar rich in hydroxyl and ester groups, which ultimately increases the formation of electron-donor–acceptor interaction between the surface of biochar and contaminated soil and water [61]. This methanol modified biochar had greater capacity for tetracycline removal. This is because of the increased electron density of O atoms on biochar surface, which plays an important role in the removal of tetracycline [61].

Coating of biochar by metal oxide, CNT, grapheme improves the properties of biochar and ultimately enhances the sorption capacity of biochar. Fe and Co coated bamboo charcoal significantly increased the hydroxyl group because of the formation of their oxides on carbon surface, which was used for the removal of heavy metals from aqueous solution [62-64] synthesized Fe_3O_4 coated bamboo charcoal which was used as a sorbent for solid-phase micro-extraction to remove PBDEs in water. A large increase of surface area(from $9 m^2 g^{-1}$ to $390 m^2 g^{-1}$ on 1% w/w CNT coating), thermal stability and pore volume was observed on CNT loaded bagasse biochar [65], was used for methylene blue sorption. Graphene coated biochar also enhances the sorption of phenol and methylene blue because of the high surface area and pore volume after coating [66]. Table 1 shows the different modification methodologies for biochar synthesis and its,applications.

3.2. Physical Modification

The methodologies involved in physical or mechanical methods are simple and economically viable, uses various oxidizing agents like carbon dioxide, steam, air. These physical methodologies do not involve any chemicals.

Table 1. Biochar modification methodologies, contaminants removal and applications

Method of modification	Feedstock	Target sorbate	Enhancement	References
Chemical modifications				
H_2O_2modification	Peanut hull	Pb^{+2},Cu^{+2}, $Ni^{+2}$and Cd^{+2}	Enhanced Pb^{2+} sorption from 0.88 to 22.82 mg g_1, comparable to commercial activated carbon	[47]
Chemical oxidation ($KMnO_4$, HNO_3, or NaOH)	Bamboo	Furfural	Chemical treatments inhibited sorption of furfural	[107]
Methanol-modified Biochar	Rice husk	Tetracycline (TC)	Approximately 45.6% enhancement ofsorption capacity in 12 h and 17.2% at equilibrium	[61]
Hybrid multi-walled carbon nanotube (CNT)-coated biochars	Hickory chips and sugarcane bagasse	MB	Highest sorption capacity among all the sorbents	[65]
Physical modifications				
Steam and KOH activation	Whitewood (Spruce)	Emission of N_2O and CH_4		[172]
Steam activation	Chicken litter, alfalfa	Cu^{+2}		[68]
Impregnation with mineral sorbents				
Clay-biochar Composites	stems, switchgrass	MB	Increase of sorption capacities by about 5 Times	[40]
Porous MgO-biochar Nanocomposites	corn cob and stover	Phosphorus and nitrates		[106]
Mg-biochar Nanocomposite	Bamboo, bagasse, and hickory chips	Phosphorus	Maximum sorption capacity >100 mg/g	[53]
Magnesium modified biochar		Phosphorus		[59]
Biochar supported nanoscale zerovalent iron (nZVI) composite	Mg enriched tomato leaves	Rice hull	Degradation efficiency of TCE (0.15 mmol/L) was 99.4% in the presence of nZVI/biochar (4.5 mmol/L, nZVI to biochar mass ratio was 1:5) and persulfate (4.5 mmol/L) within 5 min	[143]
Magnetic modifications				
Magnetic biochar/g-Fe2O3 composite	Corn	Arsenic	Maximum sorption capacity of3147mg/kg	[161]
Magnetic biochars	Rice hull, Cotton wood	Cd^{+2}		[128]

In steam activation process a conventional biochar could be converted to activated biochar having improved carbonaceous structure and high surface area. In this process after the initial pyrolysis reaction, biochar is complemented with partial gasification with steam. This helps to promote the formation of crystalline carbon in biochar and changes the properties of biochar by removing the trapped products of incomplete combustion during pyrolysis [67]. New pores are created and smaller pores were enlarged during pyrolysis [68]. A significant increase in suface area from 345 $m^2 mg^{-1}$ to 576 $m^2g^{-1}$was observed in tea waste biochar by steam activation [69]. This increase in surface area could be accredited to biochar surface corrosion and release of further syn gas from biochar, generally in the form of hydrogen. The sorption capacity of inorganic and organic contaminants was increased by activation of biochar using an oxidizing agent like steam [69].

Modification of biochar by purging gases like ammonia and carbon dioxide significantly enhances the gas sorption capacities of conventional biochar [70, 71]. In cotton stack biochar after its production it was heated upto preset temperature in a quartz reactor with nitrogen and carbon dioxide purging. The carbon dioxide treatment facilitated the pore formation and improved the microporous structure of biochar which results in the enhancement of gas sorption capacity of biochar.

3.3. Slow Pyrolysis

Pyrolysis is a thermochemical decomposition process during which biomass is heated at elevated temperature in the absence of oxygen. Biochar for many purpose can is traditionally made by slow pyrolysis: heating in the absence of oxygen. The process is characterized by slow heating rates and long residence times. Slow-pyrolysis is regarded as the main pyrolysis process for the production of biochar because of higher solid yield (25–35%) compared to the other pyrolysis processes [72, 73]. During slow-pyrolysis, biomass is heated in a temperature range of 300–650°C with long residence time (few minutes to couple of hours) and low heating rates (10–30^{0}C/min) [74]. Their action time, temperature, pressure, heating rates, and initial moisture content of raw material are considered as the main key parameters affecting percentage yield and physicochemical properties of biochar [75, 76]. Low operating temperatures and slow heating rate favor high solid product yield [77, 78] where the high operating temperature and high heating rate show significant influence on the carbon percentage, HHV and BET-surface area of biochar [79].

3.4. Fast Pyrolysis

Fast pyrolysis is somewhat new, promising technology which consequently maximizes condensable oil and also gasification. These produce predominantly heavy oil or gas, yielding only small proportions of char. Fast pyrolysis involving a high liquid yield achieved through rapid heating rates of 10 to >1000°C/s, short residence times of <2 s, and rapid quenching of the vapors [80, 81].). The peak temperature is usually set between 500 and 550°C in order to obtain the highest bio oil yield [82-85]. These operating conditions particularly favor the formation of liquid products (bio-oil), but inhibit the formation of charcoal [86].

Various technologies have been used and proposed or tested including: fixed beds, augers, ablative methods, rotating cones, fluidized beds and circulating fluidized beds. Surface charring must be continuously removed during reaction to prevent pyrolysis of particle interiors being inibited by its insulating effect.

Concerning the physical properties of charcoals obtained by fast pyrolysis, Boateng, [87] Brewer and co-workers,[88]and Mullen and co-workers[89]reported relatively low BET surfaces areas (3.10–21.6 m^2 g−1) for chars formed from switch grass and corn stover in a fluidized bed reactor. This result is expected because of the short residence time of solid particles [90]. In addition to this, Brewer and co-workers also observed a very small particle size for charcoals obtained from fast pyrolysis of switch grass. This fact is mainly due to the small particle size of the feedstock (averaging around 1 mm) usually required in fast pyrolysis systems, and, probably, to the hypothesis that fast de-volatilization might create very fragmented char structures [91]. From a chemical composition point of view, charcoals obtained at high heating rates are characterized by high oxygen content [92] and low calorific value, [93] probably as a result of the relatively short particle residence time.

The fast pyrolysis systems employs six distinct steps for conversion of biomass to transportation fuels: pre-treatment, pyrolysis, solids removal, oil recovery, heat generation, and hydro-processing. Pre-treatment consists of drying biomass to 7% moisture content and grinding it to a final particle size diameter of 3 mm. Biochar recovery employs conventional cyclones to separate 90% of solid particles from the vapor stream. The oil recovery section employs indirect heat exchangers and an electrostatic precipitator to collect condensable vapors. Non condensablegases are recycled through the heat generation unit where they are combusted to provide heat for drying;the combustion fuel gases are employed as a fluidizing agent in the pyrolysis fluid bed unit. A fraction of the biochar is combusted to provide sufficient energy to dry biomass feedstock and to sustain the pyrolysis process.

3.5. Gasification

As the name indicates, gasification transforms biomass into primarily a gaseous mixture and smaller quantities of higher hydrocarbons by supply a controlled amount of oxidizing agent under high temperatures (750–1800°C), sometimes at high pressures of 15–50 bars [94] and the presence of some oxygen, measured in equivalence ratio or the fraction of the amount of oxygen needed for stoichiometric combustion (typically around 0.25 or 25%). Air gasification produces syngas ('synthesis gas') or producer gas if it contains nitrogen, consists mostly of carbon monoxide (CO) and hydrogen (H2) with smaller amounts of carbon dioxide, methane, and other low-molecular-weight hydrocarbons with low heating values of 4–7 MJ/Nm3, while gasification with steam produces syn gas with high heating values of 10–14MJ/Nm3 [95]. The gas mixture is the key energy output and the gasification process has an application as a clean waste disposal technique [96]. Syngas may be used for electricity generation via gas or steam turbines (or both), used to manufacture chemicals and fertilizers, or further cleaned for use as liquid fuel. Overall, gasification is very similar to combustion, but due to the limited oxygen, it is not able to complete the gas-phase and solid-phase oxidation steps which would yield carbon dioxide (CO_2) and water (H_2O). An ideal gasification situation, the reaction is thermodynamically controlled. The gas composition and carbon conversion can be predicted based on temperature and pressure, and the only co-product is char. The typical char yield of gasification averages about 10 wt% of biomass [97, 98]. However, this also carries the risk of higher levels of metals and minerals that may be concentrated in biochar, with potential safety implications with regard to application to soil [99]. Although biochar from gasification has a high calorific value [100, 101] it has high value use as the basis for activated carbon production. Its potential to act as the carrier in a slow nutrient release fertilizer product has also been noted [102].

4. Synthesis of Magnetic Biochar Based Material

Despite biochar's exhibition of outstanding adsorption performance, the difficulty of separating of recovering it, especially powdered-biochar, from aqueous media without filtration and/or centrifugation at post-adsorption stage unsuitable for practical application. Hence, in order to mitigate the above obstacles and further enhance its practical utility. Biochar functioning in soil fertility enhancement, carbon sequestration and wastewater treatment applications can be improved with innovations like incorporating engineered nanoparticles into biochar production [65]. Biochar is difficult to separate from the aqueous solution, so biochar can be converted to magnetic biochar considering its difficulty to be separated from the aqueous solution, biochar can also be converted into magnetic materials by three synthesis methods, i.e., in-situ synthesis, impregnation, coating. By using these methods, nanosized magnetic iron oxide including Fe_3O_4 [103]. c-Fe_2O_3[104] and $CoFe_2O_4$ [105] particles can be coated or impregnated on the surface of biochar, resulting in the introduction of magnetism and active sites of iron oxide for contaminants removal. The magnetic biochar could formed can be easily separated from solution by an external magnetic field [28, 106].

4.1. In-situ Synthesis

Most of the magnetic material is synthesized by the in-situ synthesis which involves synthesis in the reaction mixture. Various magnetic materials are synthesized by in-situ synthesis are as follow: biochar coated silica material from rice husk is synthesized by in-situ preparation [107]. Magnetic oak bark (MOBBC) and oak wood (MOWBC) biochars were obtained by fast pyrolysis (400 and 450°C) during bio-oil production in an auger-fed reactor and used for aqueous Cd^{2+} and Pb^{2+} remediation. Mohan et al. [49] reported that aqueous biochar suspensionsupon mixing with aqueous Fe^{3+}/Fe^{2+} solutions, followed by NaOH, get magnetized causing mixed iron oxides to nucleate and bind. In situ synthesis is mostly favored over the other synthesis process because of the high purity of the magnetic biochar formed.

4.2. Impregnation

Impregnation of foreign materials onto raw biochars to get biochar-based nano-composites has become an important practice for expanding the environmental applications of biochar and nanotechnology [21, 27]. Mostly impregnation of the iron oxide particles is done on the surface of the biochar to increase its adsorption capacity. Hickory biochar surface is impregnated with the Iron (Fe) through direct hydrolysis of iron salt onto hickory biochar, and results showed that iron hydroxide particles on the biochar surface served as the main adsorption sites for As sorption [109]. Similarly Wang et al showed the impregnation of the Amorphous hydrous manganese oxide on the Peeled pine wood biochar lead to the adsorption of the Pb with 5 times more adsorption capacity than the pristine biochar [110]. Also a novel magnetic biochar is formed with colloidal or nanosized-Fe_2O_3 particles impregnated in porous biochar matrix were synthesized through thermal pyrolysis of $FeCl_3$ pretreated biomass [106]. The resulting magnetic nanocomposite showed good adsorption ability for As(V) in aqueous solution and wonderful ferromagnetic properties .

4.3. Coating

Biochar surface can be coated with the functional nanoparticles(e.g., chitosan, graphene, graphene oxide, carbon nanotubes, ZnS nanocrystals, layered double hydroxides, nanoscale zero-valent iron, graphitic C3N4) can achieve an affordable composite material capable of removing contaminants with the advantages of both biochar matrix and functional nanoparticles. The coating of these biochar materials leads to the improvement in the surface properties of the biochar which includes surface functional groups, surface area, porosity, and thermal stability of biochar, which further contribute to better performance of contaminants removal. Particularly, the main advantage of biochar-based nano-composites with different catalytic and oxidative/reductive nanoparticles (such as nanoscale zero-valent iron and graphitic C_3N_4) coated on biochar surface is the simultaneous adsorption and degradation ability for removing organic contaminants. A graphene coated biochar material showed excellent adsorption ability of methylene blue [111]. Similarly, another grapheme/biochar composite was produced from the wheat straw and it is coated to the surface of the biochar mainly through p-p interactions which results in enhanced surface properties to the higher removal efficiency of phenanthrene and mercury than the normal biochar [112, 113] also reported the Multi-walled carbon nanotube (CNT)-coated biochars had greater adsorption capacity than the pristine biochar. Besides the advantages of the coating of the nanoparticles on the biochar surface, it has some disadvantages including partial blockage of the pores of the biochar surface and the nanotoxicity of the nanomaterials used.

5. Application as Adsorbent

Over the past few years, Biochar is widely used for the removal of heavy metals, organic dyes etc due to their excellent adsorption properties. The adsorption mechanism of biochar is different for the different contaminant such as adsorption mechanism of heavy metal is quite different from the adsorption mechanism of organic dye. Also the biochar various properties like surface functional groups, surface area, porous structure and mineral components had a large effect on the adsorption mechanism [115].

5.1. Removal of Heavy Metals: Effects of Functional Groups and Mechanism

Heavy metals are stable, non-biodegradable substances and therefore liable to display long-term persistence and concentration in the soils, which may lead to the toxicity for plants, animals and humans [116]. These metals are usually acquired from the feedstock used for the manufacturing of the biochar [117, 118, 119]. To avoid the contamination of the biochar with increased level of heavy metals it is essential that the selection and analysis of the feedstock should be done carefully. During the volatilisation of the associated organic molecules, heavy metals are likely to retain in the organic molecule due to their stable nature. Most of the heavy metals are present as ash within the biochar therefore a selective manipulation is need for the removal of the heavy metals from the biochar. Many advanced spectroscopic techniques have recently been used to find out the adsorption capacities and binding mechanism of heavy metals in aqueous media [120, 121].

Biochar possess different surface functional group mainly consisting of oxygen containing groups like carboxylate ACOOH; and hydroxyl AOH on its surface. These functional groups strongly interact with the heavy metals through different interactions like electrostatic interaction, ion exchange and surface complexation. These effects can be visualized by comparing changes in functional groups of biochar before and after the metal adsorption [122, 123].

Different heavy metals interact with the surface functional group with different mechanism. The four different proposed mechanism are as follow: (i) heavy metal exchange with K^+ and Na^+ present on the outer sphere lead to electrostatic interaction, (ii) organic matter and mineral oxide of biochar lead to co-precipitation and inner sphere complexation of heavy metals with them, (iii) interaction with active carboxyl and hydroxyl functional groups of the biochar, and (iv) precipitation as lead–phosphate–silicate. These various mechanisms are shown by different biochar. Some examples are as follow: Pb^{2+} can be precipitated in water on biochar derived from dairy manure [124]. Likewise $Hg(OH)_2$ or $HgCl_2$ precipitate can be obtained from alkaline biochar surface containing high chloride content for the removal of water soluble Hg [125]. Cr(VI) is first reduced to Cr(III) due to O- containing functional group then its binding to the negatively charged biochar active sites lead to its sorption [123, 126, 127, 128]. Similar mechanisms can also be seen in the adsorption of Hg(II) onto soybean stalk-based biochar and Cd(II) on corn straw biochar, [125, 129].

In addition to surface area, the mineral components in biochar played a vital role in the adsorption process of heavy metals. [115]. compared the adsorption of the different heavy metals(Pb, Zn, Cd) on the biochar derived from different sources(rice husk, dairy manure). The results showed that the different mineral components and biochar feedstock sources had varied removal ability for the heavy metals. In another study of removal of heavy metals the researchers found that the mineral component ($PO_4{}^{3-}$, $CO_2{}^{3-}$) found in diary manure biochar provide additional adsorption sites leading to the high adsorption capacity of biochar for heavy metals [115]. The adsorption of Pb by dairy-manure derived biochar is also attributed to the mineral component present on the surface of the biochar that leads to the precipitation of the Pb with the mineral component [130]. The surface adsorption, and co-precipitation of Al with silicate particles present in the surface of the biochar lead to Al adsorption onto biochar [98]. Biochars possess different magnitudes of pores and surface areas within the particles, which enable them easily accessible to metals but it has less effect on heavy metal adsorption as compared to the oxygen containing functional group [122]. The adsorption of Pb onto low temperature (250 and 400°C) is due to the oxygen functional group while the adsorption of Pb onto high temperature (500 and 600°C) is due to the intraparticle diffusion.Samsuri et al. (2014) [131] also showed that the oxygen containing functional groups are more important that the surface area through adsorption of heavy metal using oil palm biochar and rice husk biochar. The former having less surface area exhibit higher adsorption capacity than the latter due to oxygen containing groups. Table 2 presents removal of heavy metals by magnetic biochar

Table 2. Removal of heavy metals by magnetic biochar

Biochar	Removal of organic contaminant	References
Magnetic Kans grass biochar	As(III, V) adsorption from aqueous solutions	[103]
Hickory wood Biochar covered with MnOx ultrafine Particles	Pb(II), Cu(II), and Cd(II)	[132-136]
Sugarcane bagasse biochar	Cr(VI)	[137]
Cotton Wood/c-Fe_2O_3 particles	As(V)	[106]
Hickory chips/Iron hydroxide particles	As(V)	[109]
Kans grass/ Fe_3O_4	As(V)	[103]
Loblolly pine wood/c-Fe_2O_3 particles	As(V)	[135]
Pine wood/Manganosite	As(V)	[134]
Pine wood/Birnessite	As(V)	[133]
Rice husks/Iron oxide	As(V)	[138]
Rice husks/Iron oxide	As(V)	[138]
Hickory wood/ MnOx	Cd(II)	[132]
Pine bark waste/Magnetic $CoFe_2O_4$	Cd(II)	[105]
Eichhornia crassipes/Chitosan and c-Fe_2O_3	Cr(VI)	[139]
Sugarcane bagasse/ZnO	Cr(VI)	[137]
Walnut shell/Chitosan	Cr(VI)	[140]
Corn straws/MnOx	Cu(II)	[141]
Hickory wood/MnOx	Cu(II)	[135]
Elovich Wheat straw/Graphene	Hg(II)	[112]
Bamboo/Chitosan	Pb(II)	[142]
Hickory chips/Carbon nanotubes	Pb(II)	[113]
Hickory wood/MnOx	Pb(II)	[132]
Peeled pine wood/Hydrous-manganeseoxide	Pb(II)	[133]
Pine bark waste/Magnetic $CoFe_2O_4$	Pb(II)	[105]
Pine wood/Manganosite	Pb(II)	[134]
Pine wood/Birnessite	Pb(II)	[134]
Sugarcane bagasse /Carbon nanotubes (CNTs)	Pb(II)	[113]

5.2. Removal of Dyes and Organic Pollutants: Factors and Mechanisms

The adsorption mechanism of binding of the organic contaminants to the biochar is also based upon different kinds of interactions. Mainly these interactions involve electrostatic interaction, hydrophobic effect, hydrogen bonds, and pore-filling for the adsorption of organic contaminants onto biochar. The mechanism of the adsorption of different organic contaminant is different and co-related to the properties of the properties of the biochar. The major role in the adsorption of organic contaminant is played by the surface properties of the biochar which includes surface area, oxygen containing functional groups, surface pores and co-existing carbonized and non-carbonized phases. The adsorption of organic compound was both decided by the non-carbonized organic matter and by adsorption onto the carbonized matter [22, 130, 65] studies reveals that the electrostatic attraction was the main dominant mechanism for the adsorption of the organic contaminant while other mechanism works as the supporting mechanisms. The mechanism of the adsorption of the dye on the straw-based biochar is due the p-p interaction between dye molecules and graphene layers of the biochar [143] In the further study, Xu et al. (2013) [115] studied that the adsorption of methyl violet is due to the specific interaction between the dye and the oxygen containing group's of the biochar surfaces. The adsorption of the carbaryl and atrazine on pig manure derived biochar is attributed to the various processes that are hydrophobic effect, pore-filling and p–p electron donor–acceptor interactions [106] in the adsorption of sulfonamides by different biochars. Xie et al. (2014) reported that the adsorption of sulfonamides by different biochars is dependent on the degree of graphitization, signifying that p–p electron–donor–acceptor interaction between the sulfonamides molecules and the graphitic surfaces of the biochars was the major adsorption mechanism [144]. Chen et al., 2012c reveals that pore-filling was another important mechanism to organic compounds adsorption on biochars and the adsorption capability is directly proportional to the micropore surface area. The adsorption of phenol and methylene blue by biochars from water is due to the prominent pore-filling effect [142, 65]. Table 3 shows degradation of organic pollutants by using magnetic biochar.

6. Soil Enrichment and Detoxification

Biochar when applied to soil it adsorbed the organic contaminants, so that altering their susceptibility to biodegradation. For example, biochar strongly sorbs phenanthrene [147], hexachlorobenzene [148] sulfamethoxazole [149] and substantially reduced the availability of PCDD/Fs measured by polyoxymethylene (POM). Addition of biochar to sewage sludge caused significant decreases in dissolved PAH concentration [150]. It was also found that biochar application to soil affected the persistence and efficacy of pesticides of isoproturon (IPU) [151] and benzonitrile [152], and herbicides such as (4-chloro- 2-methylphenoxy) acetic acid (MCPA), atrazine and trifluralin in soil [153-155]. In addition to the impact on chemical sorption, biochar amendments also affected the fate and transport of the chemicals' products.

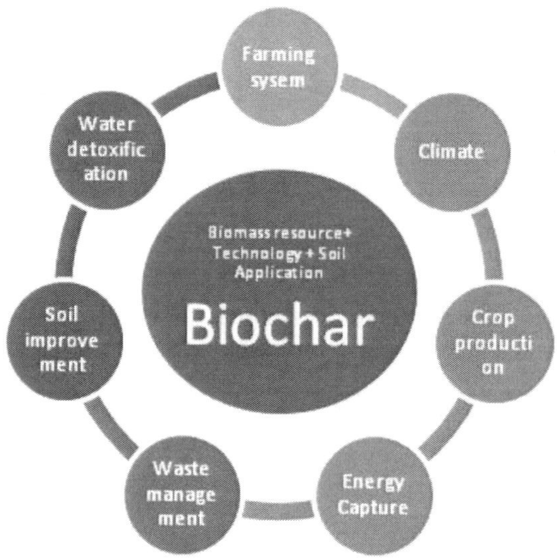

Biochar properties significantly affect the sorption of organic contaminants. Biochars perform better in terms of organic compound sorption and removal, due to the effective particle size and a larger surface area Recent studies have shown appreciable adsorption of PCDD/Fs onto biochar, especially when in a powdered form [156]. Relative to biochars with larger particle sizes, biochar with small particle size (i.e., <0.075 mm) needed less time to reach sorption equilibrium for both atrazine and simazine [157]. In different study by Ahmad et al. investigating TCE adsorption, biochar produced at higher pyrolysis temperature had higher adsorption capacity. The high trichloroethylene adsorption capacity of biochar produced at 700°C was attributed to their high aromaticity and low polarity. Similarly, compared to biochar produced at 450°C, the biochar produced at 850°C was particularly effective in reducing plant availability of both pesticides, due to its high affinity to sorb pesticide residues.

The kinetics for pollutants adsorption followed differential models. The sorption isotherms of atrazine and simazine on the biochar are nonlinear and follow a Freundlich model [157], which is similar to the adsorption of PCBs [158]. For the adsorption of pentachlorophenol (PCP) in biochar, the Freundlich and dual-site models could describe all the sorption isotherm data well [159]. The Freundlich model also indicated the competition between different contaminants during sorption. When atrazine and simazine co-existed, a competitive sorption occurred between these two pesticides on the biochar, reflecting a decrease in sorption capacity of both pesticides [160].

Table 3. Degradation of organic pollutants by magnetic biochar

Biochar	Removal of inorganic contaminant	References
Hickory chips and sugarcane Bagasse/CNT	Methylene blue	[65]
Hickory chips and sugarcane Bagasse/(CNT) suspensions with sodium	Sulfapyridine	[113]
Wheat straw/graphene	Phenanthrene	[112]
Bamboo/ZVI particles	methylene blue	[142]
Sewage sludge/citric acid–$ZnCl_2$ mixed	Benzene Derivatives	[125]
Chicken droppings, wood, and old car tire/neutral hydrogel matrix	Phenol	[145]
Paper mill sludge/Zero-valent iron (ZVI))	Pentachlorophenol	[46]
Paper mill sludge/Zero-valent iron (ZVI))/doping of Ni	Pentachlorophenol	[146]
Corn stalks/Fe_3O_4 nanoparticles	Crystal violet	[129]
Cotton Wood/ AlOOH particles/flakes	Methylene Blue	[106]
Hickory chips/ Hybrid multi-walled carbon nanotube (CNT)	Methylene blue	[113]
Milled cotton wood/Graphene	Methylene blue	[111]
Sugarcane bagasse/Hybrid multi-walled carbon nanotube (CNT)	Methylene blue	[65]
Pecan nut shells/calcium salts	Dyes (acid blue 74 and reactive blue 4)	[147]

The soil conditions normally affect the degradation and adsorption of contaminants. The sorption of the biochar for pesticides was generally favored by low pH [158]. However, Zheng et al. (2010) demonstrated that the benzonitrile degradation was dependent on the soil pH only in the un-amended soil, while the degradation was not affected by soil pH after biochar amendment. Thus, it is likely the presence of crop residue-derived chars may alter pesticide degradation patterns normally observed in soils and thus significantly affect their environmental fate [160].

As expected, surface activation increases the sorption of organic contaminants due to the higher surface area of the activated char relative to an untreated biochar of similar composition and production method [158, 161].

Adsorption is likely the main mechanism for reduction of organic contaminants in soil and plants, combined with the immobilization and degradation of organic matter. Like other carbonaceous sorbents, biochars have also been found to decrease the bioavailability of PAHs. Inaddition to reducing the opportunity for soil-pore water-plant transfer of PAHs, biochar's influence upon PAHs partition will also reduce the opportunity for PAHs transfer from soil to air to plant [162]. As nutrient rich media, biochar is effective in enhancing the cometabolic biodegradation of PAHs, such as phenanthrene, despite the weaker adsorption effect than activated carbon [163]. Black carbon amendments have been shown to reduce pesticide degradation in soils because adsorbed pesticides are less available for microbial degradation compared to pesticides in solution [164].

7. Porosity and Surface Area

The tough pyrolysis conditions and increased pyrolysis temperature result in a growing proportion of biochar particles with smaller particle size distributions [165]. Based on Scanning Electron Microscope (SEM) images. Identified different size ranges of pores in biochar particles. The biochar of casein was characterizing with mostly meso pores ($3.7 < Ø < 41.7$ nm) and macro pores ($Ø >$ 88.04-6811.04 nm) [166]. Temperature is clearly an important factor on the char formation and it was observed that increasing temperature favored the rise in both, porosity and surface area [167-168]. As the surface area (SA) is a main indicator of the char uptake ability thus SA values vary significantly among biochar from digested and undigested biomass. While the biochar from digested biomass had significant N2 SA (336 m^2/g, indicating the presence of mesopores), the N2 SA of the biochar from undigested biomass was very small (2.6 m^2/g), indicating that its surface was dominated by the microporesonly. In some studies, SA values for wood char (347m^2/g) were significantly higher than those of grass char (140 m^2/g) [168]. However, the chemical composition of these surfaces are also very important for determining their interfaces in the environment [169-170]. As a commercial product, activated carbons (AC) possesses a large surface area, which reflects the AC's fine-pore structure created through a well-controlled activation process. Considering that biochar samples were produced without an activation process, it is not surprising that their surface areas and thus porosities were significantly lower [171]). Alternatively, activation of the biochar during production may be employed to increase the porosity and surfacearea of biochar to values comparable with AC. In one study, physical (steam) and chemical (KOH) activation of feedstock increased total pore volume of activated carbons prepared fromcoal and biochar by ~43% and 11%, respectively [171].

8. Cation Exchange Capacity

Overall, maximum of biochars have comparatively high cation exchange capacity due to their negative surface charge and resultant affinity for soil cations, including most heavy metals (e.g., Pb^{2+}, Cr^{3+}). For this reason, several studies have been conducted investigating the adsorption of cations onto biochar [32, 172], which is the primary mechanism by which biochar can be used for remediation of heavy metal contamination. CEC varies significantly between terrestrial-derived biomass from different feedstocks, ranging from 4.5 to 40 cmol/kg [173-174]. Compared to many terrestrial biomass biochar, the CEC of biochar from algae is relatively high (29~41 cmol/kg) with extractable Ca, Mg and K ranging from 27 to 485 cmol/kg. The CEC of biochar-amended soil is strongly dependent on the age and surface functional properties and charge of the applied char [175].

9. Other Applications

Other than the adsorption of organic and inorganic contaminants, biochar is also widely used in other fields like the introduction of biochar to the compositing mixture lead to the increased carbon content in the humic moietyand also enhances stability of chemical structure of humic substances [176]. Also biochar made from the rice husk can be used for the removal of TNT and RDX from groundwater [177]. For the introduction of bacteria into the soil for its enrichment as well as for the detoxification, biochar serve as a carrier for those bacteria's [178]. Transportation and retention of bacterias like *Escherichia coli* and *bacteriophages* (PRD1, MS2 and ΦX174), as surrogates for human pathogenic bacteria and viruses, respectively, were studied in the sand that was amended with several types of biochar produced from various feedstocks [179]. Fe_3O_4 nanoparticles impregnated on the biochar material can be used as excellent anode materials for the Li-ion batteries [180].

Conclusion

In this chapter, we have done a detailed study of biochar by giving information about the biochar magnetization, modification synthesis techniques and properties. Magnetic biochar enhances the ability of biochar by altering its properties. Biochars have tremendous applications as adsorbents, detoxifying agents, soil enrichment agents, photocatalysts, etc. Biochar production helps with waste management because it is produced from various wastes like industrial waste, household waste and agriculture waste. The overall study showed that biochar is an economic, ecofreindly material that has wide range of application that needs attention for further implications.

References

[1] Sohi, S. P., (2012). Carbon storage with benefits. *Science*, 338, 1034-1035.

[2] Spokas, K. A., (2010). Review of the stability of biochar in soils: predictability of O: C molar ratios. *Carbon Manag.* 1(2), 289-303.

[3] Fowles, M., (2007). Black carbon sequestration as an alternative to bioenergy. *Biomass and Bioenergy*, 31(6), 426-432.

[4] Goldberg, E., (1985). Black carbon in the environment.

[5] Glaser, B., Lehmann, J., & Zech, W., (2002). Ameliorating physical and chemical properties of highly weathered soils in the tropics with charcoal-a review. *Bio. fertility of soils* 35(4), 219-230.

[6] Laird, D. A., (2008). The charcoal vision: a win–win–win scenario for simultaneously producing bioenergy, permanently sequestering carbon, while improving soil and water quality. *Agronomy J* 100, 178-181.

[7] Lehmann, J., (2007). Bio- energy in the black. *Frontiers in Ecology Env.* 5(7), 381-387.

[8] Haberstroh, P. R., Brandes, J. A., Gélinas, Y., Dickens, A. F., Wirick, S., and Cody, G., (2006). Chemical composition of the graphitic black carbon fraction in riverine and marine sediments at sub-micron scales using carbon X-ray spectromicroscopy. *Geochimica et Cosmochimica Acta*. 70(6), 1483-1494.

[9] Kuhlbusch, T. A., Zepp, R. G., Miller, W. L., and A BURKE, R., (1998). Carbon monoxide fluxes of different soil layers in upland Canadian boreal forests. *Tellus B*, 50(4), 353-365.

[10] Masiello, C. A., and Druffel, E. R. M., (1998). Black carbon in deep-sea sediments. *Science*, 280(5371), 1911-1913.

[11] Masiello, C. A., Druffel, E. R. M., and Currie, L. A., (2002). Radiocarbon measurements of black carbon in aerosols and ocean sediments. *Geochimica et Cosmochimica Acta.* 66, 1025-1036.

[12] Kuhlbusch, T. A. J., and Crutzen, P. J., (1995). Toward a global estimate of black carbon in residues of vegetation fires representing a sink of atmospheric CO2 and a source of O2. *Global Biogeochemical Cycles* 9, 491-501.

[13] Whitman, T., Nicholson, C. F., Torres, D., and Lehmann, J., (2011). Climate change impact of biochar cook stoves in Western Kenyan farm households: system dynamics model analysis. *Env. Sci. tech.* 45(8), 3687-3694.

[14] Abit, S. M., Bolster, C. H., Cai, P., & Walker, S. L., (2012). Influence of feedstock and pyrolysis temperature of biochar amendments on transport of Escherichia coli in saturated and unsaturated soil. *Env. Sci. tech.* 46(15), 8097-8105.

[15] Mao, J. D., Johnson, R. L., Lehmann, J., Olk, D. C., Neves, E. G., Thompson, M. L., and Schmidt-Rohr, K., (2012). Abundant and stable char residues in soils: implications for soil fertility and carbon sequestration. *Env. Sci. techn.* 46(17), 9571-9576.

[16] Khare, P., and Goyal, D. K., (2013). Effect of high and low rank char on soil quality and carbon sequestration. *Ecological engineering*, 52, 161-166.

[17] Verheijen, F. G. A., Graber, E. R., Ameloot, N., Bastos, A. C., Sohi, S., and Knicker, H., (2014). Biochars in soils: new insights and emerging research needs. *Europ. J Soil Sci.* 65(1), 22-27.

[18] Huggins, T., Wang, H., Kearns, J., Jenkins, P., and Ren, Z. J., (2014). Biochar as a sustainable electrode material for electricity production in microbial fuel cells. *Bioresource techn.* 157, 114-119.

[19] Kacprzak, A., Kobyłecki, R., Włodarczyk, R., and Bis, Z., (2014). The effect of fuel type on the performance of a direct carbon fuel cell with molten alkaline electrolyte. *J Power Sourc.* 255, 179-186.

[20] Titirici, M. M., White, R. J., Falco, C., and Sevilla, M., (2012). Black perspectives for a green future: hydrothermal carbons for environment protection and energy storage. *Energy Env. Sci.* 5(5), 6796-6822.

[21] Chen, B., Chen, Z., and Lv, S., (2011). A novel magnetic biochar efficiently sorbs organic pollutants and phosphate. *Bioresource techn. 102*(2), 716-723.

[22] Chen, Y. N., Chai, L. Y., and Shu, Y. D., (2008). Study of arsenic (V) adsorption on bone char from aqueous solution. *J Hazardous Mat. 160*(1), 168.

[23] Zhang, J., Fu, H., Lv, X., Tang, J., and Xu, X., (2011). Removal of Cu (II) from aqueous solution using the rice husk carbons prepared by the physical activation process. *Biomass and bioenergy*, *35*(1), 464-472.

[24] Mohan, D., Pittman, C. U., Bricka, M., Smith, F., Yancey, B., Mohammad, J., and Gong, H., (2007). Sorption of arsenic, cadmium, and lead by chars produced from fast pyrolysis of wood and bark during bio-oil production. *J colloid and interface sci. 310*(1), 57-73.

[25] Mohan, D., Sarswat, A., Ok, Y. S., and Pittman, C. U. (2014). Organic and inorganic contaminants removal from water with biochar, a renewable, low cost and sustainable adsorbent–a critical review. *Bioresource techn. 160*, 191-202.

[26] Tang, W., Su, Y., Li, Q., Gao, S., and Shang, J. K., (2013). Superparamagnetic magnesium ferrite nanoadsorbent for effective arsenic (III, V) removal and easy magnetic separation. *Water research 47*(11), 3624-3634.

[27] Zhang, M., Gao, B., Varnoosfaderani, S., Hebard, A., Yao, Y., and Inyang, M. (2013). Preparation and characterization of a novel magnetic biochar for arsenic removal. *Bioresource techn. 130*, 457-462.

[28] Baig, S. A., Sheng, T., Sun, C., Xue, X., Tan, L., and Xu, X. (2014). Arsenic removal from aqueous solutions using Fe 3 O 4-HBC composite: effect of calcination on adsorbents performance. *PloS one 9*(6), e100704.

[29] Lim, S. F., Zheng, Y. M., and Chen, J. P.,(2009). Organic arsenic adsorption onto a magnetic sorbent. *Langmuir 25*(9), 4973-4978.

[30] Zhou, Y., Gao, B., Zimmerman, A. R., Chen, H., Zhang, M., and Cao, X., (2014). Biochar-supported zerovalent iron for removal of various contaminants from aqueous solutions. *Bioresource techn. 152*, 538-542.

[31] Devi, P., & Saroha, A. K., (2014). Synthesis of the magnetic biochar composites for use as an adsorbent for the removal of pentachlorophenol from the effluent. *Bioresource techn. 169*, 525-531.

[32] Tian, Y., Wu, M., Lin, X., Huang, P., and Huang, Y. (2011). Synthesis of magnetic wheat straw for arsenic adsorption. *J hazardous mat. 193*, 10-16.

[33] Barrow, C. J., (2012). Biochar: potential for countering land degradation and for improving agriculture. *Applied Geography*, 34, 21-28.

[34] Brown, R. A., Kercher, A. K., Nguyen, T. H., Nagle, D. C., and Ball, W. P., (2006). Production and characterization of synthetic wood chars for use as surrogates for natural sorbents. *Organic Geochemistry*, 37(3), 321-333.

[35] Abdullah, H., and Wu, H. (2009). Biochar as a fuel: 1. Properties and grindability of biochars produced from the pyrolysis of mallee wood under slow-heating conditions. *Energy & Fuels*, 23(8), 4174-4181.

[36] Lucchini, P., Quilliam, R. S., DeLuca, T. H., Vamerali, T., and Jones, D. L. (2014). Increased bioavailability of metals in two contrasting agricultural soils treated with waste wood-derived biochar and ash. *Envi. Sci. Pollution Research* 21(5), 3230-3240.

[37] Feng, Y., Xu, Y., Yu, Y., Xie, Z., and Lin, X., (2012). Mechanisms of biochar decreasing methane emission from Chinese paddy soils. *Soil Biology and Biochem.* 46, 80-88.

[38] Chun, Y., Sheng, G., Chiou, C. T., and Xing, B., (2004). Compositions and sorptive properties of crop residue-derived chars. *Env. Sci. techn.* 38(17), 4649-4655.

[39] Peng, X., Ye, L. L., Wang, C. H., Zhou, H., and Sun, B., (2011). Temperature-and duration-dependent rice straw-derived biochar: Characteristics and its effects on soil properties of an Ultisol in southern China. *Soil and Tillage Research,* 112(2), 159-166.

[40] Yao, Y., Gao, B., Inyang, M., Zimmerman, A. R., Cao, X., Pullammanappallil, P., and Yang, L., (2011). Biochar derived from anaerobically digested sugar beet tailings: characterization and phosphate removal potential. *Bioresource techn.* 102(10), 6273-6278.

[41] Rondon, M. A., Lehmann, J., Ramírez, J., and Hurtado, M., (2007). Biological nitrogen fixation by common beans (Phaseolus vulgaris L.) increases with bio-char additions. *Bio. fertility of soils*, 43(6), 699-708.

[42] Inyang, M., Gao, B., Pullammanappallil, P., Ding, W., and Zimmerman, A. R., (2010). Biochar from anaerobically digested sugarcane bagasse. *Bioresource Techn.* 101(22), 8868-8872.

[43] Yu, X. Y., Ying, G. G., and Kookana, R. S., (2009). Reduced plant uptake of pesticides with biochar additions to soil. *Chemosphere* 76(5), 665-671.

[44] Lehmann, J., Czimczik, C., Laird, D., and Sohi, S., (2009). Stability of biochar in soil. *Biochar for env. manag: sci techn.* 183-206.

[45] Lu, H., Zhang, W., Yang, Y., Huang, X., Wang, S., and Qiu, R. (2012). Relative distribution of Pb 2+ sorption mechanisms by sludge-derived biochar. *Water research* 46(3), 854-862.

[46] Devi, P., and Saroha, A. K., (2014). Synthesis of the magnetic biochar composites for use as an adsorbent for the removal of pentachlorophenol from the effluent. *Bioresource techn. 169*, 525-531.

[47] Xu, X., Cao, X., and Zhao, L., (2013). Comparison of rice husk-and dairy manure-derived biochars for simultaneously removing heavy metals from aqueous solutions: role of mineral components in biochars. *Chemosphere* 92, 955–961.

[48] Inyang, M., Gao, B., Zimmerman, A., Zhang, M., and Chen, H., (2014). Synthesis, characterization, and dye sorption ability of carbon nanotube–biochar nanocomposites. *Chem. Eng. J 236*, 39-46.

[49] Mohan, D., Kumar, H., Sarswat, A., Alexandre-Franco, M., and Pittman, C. U., (2014). Cadmium and lead remediation using magnetic oak wood and oak bark fast pyrolysis bio-chars. *Chem. Eng. J 236*, 513-528.

[50] Demiral, İ., and Kul, Ş. Ç., (2014). Pyrolysis of apricot kernel shell in a fixed-bed reactor: characterization of bio-oil and char. *J Analyt. Applied Pyrolysis 107*, 17-24.

[51] Sewu, D. D., Boakye, P., and Woo, S. H., (2017). Highly efficient adsorption of cationic dye by biochar produced with Korean cabbage waste. *Bioresource Techn. 224*, 206-213.

[52] Liu, P., Liu, W. J., Jiang, H., Chen, J. J., Li, W. W., and Yu, H. Q., (2012). Modification of bio-char derived from fast pyrolysis of biomass and its application in removal of tetracycline from aqueous solution. *Bioresource techn. 121*, 235-240.

[53] Xue, Y., Gao, B., Yao, Y., Inyang, M., Zhang, M., Zimmerman, A. R., and Ro, K. S., (2012). Hydrogen peroxide modification enhances the ability of biochar (hydrochar) produced from hydrothermal carbonization of peanut hull to remove aqueous heavy metals: batch and column tests. *Chem. Eng. J 200*, 673-680.

[54] Gutiérrez-Valero, M. D., Godino-Salido, M. L., Arranz-Mascarós, P., López-Garzón, R., Cuesta, R., and García-Martín, J., (2007). Adsorption of designed pyrimidine derivative ligands on an activated carbon for the removal of Cu (II) ions from aqueous solution. *Langmuir 23*(11), 5995-6003.

[55] Azargohar, R., and Dalai, A. K. (2008). Steam and KOH activation of biochar: Experimental and modeling studies. *Microporous and Mesoporous Mater. 110*(2), 413-421.

[56] Qian, K., Kumar, A., Zhang, H., Bellmer, D., and Huhnke, R., (2015). Recent advances in utilization of biochar. *Renewable and Sustainable Energy Reviews*, *42*, 1055-1064.

[57] Lin, Y., Munroe, P., Joseph, S., Henderson, R., and Ziolkowski, A. (2012). Water extractable organic carbon in untreated and chemical treated biochars. *Chemosphere 87*(2), 151-157.

[58] Vithanage, M., Rajapaksha, A. U., Zhang, M., Thiele-Bruhn, S., Lee, S. S., and Ok, Y. S., (2015). Acid-activated biochar increased sulfamethazine retention in soils. *Env. Sci. Pollu. Research 22*(3), 2175-2186.

[59] Fan, Y., Wang, B., Yuan, S., Wu, X., Chen, J., and Wang, L., (2010). Adsorptive removal of chloramphenicol from wastewater by NaOH modified bamboo charcoal. *Bioresource techn. 101*(19), 7661-7664.

[60] Yang, G. X., and Jiang, H., (2014). Amino modification of biochar for enhanced adsorption of copper ions from synthetic wastewater. *Water research 48*, 396-405

[61] Jing, X. R., Wang, Y. Y., Liu, W. J., Wang, Y. K., and Jiang, H., (2014). Enhanced adsorption performance of tetracycline in aqueous solutions by methanol-modified biochar. *Chem. Eng. J 248*, 168-174.

[62] Wang, Y., Wang, X. J., Liu, M., Wang, X., Wu, Z., Yang, L. Z., and Zhao, J. F., (2012). Cr (VI) removal from water using cobalt-coated bamboo charcoal prepared with microwave heating. *Industrial crops and products, 39*, 81-88.

[63] Zhang, Z., Wang, X., Wang, Y., Xia, S., Chen, L., Zhang, Y., and Zhao, J., (2013). Pb (II) removal from water using Fe-coated bamboo charcoal with the assistance of microwaves. *J Env. Sci. 25*(5), 1044-1053.

[64] Zhao, R. S., Liu, Y. L., Chen, X. F., Yuan, J. P., Bai, A. Y., and Zhou, J. B., (2013). Preconcentration and determination of polybrominated diphenyl ethers in environmental water samples by solid-phase microextraction with Fe 3 O 4-coated bamboo charcoal fibers prior to gas chromatography–mass spectrometry. *Analytica chimica acta. 769*, 65-71.

[65] Inyang, M., Gao, B., Zimmerman, A., Zhang, M., and Chen, H. (2014). Synthesis, characterization, and dye sorption ability of carbon nanotube–biochar nanocomposites. *Chem. Eng. J 236*, 39-46.

[66] Ghaffar, A., and Younis, M. N., (2014). Adsorption of organic chemicals on graphene coated biochars and its environmental implications. *Green Processing and Syn. 3*(6), 479-487.

[67] Manyà, J. J., (2012). Pyrolysis for biochar purposes: a review to establish current knowledge gaps and research needs. *Env. Sci. techn. 46*(15), 7939-7954.

[68] Lima, I. M., Boateng, A. A., and Klasson, K. T., (2010). Physicochemical and adsorptive properties of fast- pyrolysis bio- chars and their steam activated counterparts. *J chem.. techn. Biotechn. 85*(11), 1515-1521.

[69] Rajapaksha, A. U., Vithanage, M., Lim, J. E., Ahmed, M. B. M., Zhang, M., Lee, S. S., and Ok, Y. S. (2014). Invasive plant-derived biochar inhibits sulfamethazine uptake by lettuce in soil. *Chemosphere 111*, 500-504.

[70] Xiong, Z., Shihong, Z., Haiping, Y., Tao, S., Yingquan, C., and Hanping, C. (2013). Influence of NH3/CO2 Modification on the Characteristic of Biochar and the CO2 Capture. *BioEnergy Research, 6*(4), 1147-1153.

[71] Zhang, X. N., Mao, G. Y., Jiao, Y. B., Shang, Y., and Han, R. P. (2014). Adsorption of anionic dye on magnesium hydroxide-coated pyrolytic bio-char and reuse by microwave irradiation. *I J Env. Sci. Techn 11*(5), 1439-1448.
[72] Mohan, D., Pittman, C. U., and Steele, P. H., (2006). Pyrolysis of wood/biomass for bio-oil: a critical review. *Energy & fuels* 20(3), 848-889.
[73] Brownsort, P. A., (2009). Biomass pyrolysis processes: performance parameters and their influence on biochar system benefits.
[74] Onay, O., and Kockar, O. M. (2003). Slow, fast and flash pyrolysis of rapeseed. *Renewable energy,* 28(15), 2417-2433.
[75] Laird, D. A., Brown, R. C., Amonette, J. E., and Lehmann, J. (2009). Review of the pyrolysis platform for coproducing bio- oil and biochar. Biofuels, *Bioproducts and Biorefining*, 3(5), 547-562.
[76] Şensöz, S., Angın, D., & Yorgun, S., (2000). Influence of particle size on the pyrolysis of rapeseed (Brassica napus L.): fuel properties of bio-oil. *Biomass and Bioenergy,* 19(4), 271-279.
[77] Onay, O., (2007). Influence of pyrolysis temperature and heating rate on the production of bio-oil and char from safflower seed by pyrolysis, using a well-swept fixed-bed reactor. *Fuel Processing Techn.* 88(5), 523-531.
[78] Karaosmanoğlu, F., Tetik, E., and Göllü, E. (1999). Biofuel production using slow pyrolysis of the straw and stalk of the rapeseed plant. *Fuel processing techn.* 59(1), 1-12.
[79] Ronsse F, Van Hecke Dickinson D., and Prins W., (2012) Production and characterization of slow pyrolysis biochar: influence of feedstock type and pyrolysis conditions. *GCBBioenergy*; 5(2):104–15.
[80] Balat, M., Balat, M., Kırtay, E., and Balat, H. (2009). Main routes for the thermo-conversion of biomass into fuels and chemicals. Part 1: Pyrolysis systems. *Energy Conversion and Manag.* 50(12), 3147-3157.
[81] Bridgwater, A. V., and Peacocke, G. V. C., (2000). Fast pyrolysis processes for biomass. *Renewable and sustainable energy reviews,* 4(1), 1-73.
[82] Maschio, G., Koufopanos, C., and Lucchesi, A., (1992). Pyrolysis, a promising route for biomass utilization. *Bioresource techn.* 42(3), 219-231.
[83] Yanik, J., Kornmayer, C., Saglam, M., and Yüksel, M., (2007). Fast pyrolysis of agricultural wastes: Characterization of pyrolysis products. *Fuel Processing Techn.* 88(10), 942-947.
[84] Onay, Ö., Beis, S. H., and Koçkar, Ö. M., (2001). Fast pyrolysis of rape seed in a well-swept fixed-bed reactor. *J analytical and applied pyrolysis*, 58, 995-1007.
[85] Uzun, B. B., Pütün, A. E., and Pütün, E., (2007). Composition of products obtained via fast pyrolysis of olive-oil residue: effect of pyrolysis temperature. *J Analytical and Applied Pyrolysis*, 79(1), 147-153.

[86] Zhang, L., Xu, C. C., and Champagne, P., (2010). Overview of recent advances in thermo-chemical conversion of biomass. *Energy Conversion and Manag.* 51(5), 969-982.

[87] Boateng, A. A., (2007). Characterization and thermal conversion of charcoal derived from fluidized-bed fast pyrolysis oil production of switchgrass. *Industrial & eng. Chem.. research* 46(26), 8857-8862.

[88] Brewer, C. E., Schmidt- Rohr, K., Satrio, J. A., and Brown, R. C. (2009). Characterization of biochar from fast pyrolysis and gasification systems. *Env. Progress & Sustainable Energy*, 28(3), 386-396.

[89] Mullen, C. A., Boateng, A. A., Goldberg, N. M., Lima, I. M., Laird, D. A., and Hicks, K. B. (2010). Bio-oil and bio-char production from corn cobs and stover by fast pyrolysis. *Biomass and bioenergy*, 34(1), 67-74.

[90] Day, D., Evans, R. J., Lee, J. W., and Reicosky, D., (2005). Economical CO 2, SO x, and NO x capture from fossil-fuel utilization with combined renewable hydrogen production and large-scale carbon sequestration. *Energy* 30(14), 2558-2579.

[91] Scala, F., Chirone, R., and Salatino, P., (2006). Combustion and attrition of biomass chars in a fluidized bed. *Energy & Fuels*, 20(1), 91-102.

[92] Yanik, J., Kornmayer, C., Saglam, M., and Yüksel, M. (2007). Fast pyrolysis of agricultural wastes: Characterization of pyrolysis products. *Fuel Processing Techn.* 88(10), 942-947.

[93] Duman, G., Okutucu, C., Ucar, S., Stahl, R., and Yanik, J., (2011). The slow and fast pyrolysis of cherry seed. *Bioresource Techn.* 102(2), 1869-1878.'

[94] Bridgwater, T., (2006). Biomass for energy. *J Sci Food and Agric.* 86(12), 1755-1768.

[95] Kumar, A., Jones, D. D., and Hanna, M. A., (2009). Thermochemical biomass gasification: a review of the current status of the technology. *Energies* 2(3), 556-581.

[96] Bapat, H. D., and Manahan, S. E. (1998, April). Chemchar gasification of hazardous wastes and mixed wastes on a biochar matrix. In: *Abstracts of Papers of The American Chemical Society* (Vol. 215, pp. U571-U571). 1155 16th St, NW, Washington, DC 20036 USA: Amer Chemical Soc.

[97] Meyer, S., Glaser, B., and Quicker, P. (2011). Technical, economical, and climate-related aspects of biochar production technologies: a literature review. *Env. Sci. techn.* 45(22), 9473-9483.

[98] Qian, K., Kumar, A., Patil, K., Bellmer, D., Wang, D., Yuan, W., and Huhnke, R. L., (2013). Effects of biomass feedstocks and gasification conditions on the physiochemical properties of char. *Energies* 6(8), 3972-3986.

[99] Fernandes, M. B., and Brooks, P. (2003). Characterization of carbonaceous combustion residues: II. Nonpolar organic compounds. *Chemosphere* 53(5), 447-458.

[100] Demirbas, A., (2006). Production and characterization of bio-chars from biomass via pyrolysis. *Energy Sources, Part A,* 28(5), 413-422.

[101] Demirbas, A., Pehlivan, E., and Altun, T., (2006). Potential evolution of Turkish agricultural residues as bio-gas, bio-char and bio-oil sources. *Int. J Hydrogen Energy* 31(5), 613-620.

[102] Ueno, M., Kawamitsu, Y., Komiya, Y., and Sun, L., (2007). Carbonisation and gasification of bagasse for effective utilisation of sugarcane biomass. In *Proc. Int. Soc. Sugar Cane Technol* (Vol. 26).

[103] Baig, S. A., Zhu, J., Muhammad, N., Sheng, T., and Xu, X. (2014). Effect of synthesis methods on magnetic Kans grass biochar for enhanced As (III, V) adsorption from aqueous solutions. *Biomass and Bioenergy*, *71*, 299-310.

[104] Liu, F., Zuo, J., Chi, T., Wang, P., and Yang, B. (2015). Removing phosphorus from aqueous solutions by using iron-modified corn straw biochar. *Frontiers of Environ. Sci. & Engin. 9*(6), 1066-1075.),

[105] Reddy, D. H. K., and Lee, S. M. (2014). Magnetic biochar composite: facile synthesis, characterization, and application for heavy metal removal. *Colloids and Surfaces A: Physicochemical and Engineering Aspects*, *454*, 96-103.)

[106] Zhang, X., Wang, H., He, L., Lu, K., Sarmah, A., Li, J., and Huang, H. (2013). Using biochar for remediation of soils contaminated with heavy metals and organic pollutants. *Env. Sci. Pollution Research*, *20*(12), 8472-8483.).

[107] Li, Y., Wang, X., Zhu, Y., Wang, L., and Wang, Z. (2012). In situ preparation of biochar coated silica material from rice husk. *Colloids and Surfaces A: Physicochemical and Engin. Aspects*, *395*, 157-160.

[108] Li, Y., Wang, X., Zhu, Y., Wang, L., and Wang, Z. (2012). In situ preparation of biochar coated silica material from rice husk. *Colloids and Surfaces A: Physicochemical and Engin. Aspects*, *395*, 157-160.

[109] Hu, X., Ding, Z., Zimmerman, A.R., Wang, S., and Gao, B., (2015). Batch and column sorption of arsenic onto iron-impregnated biochar synthesized through hydrolysis. *Water Res*. 68, 206–216.).

[110] Wang, M., Sheng, G., and Qiu, Y., (2015). A novel manganese-oxide/biochar composite for efficient removal of lead (II) from aqueous solutions. *Int. J. Environ. Sci. Technol*. 12, 1719–1726.

[111] Zhang, M., Gao, B., Yao, Y., Xue, Y., & Inyang, M. (2012). Synthesis, characterization, and environmental implications of graphene-coated biochar.*Science of the total environment*, *435*, 567-572.).

[112] Tang, J., Lv, H., Gong, Y., & Huang, Y. (2015). Preparation and characterization of a novel graphene/biochar composite for aqueous phenanthrene and mercury removal. *Bioreso. Technol.*, *196*, 355-363.).

[113] Inyang, M., & Dickenson, E. (2015). The potential role of biochar in the removal of organic and microbial contaminants from potable and reuse water: a review. *Chemosphere*, *134*, 232-240.)

[114] Tan, X., Liu, Y., Zeng, G., Wang, X., Hu, X., Gu, Y., & Yang, Z. (2015). Application of biochar for the removal of pollutants from aqueous solutions.*Chemosphere*, *125*, 70-85.

[115] Xu, X., Cao, X., Zhao, L., 2013. Comparison of rice husk-and dairy manure-derived biochars for simultaneously removing heavy metals from aqueous solutions: role of mineral components in biochars. *Chemosphere* 92, 955–961.

[116] (Ahmad, M., Rajapaksha, A. U., Lim, J. E., Zhang, M., Bolan, N., Mohan, D., ... & Ok, Y. S. (2014). Biochar as a sorbent for contaminant management in soil and water: a review. *Chemosphere*, *99*, 19-33.).

[117] Helsen, L., Van den Bulck, E., Van den Broeck, K., & Vandecasteele, C. (1997). Low-temperature pyrolysis of CCA-treated wood waste: chemical determination and statistical analysis of metal input and output; mass balances. *Waste Management*, *17*(1), 79-86.

[118] Molinski, T. F., Dalisay, D. S., Lievens, S. L., & Saludes, J. P. (2009). Drug development from marine natural products. *Nature reviews Drug discovery*,*8*(1), 69-85.

[119] Vassilev, S. V., Braekman-Danheux, C., Laurent, P., Thiemann, T., & Fontana, A. (1999). Behaviour, capture and inertization of some trace elements during combustion of refuse-derived char from municipal solid waste. *Fuel*, *78*(10), 1131-1145.

[120] (Ippolito, J.A., Strawn, D.G., Scheckel, K.G., Novak, J.M., Ahmedena, M., Niandou, M.A.S., 2012. Macroscopic and molecular investigations of copper sorption by a steam-activated biochar. *J. Environ. Qual*. 41, 150–156.).

[121] Bolan, N., Kunhikrishnan, A., Thangarajan, R., Kumpiene, J., Park, J., Makino, T., Kirkham, M.B., Scheckel, K., 2014. Remediation of heavy metal(loid)s contaminated soilseto mobilize or to immobilize? *J. Hazard. Mater*. 266, 141-166.

[122] Khare, P., Goyal, D.K., 2013. Effect of high and low rank char on soil quality and carbon sequestration. *Ecol. Eng*. 52, 161–166.).

[123] Dong, X., Ma, L.Q., Li, Y., 2011. Characteristics and mechanisms of Hexavalent chromium removal by biochar from sugar beet tailing. J. Hazard. *Mater*. 190, 909–915.

[124] Cao, X., Harris, W., 2010. Properties of dairy-manure-derived biochar pertinent to its potential use in remediation. *Bioresour. Technol*. 101, 5222–5228.).

[125] Kong, H., He, J., Gao, Y., Wu, H., Zhu, X., 2011. Cosorption of phenanthrene and mercury(II) from aqueous solution by soybean stalk-based biochar. *J. Agric. Food Chem*. 59, 12116–12123.

[126] Choppala, G.K., Bolan, N.S., Mallavarapu, M., Chen, Z., Naidu, R., 2012. The influence of biochar and black carbon on reduction and bioavailability of chromate in soils. *J. Environ. Qual.* 41, 1–10.

[127] Bolan, N.S., Choppala, G., Kunhikrishnan, A., Park, J.H., Naidu, R., 2013. Microbial transformation of trace elements in soils in relation to bioavailability and remediation. *Rev. Environ. Contaminat. Toxicol.* 225, 1–56.).

[128] Mohan, D., Rajput, S., Singh, V.K., Steele, P.H., Pittman Jr., C.U., 2011. Modeling and evaluation of chromium remediation from water using low cost bio-char, a green adsorbent. *J. Hazard. Mater.* 188, 319–333.

[129] Sun, J., Lian, F., Liu, Z., Zhu, L., Song, Z., 2014. Biochars derived from various crop straws: characterization and Cd (II) removal potential. *Ecotoxicol. Environ. Saf.* 106, 226–231.).

[130] Cao, X., Ma, L., Gao, B., & Harris, W. (2009). Dairy-manure derived biochar effectively sorbs lead and atrazine. *Environ.sci. & tech.,43*(9), 3285-3291.

[131] Samsuri, A. W., Sadegh-Zadeh, F., & Seh-Bardan, B. J. (2014). Characterization of biochars produced from oil palm and rice husks and their adsorption capacities for heavy metals. *Inter.J. Env. Sci. Tech.*, *11*(4), 967-976.

[132] Wang, H., Gao, B., Wang, S., Fang, J., Xue, Y., Yang, K., 2015a. Removal of Pb (II), Cu(II), and Cd (II) from aqueous solutions by biochar derived from KMnO4 treated hickory wood. *Bioresour. Technol.* 197, 356–362.

[133] Wang, M., Sheng, G., Qiu, Y., 2015b. A novel manganese-oxide/biochar composite for efficient removal of lead (II) from aqueous solutions. *Int. J. Environ. Sci. Technol.* 12, 1719–1726.

[134] Wang, S., Gao, B., Li, Y., Mosa, A., Zimmerman, A.R., Ma, L.Q., Harris, W.G., Migliaccio, K.W., 2015d. Manganese oxide-modified biochars: preparation, characterization, and sorption of arsenate and lead. *Bioresour. Technol.* 181,13–17.

[135] Wang, S., Gao, B., Zimmerman, A. R., Li, Y., Ma, L., Harris, W.G., Migliaccio, K.W., 2015e. Removal of arsenic by magnetic biochar prepared from pinewood and natural hematite. *Bioresour. Technol.* 175, 391–395.

[136] Wang, S. Y., Tang, Y.K., Chen, C., Wu, J.T., Huang, Z., Mo, Y.Y., Zhang, K.X., Chen, J. B., 2015c. Regeneration of magnetic biochar derived from eucalyptus leaf residue for lead (II) removal. *Bioresour. Technol.* 186, 360–364.

[137] Gan, C., Liu, Y.G., Tan, X.F., Wang, S., Zeng, G., Zheng, B., Li, T., Jiang, Z., Liu, W., 2015. Effect of porous zinc–biochar nanocomposites on Cr (vi) adsorption from aqueous solution. *RSC Adv.* 5, 35107–35115.

[138] Cope, C. O., Webster, D. S., & Sabatini, D. A. (2014). Arsenate adsorption onto iron oxide amended rice husk char. *Sci. Total Env.*, *488*, 554-561.

[139] Zhang, F., Wang, X., Yin, D., Peng, B., Tan, C., Liu, Y., & Wu, S. (2015). Efficiency and mechanisms of Cd removal from aqueous solution by biochar derived from water hyacinth (Eichornia crassipes). *J. env.mgt.*, *153*, 68-73.

[140] Huang, X., Liu, Y., Liu, S., Tan, X., Ding, Y., Zeng, G., & Zheng, B. (2016). Effective removal of Cr (vi) using β-cyclodextrin–chitosan modified biochars with adsorption/reduction bifuctional roles. *RSC Advances*, *6*(1), 94-104.

[141] Song, Z., Lian, F., Yu, Z., Zhu, L., Xing, B., & Qiu, W. (2014). Synthesis and characterization of a novel MnOx-loaded biochar and its adsorption properties for Cu 2+ in aqueous solution. *Chem.l eng. J.*, *242*, 36-42.

[142] Zhou, Y., Gao, B., Zimmerman, A. R., Fang, J., Sun, Y., & Cao, X. (2013). Sorption of heavy metals on chitosan-modified biochars and its biological effects. *Chem.l eng. J.*, *231*, 512-518.

[143] Qiu, Y., Zheng, Z., Zhou, Z., & Sheng, G. D. (2009). Effectiveness and mechanisms of dye adsorption on a straw-based biochar. *Biores.tech.*, *100*(21), 5348-5351.

[144] Xie, T., Reddy, K. R., Wang, C., Yargicoglu, E., & Spokas, K. (2015). Characteristics and applications of biochar for environmental remediation: a review. *Critical Reviews in Environmental Science and Technology,* 45(9), 939-969.

[145] Karakoyun, N., Kubilay, S., Aktas, N., Turhan, O., Kasimoglu, M., Yilmaz, S., & Sahiner, N. (2011). Hydrogel–Biochar composites for effective organic contaminant removal from aqueous media. *Desalination*, *280*(1), 319-325.

[146] Devi, P., & Saroha, A. K. (2015). Simultaneous adsorption and dechlorination of pentachlorophenol from effluent by Ni–ZVI magnetic biochar composites synthesized from paper mill sludge. *Chem. Eng.J.*, *271*, 195-203.

[147] Rhodes, A. H., Carlin, A., & Semple, K. T. (2008). Impact of black carbon in the extraction and mineralization of phenanthrene in soil. *Environ. Sci. & Tech.*, 42(3), 740-745.

[148] Song, Y., Wang, F., Bian, Y., Kengara, F. O., Jia, M., Xie, Z., & Jiang, X. (2012). Bioavailability assessment of hexachlorobenzene in soil as affected by wheat straw biochar. *J. of hazard. Mat.*, 217, 391-397.

[149] Yao, Y., Gao, B., Chen, H., Jiang, L., Inyang, M., Zimmerman, A. R., & Li, H. (2012). Adsorption of sulfamethoxazole on biochar and its impact on reclaimed water irrigation. *J. of hazard. Mat.*, 209, 408-413.

[150] Oleszczuk, P., Hale, S. E., Lehmann, J., & Cornelissen, G. (2012). Activated carbon and biochar amendments decrease pore-water concentrations of polycyclic aromatic hydrocarbons (PAHs) in sewage sludge. *Biores.tech.*, 111, 84-91.

[151] Sopeña, F., Semple, K., Sohi, S., & Bending, G. (2012). Assessing the chemical and biological accessibility of the herbicide isoproturon in soil amended with biochar. *Chemosphere*, 88(1), 77-83.

[152] Zhang, P., Sheng, G. Y., Feng, Y. H., and Miller, D. M. (2006). Predominance of char sorptionover substrate concentration and soil ph in influencing biodegradation of benzonitrile.*Biodegradation*, 17(1), 1-8.

[153] Martin, S. M., Kookana, R. S., Van Zwieten, L., & Krull, E. (2012). Marked changes in herbicide sorption–desorption upon ageing of biochars in soil. *J. of hazard. Mat.*, 231, 70-78.

[154] Nag, S. K., Kookana, R., Smith, L., Krull, E., Macdonald, L. M., & Gill, G. (2011). Poor efficacy of herbicides in biochar-amended soils as affected by their chemistry and mode of action. Chemosphere, 84(11), 1572-1577.

[155] Tatarková, V., Hiller, E., & Vaculík, M. (2013). Impact of wheat straw biochar addition to soil on the sorption, leaching, dissipation of the herbicide (4-chloro-2-methylphenoxy) acetic acid and the growth of sunflower (Helianthus annuus L.). *Ecotoxicology and environmental safety*, 92, 215-221.

[156] Chai, Y., Davis, J. W., Currie, R. J., Fishman, V., Martin, G., Wilken, M., Lucas, S., Wandor, D.,and Ghosh, U. (2012). A field pilot study using activated carbon and biochar to reduce theavailability of PCDD/Fs/fs in soils/sediments. In: SETAC North America 33rd AnnualMeeting, Long Beach, California.

[157] Zheng, W., Guo, M., Chow, T., Bennett, D. N., & Rajagopalan, N. (2010). Sorption properties of greenwaste biochar for two triazine pesticides. *J. Hazard. Mat.*, 181(1), 121-126.

*[158]*Ghosh, U., Gomez-Eyles, J., Yupanqui, C., Xia, H., & Beckingham, B. (2012). Activated Biochars with Iron for In-Situ Sequestration of Organics, Metals and Carbon. *MARYLAND UNIV BALTIMORE COUNTY BALTIMORE MD DEPT OF CHEMICAL BIOCHEMICAL AND ENVIRONMENTAL ENGINEERING.*

[159] Lou, L., Wu, B., Wang, L., Luo, L., Xu, X., Hou, J., & Chen, Y. (2011). Sorption and ecotoxicity of pentachlorophenol polluted sediment amended with rice-straw derived biochar. *Biores. Techn.*, 102(5), 4036-4041.

[160] Zhang, M., Gao, B., Yao, Y., Xue, Y., & Inyang, M. (2012). Synthesis of porous MgO-biochar nanocomposites for removal of phosphate and nitrate from aqueous solutions. *Chem. Eng. J.*, 210, 26-32.

[161] Zhang, M., Gao, B., Varnoosfaderani, S., Hebard, A., Yao, Y., and Inyang, M. (2013). Preparation and characterization of a novel magnetic biochar for arsenic removal. *Biores. Techn.*, 130, 457-462.

[162] Khan, S., Wang, N., Reid, B. J., Freddo, A., and Cai, C. (2013). Reduced bioaccumulation ofPAHs by Lactucasatuva L. grown in contaminated soil amended with sewage sludge andsewage sludge derived biochar. *Environ Pollut*, 175, 64-68.

[163] Marchal, G., Smith, K., Rein, A., Winding, A., Trapp, S., and Karlson, U. G. (2013). Comparingthe desorption and biodegradation of low concentrations of phenanthrene sorbed toactivated carbon, biochar and compost. *Chemosphere*, 90(6), 1767-1778.

[164] Larsbo, M., Lofstrand, E., de Veer, D. V., and Ulen, B. (2013). Pesticide leaching from two Swedish top soils of contrasting texture amended with biochar. J *ContamHydrol*, 147, 73-81.

[165] Kim, K. H., Kim, J. Y., Cho, T. S., & Choi, J. W. (2012). Influence of pyrolysis temperature on physicochemical properties of biochar obtained from the fast pyrolysis of pitch pine (Pinus rigida). *Biores. Tech.,* 118, 158-162.

[166] Purevsuren, B., Avid, B., Tesche, B., & Davaajav, Y. A. (2003). A biochar from casein and its properties. *J. Mat. Sci.*, 38(11), 2347-2351.

[167] Bird, M. I., Wurster, C. M., de Paula Silva, P. H., Bass, A. M., & De Nys, R. (2011). Algal biochar–production and properties. *Biores. Tech.*, 102(2), 1886-1891.

[168] Keiluweit, M., Nico, P. S., Johnson, M. G., & Kleber, M. (2010). Dynamic molecular structure of plant biomass-derived black carbon (biochar). *Environ. Sci. tech.*, 44(4), 1247-1253.

[169] Boehm, H. P. (1994). Some aspects of the surface chemistry of carbon blacks and other carbons. *Carbon*, 32(5), 759-769.

[170] Uchimiya, M., Orlov, A., Ramakrishnan, G., & Sistani, K. (2013). In situ and ex situ spectroscopic monitoring of biochar's surface functional groups. *J. Anal. App.Pyrolysis*, 102, 53-59.

[171] Chun, Y., Sheng, G., Chiou, C. T., & Xing, B. (2004). Compositions and sorptive properties of crop residue-derived chars. *Env. Sci. tech.*, 38(17), 4649-4655.

[172] Jiang, J., Xu, R. K., Jiang, T. Y., & Li, Z. (2012). Immobilization of Cu (II), Pb (II) and Cd (II) by the addition of rice straw derived biochar to a simulated polluted Ultisol. *J. hazard. Mat.*, 229, 145-150.

[173]]Rondon, M. A., Lehmann, J., Ramírez, J., & Hurtado, M. (2007). Biological nitrogen fixation by common beans (Phaseolus vulgaris L.) increases with bio-char additions. *Bio. fertility of soils*, 43(6), 699-708.

[174] Uzoma, K. C., Inoue, M., Andry, H., Fujimaki, H., Zahoor, A., & Nishihara, E. (2011). Effect of cow manure biochar on maize productivity under sandy soil condition. *Soil use manag.*, 27(2), 205-212.

[175] Kookana, R. S., Sarmah, A. K., Van Zwieten, L., Krull, E., & Singh, B. (2011). 3 biochar application to soil: agronomic and environmental benefits and unintended consequences. *Adv. in agro.*, 112(112), 103-143.

[176] Jindo, K., Sonoki, T., Matsumoto, K., Canellas, L., Roig, A., & Sanchez-Monedero, M. A. (2016). Influence of biochar addition on the humic substances of composting manures. *Waste Manag.*, *49*, 545-552.

[177] Lingamdinne, L. P., Roh, H., Choi, Y. L., Koduru, J. R., Yang, J. K., & Chang, Y. Y. (2015). Influencing factors on sorption of TNT and RDX using rice husk biochar. *J. Indus. Eng.Chem.*, *32*, 178-186.

[178] Hale, L., Luth, M., & Crowley, D. (2015). Biochar characteristics relate to its utility as an alternative soil inoculum carrier to peat and vermiculite. *Soil Bio. Biochem.*, *81*, 228-235.

[179] Sasidharan, S., Torkzaban, S., Bradford, S. A., Kookana, R., Page, D., & Cook, P. G. (2016). Transport and retention of bacteria and viruses in biochar-amended sand. *Sci. Total Envi.r*, *548*, 100-109.

[180] Li, T., Bai, X., Qi, Y. X., Lun, N., & Bai, Y. J. (2016). Fe 3 O 4 nanoparticles decorated on the biochar derived from pomelo pericarp as excellent anode materials for Li-ion batteries. *Electrochimica Acta*, *222*, 1562-1568.

In: Modified Biopolymers: Challenges and Opportunities ISBN: 978-1-53612-116-2
Editors: D. Pathania, G. Sharma and A. Kumar

Chapter 6

NATURALLY OCCURRING BIODEGRADABLE POLYMERS

Sujata Bhattacharya[1,*], Sunil Puri[1] and Ashok Kumar[2]
[1]School of Biological and Environmental Sciences,
Faculty of Basic Sciences, Shoolini University, Solan, India
[2]Department of Biotechnology,
Himachal Pradesh University, Summer Hill Shimla, India

ABSTRACT

Expeditious diminishing natural resources are driving forces for researchers to identify innovative methodologies and exploit renewable materials alternatives to natural-based products and processes. Due to concerns over the depletion of natural resources, biodegradable biopolymers have been considered as the most promising materials. With their biodegradability, eco-friendly manufacturing processes and vast range of applications, the development of biopolymers has been driven largely in response to the growing concerns regarding the sustainability of conventional polymers. Natural polymers can be regarded as the first biodegradable biomaterials used since ancient times for nonfood and clinical uses. Several ages of human history are using natural polymers in their day today life such as skin and bone parts of animals, plant fibers, starch, silk, etc. Natural polymers possess numerous inherent advantages such as bioactivity, the ability to present receptor-binding ligands to cells, susceptibility to cell-triggered proteolytic degradation and natural remodeling. This chapter deals with an overview of development of natural polymers such as with starch, cellulose, guar gum, chitin, chitosan, pectin etc novel processing characteristics and properties.

Keywords: biopolymers, starch, cellulose, chitosan, chitin, pectin

---

*Corresponding Author Email:sujata0685@gmail.com.

1. INTRODUCTION

Natural resources bestowed by nature are vital for the survival and wellbeing of living beings including humans. The exploitation of diverse natural resources is indispensable for the sustenance of mankind and for the development of a nation, consumption of energy, soil, water, coal, oil, gas and nuclear energy is very important. Globalization leads to rapid depletion of natural resources and the growing adverse effects on the environment accompanied by their increasing rates of consumption. Resource depletion is not the only one problem to be faced, since adversities related to global warming could have impending disturbing social, economic and environmental consequences in the future, if not addressed [1]. Efficient use of resources has become a main concern for sustainable economies. Fast depleting natural resources are driving forces for researchers to identify innovative methodologies and exploit renewable materials alternatives to natural-based products and processes.

In this context, the concept of eco-friendly materials has been receiving the attention of a broad research community at the global level and led to development of new advanced biodegradable polymers. Since 1970s, biodegradable polymers have undergone extensive investigation from academia and industry and experienced several important stages of development. Natural polymers are derived from renewable and synthetic polymers are derived from nonrenewable resources. Enhanced environmental pollution and the depletion of fossil oils, intensive research is being conducted for developing biodegradable polymers. Biopolymers have been considered as the most promising materials to non-sustainable products in a cost-effective manner, with biodegradable materials offering competitive mechanical properties [2]. With their biodegradability, eco-friendly manufacturing processes and vast range of applications, the development of biopolymers has been driven largely in response to the growing concerns regarding the sustainability of conventional polymers [3]. Biodegradable polymers mainly include aliphatic polyesters such as poly (butylenes succinate-*co*-adipate) (PBSA), polycaprolactone (PCL), poly (butylene succinate) (PBS), and other aliphatic co-polyesters and aliphatic–aromatic copolyesters such as poly (butylene adipate-*co*-terephalate) (PBAT). Globally, researchers also made extensive efforts in developing biodegradable from renewable natural resources such as starch, cellulose, chitosan, chitin and soy protein [4]. Biopolymers can be obtained from different types of feedstock including agricultural products, such as corn or soybeans and from other sources like algae or food waste and offer a renewable substitute to traditional petroleum based plastics [1]. For instance, biopolymers with their eco-friendly manufacturing processes presenting similar quality level as traditional materials and are quite suitable to replace unsustainable products [5]. The use of biopolymers would also enable reduced exploitation of non-renewable primary-energy resources [6].Current attention in biopolymers is motivated diversely by both government and corporate commitments to reduced energy use and carbon dioxide emissions which directed the industries to emphasize on biologically-produced, renewable raw materials. Moreover, waste-disposal problems associated with synthetic plastics can be accomplished by biodegradable polymers. This chapter summarized current research achievements in biodegradable and biobased polymers such as their classification, sources, applications etc.

1.1. Biodegradable Polymers

Biodegradable polymeric biomaterials have been extensively explored as both synthetic polymers and biologically derived (or natural) polymers. Cleavage of hydrolytically sensitive bonds in the polymer leading to polymer erosion involves in the biodegradation of polymeric biomaterials [7]. Polymeric biomaterials can be further classified into hydrolytically degradable polymers and enzymatically degradable polymers, depending on the mode of degradation (Figure 1) [8]. Biodegradable polymers could be derived from either renewable or nonrenewable resources and depending on the source of derivation they can be further classified into naturally occurring biodegradable polymers, biodegradable polymers derived from renewable resources, biodegradable polymers based on petroleum and biobased polymers (Figure 2) [1].

1.2. Naturally Occurring Biodegradable Polymers

Natural polymers can be regarded as the first biodegradable biomaterials used since ancient times for nonfood and clinical uses. Several ages of human history are using natural polymers in their day today life such as skin and bone parts of animals, plant fibers, starch, silk, etc. Natural polymers possess numerous inherent advantages such as bioactivity, the ability to present receptor-binding ligands to cells, susceptibility to cell-triggered proteolytic degradation and natural remodeling [8]. But the advent of low-cost petrochemical polymers delayed the development of natural polymers. It was only about two decades ago that intensive research on natural polymers was revived, primarily due to the issues of environment pollution and the depletion of fossil oils [4]. Synthesis, structures, and properties of the natural polymers are constantly explored by modern technologies which lead to development of natural polymers with novel processing characteristics and properties. This section deals with three major natural polymers: starch, cellulose, guar gum, chitin, chitosan, pectin etc.

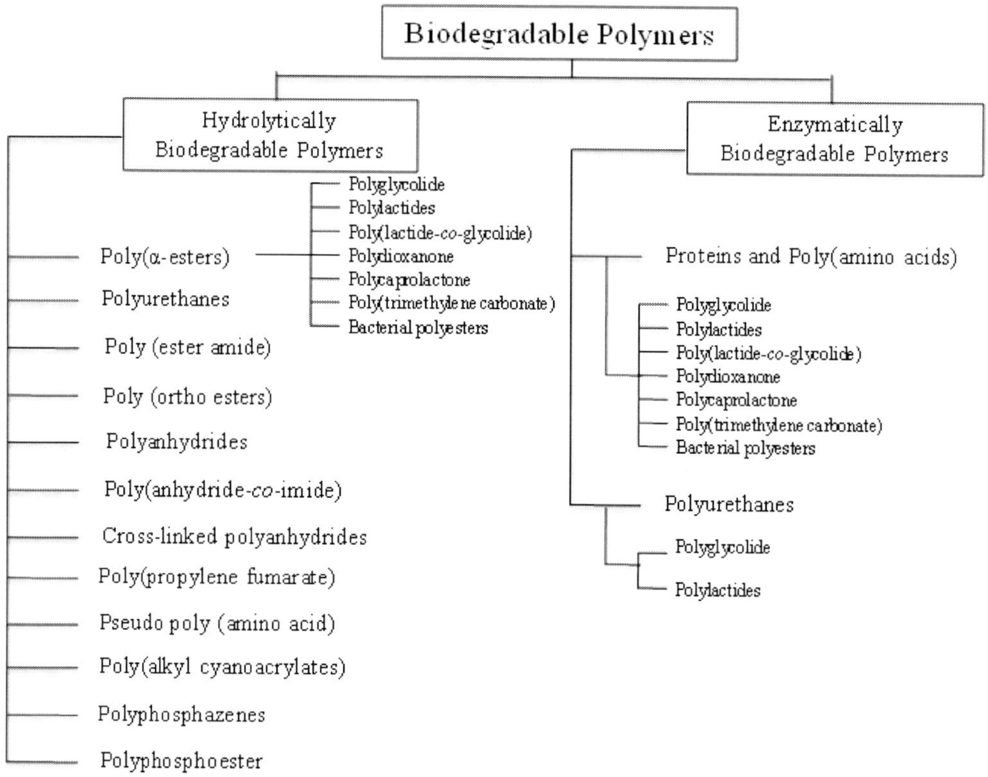

Figure 1.Classification of biopolymers on the basis of mode of degradation.

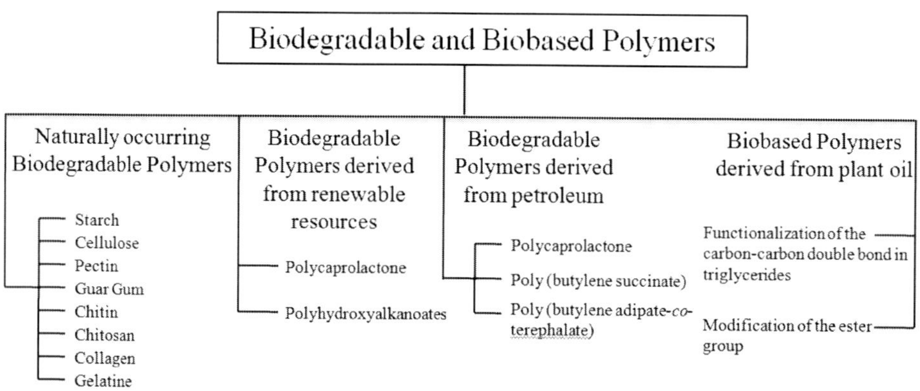

Figure 2.Classification of biopolymers on the basis of source of derivation.

1.2.1. Starch

Starch is the most important renewable storage carbohydrate for food and feed [9]. It is a major component contributing to yield in a variety of crop plants and also a sustainable feedstock for many industrial applications. A seed at maturity has about 50 to 80% starch content, depending upon genotype and environmental conditions [10]. It is a distinct, semi-crystalline granule, composed of two kinds of water-insoluble polymers of glucose molecules, i.e., amylose (one-quarter) and amylopectin (three-quarter) [11]. Amylose is predominately a linear α-1,4 linked glucan polymer of 500-3,000 glucose moieties, has <1% α-1,6 linkages, while amylopectin is a highly branched glucan polymer with α-1,4 linkages and 5-6% α-1,6 linkages (Figure 3) [12]. Besides these polymers, phosphate esters are covalently bound to starch granules, providing specific functionalities to starches and thus making these more suitable for different industrial applications.

The well-organized structure and shape of starch granules was provided by the chains of amylose and amylopectin subunits, with size varying from 1 to 50 μm, and grouped into large (lenticular) and small (round) sized starch granules [13, 14]. This variation in the starch granule size, structure and has important impact on various starch physico-chemical properties [15, 16]. The texture of cooked food, swelling, solubility, gelatinization, retrogradation and pasting properties influenced by both high or low amylose starch and thus, affecting the processing, cooking, and organoleptic qualities and digestibility of wheat-based food products [17]. The phosphate esters are mainly present in the amylopectin chains and monoesterified at both the C-6 (70%) and C-3 (30%) positions of glucose units [18]. Starch, the major biopolymer of plants, viz., grains, seeds and tubers, occurs in characteristic granular forms of various shapes and sizes [19].

Glucose-1-phosphate synthesize starch in amyloplast of seed endosperm [20]. The complex and coordinated actions of metabolic enzymes help in the catalysis of this process: ADP-glucose pyrophosphorylase (AGPase), starch synthases [granule-bound (GBSS and soluble starch synthase (SS)], branching enzyme (BE), debranching enzymes (DBEs) [Isoamylase (ISA) and pullulanase (PUL)], and starch degrading enzymes [starch phosphorylase (Pho), disproportionating enzyme (DPE), and amylases (α-amylase and β-amylase)] [21]. Other transcription factors involved in starch biosynthesis and its gene regulation are: rice starch regulator (RSR1) [22] and storage protein activator (SPA) [23] and transporters, such as sucrose transporter (SUT) [24], ADP-Glc transporter (brittle 1 transporter, BT1) and triose phosphate translocator (TPT). The level of starch phosphorylation depends on plant organ and species. In potato tubers, storage starches contain up to 33 nmol phosphate/mg starch, whereas the phosphate content in transitory starch does not exceed 0.1% [18]. The presence of high phosphate content in potato starch is in part responsible for the high swelling power and stable-paste properties, thereby broadening the end-use of potato starches to a wide variety of applications in food and non-food industries [25].

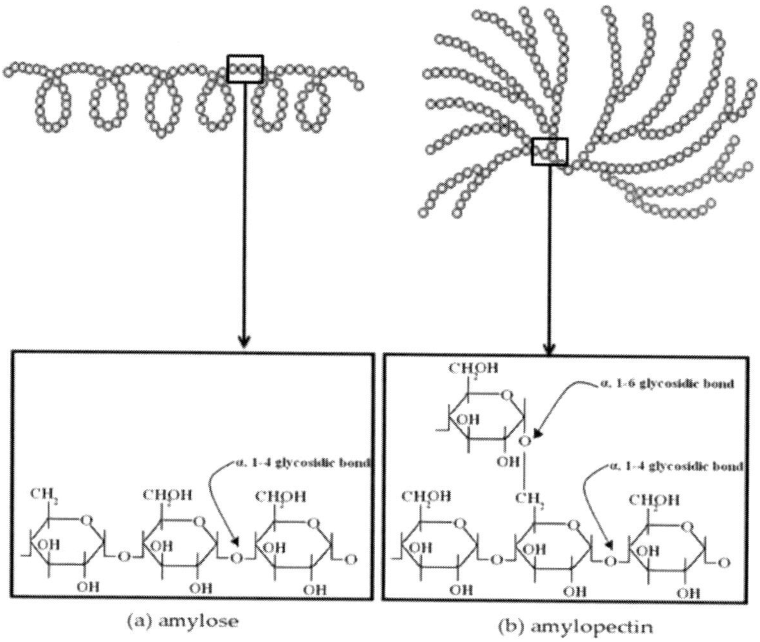

Figure 3.Structure of starch (a) Amylose (b) Amylopectin.

Starch is widely used as an ingredient in various food and nonfood products. Starch can be modified to develop desirable functional properties, such as solubility, texture, adhesion, dispersion and heat tolerance, in order to be suitable for industrial applications [26]. Starch, composed of repeating D-glucopyranosyl units, has been recognized as one of the most promising natural feedstocks to substitute for petrochemical plastics in a variety of applications [27]. The biodegradability of starch together with its low cost and wide availability offers several advantages [28], but its applications are limited by poor mechanical strength, hydrophilic characteristics and susceptibility to microbial attack. These can be managed by blending starch with other synthetic polymers [29]. Amongst the starch-based composites, starch/ Polyvinyl Alcohol (PVOH) blends have attracted great interest since the 1980s because of their excellent compatibility and processability, and the improved properties of the blends [28, 29]. Starch/PVOH blends have developed rapidly in the last decades and are widely applied as packaging or agricultural mulch film. For instance, the starch/PVOH loosefills have been commercialized under the Mater-Bi trademark since the 1990s [30], and nowadays approximately 25,000 tonnes of starch–PVOH loosefills are estimated to be consumed annually worldwide (data from Greenlight Products Ltd.). The increasing consumption of starch/PVOH based products brings concern about their waste treatment. Some research has been carried out on the biodegradability of PVOH and PVOH/starch blends under various environmental conditions including anaerobic digestion [28, 31].

Starch products are the most frequently used biomaterials in many industries because they are of low cost and biodegradable nature. Starch is one of the most widely used biopolymers in the manufacture of food industry, solid dosage, can be formulated as a filler, disintegrant, or binder. Besides, starch-based materials with high amylose contents exhibit good oxygen barrier functions. Amylopectin used as subcoating material improves the acidic resistance of enteric-coated pellets. Physically or chemically modified starches are used as a forming-matrix excipient and a coating material for pellets and tablets. By means of temperature modification fully or partially gelatinized starch is of great interest for the formulation of sustained release products owing to its coldwater-swelling capacity and gel barrier formation. Behaviour of the layer, formed around matrices after contact with water is of significant importance also for the drug delivery profiles. Recently pregelatinized and modified starches have been described as possessing mucoadhesive properties [32].

1.2.2. Cellulose

Cellulose is the most abundant renewable organic resource on this planet and is widespread in higher plants, bacteria, marine algae and other biomass. Cellulose is the main component of the plant cell. Some animals (such as tunicates) and some bacteria also contain cellulose, but the amount is negligible as compared to plants. Cellulose is the most abundant resource biosynthesized on earth with an annual production over 7.5 × 1010 tons [33-36].

Cellulose can be characterized as highly polymerized linear macromolecule consisting solely of 1–4- β-linked anhydro-d-glucose [5-7]. Particularly, the β-configuration at the anomeric carbon of glucose unit gives rise to a stretched chain conformation. Then, these chains are linked via hydrogen-bonds to make flat sheets. The linear alignment of molecular chain enables the compact packing of numerous cellulose strands into crystalline fibrils, which makes it challenging to separate individual molecular chain. Moreover, many other structural aspects have been identified, including the high molecular weight, the comparatively low flexibility of polymer chain, together with the hydrophobic surface attributed to the inferior solubility of cellulose in most solvents. Correspondingly, valorization of natural cellulose via physical and chemical conversion was severely restricted.

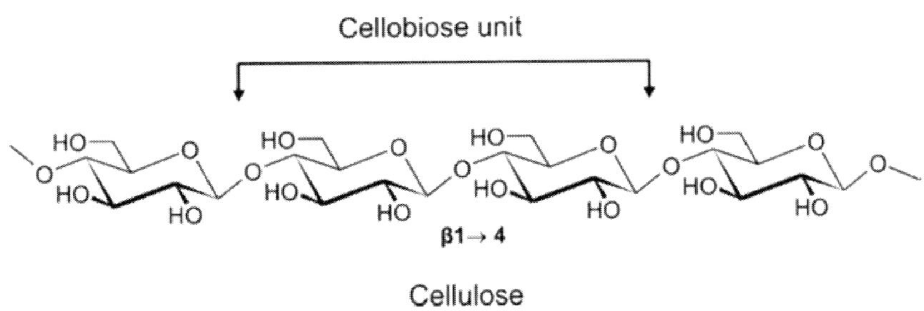

Figure 4.Structure of cellulose.

Cellulose is an organic compound with the molecular formula $(C_6H_{10}O_5)_n$, and consisting of a linear chain of several hundred to over ten thousand β - 1, 4 linkedD-glucose units (Figure 4). The -OH groups in the linear cellulose structure form hydrogen bonds with neighboring chains, resulting in highly crystalline structures called microfibrils. Cellulose forms the basic structural foundation of the cell wall of eukaryotic plants, some algae and fungi [40]. In total plant matter, about 33% is cellulose. However, the cellulose content of cotton is 90% and that of wood is 40 – 50%. Together, cotton and wood are the major resources for all cellulose products such as paper, textiles, construction materials, cardboard, as well as cellulose derivatives as cellophane, rayon, and cellulose acetate [41-44].

Cellulose is a natural polymer mainly derived from wood taken from trees, such as acacia, eucalyptus and pines. The cellulosic waste generated in cereal crop such as corn, wheat, sorghum, oat and rice are discarded in the soil used as protection and coverage for the next planting. In the industrial process, the rice and oat grains generate a high amount of husks that have high cellulose content. According to [46], global rice annual production is 741 million tons and generates approximately 148 million tons of husks, which represent 20% of the grain weight [47]. Plant cellulose is synthesized in the presence of hemicelluloses (xylan, xyloglucan and glucomannan) and pectin, which impact its structure. Xyloglucan is the most abundant hemicellulose in the plant primary wall and has a β-1,4-linked-glucose backbone with frequent substitution of β -d-xylosyl residues at C-6 positions [48]. Another hemicellulose, glucomannan, is present in the secondary cell wall and composed of β-(1 → 4)-linked d-mannose and d-glucose [49].

Cellulose, the most abundant biopolymer on Earth is frequently obtained from plant sources. Some bacterial strains such as *Gluconacetobacter*, *Agrobacterium*, *Rhizobium*, *Rhodobacter*, and *Sarcina* can produce cellulose extracellularly [50-51].Biopolymers are natural polymers produced by living organisms. These contain monomeric subunits covalently linked to form larger structures [51]. In nature, biopolymers and their derivatives are diverse, abundant and are important for life. These exhibit many properties making them important for different applications. Biopolymers are found in all living organisms and contribute to the major fraction of the cellular dry matter. They carry out a wide range of different essential and beneficial functions such as catalysis of reactions, conservation and expression of genetic information, storage of carbon, nitrogen, phosphorus and other nutrients, defence and protection against attacks by other cells or hazardous environmental or intrinsic factors. These biopolymers also act as sensors of biotic and abiotic factors, communication with the environment and other organisms or with non-living matter and many more. In addition, many of these are structural components of cells and tissues of organisms [52-53]. Cellulose is one of the most abundant biopolymerpresent on the planet earth. It is present all around us: we wear it, eat it,build with it, print on it and burn it. Thus, cellulose is of great economic importance globally [54-58]. A major portion of the world's economy depends directly upon this. As a chemical raw material, cellulose and its derivatives have been widely utilized in many fields, such as papermaking, textile and pharmaceuticals [59-61]. Because of its inherent biodegradability and biocompatibility, bacterial cellulose (BC) has several potential applications in various biomedical fields, including artificial blood vessel manufacturing, wound dressing, dialysis membranes, and biosensors [62]. The structural features of BC are superior to those of plant cellulose, as BC has unique properties such as high purity, high crystallinity, ultrafine networks, and remarkable mechanical strength [63].

1.2.3. Pectin

Pectin is one of the main constituent of the primary cell wall of most plants; it is present in cells during growth and development and between adjacent cells (middle lamella). Pectins are complex polymers which consist mainly of homogalacturonan (HG), rhamnogalacturonan I (RGI) and rhamnogalacturonan II (RGII) blocks covalently linked to one another [64]. In the most common representation of the primary plant cellwall of growing plants, pectin is depicted as an independent network, embedded in a load-bearing cellulose-hemicellulose net-work, with some structural proteins [65]. To date the major hemicellulose thought to be implicated in this net-work has been xyloglucan (XG) pectin is not only present as an independent network butcan also be in close contact with cellulose, through interactions with e.g., calcium-deficient regions of homogalacturonan [66]and/or the arabinan and galactan side chains characteristic of rhamnogalacturonan I [66,67].We have recently demonstrated that pectins contribute to the mechanical properties of these types of cellulose networks when they are present as a solution in the absence of Ca^{2+}.

Pectin was first discovered in apple juice by [68] and named by [69] borrowing from the Greekword: pektikos which means to congeal or solidify. Since then, pectin has been studied extensively by scientists from different branches of science. Chemically, pectin is defined as an anionic polysaccharide made up largely by covalently -1,4-linked d-galacturonic acid (also known as galactosyluronic acid) (GalA) units (Figure 5) . It is derived naturally from plants, in which pectin makes up approximately one third of the cell walls in dicotyledonous and some monocotyledonous plants [69]. Pectin can also be found in small proportions of cell walls in grass (2–10%) and wood tissue (5%) [70]. Biologically, pectin has specific functions depending on its location and molecular structure. Pectin can function as a gel and there by assisting with cell adhesion and softening cell walls for cellelongation [71-73]. This polysaccharide lends strength and support to plants by maintaining cell consistency and mechanical resistance [70]. Additionally, pectin influences various cellwall properties such as porosity, surface charge, pH, and ion balance by forming networks and trapping solute molecules that enable ion transportation [70, 74]. Pectin also activates plant defenses by stimulating the accumulation of phytoalexins which have a wide spectrum of anti-microbial activity [75, 76].

Although it has been more than 210 years since the discovery of pectins, their chemical and structural properties are still the subject of investigation due to the in homogeneity of this polymer family. To begin with basic structure, pectin is assembled by at least 17 different monosaccharides, of which GalA is the most plentiful, followed by l-arabinose, d-galactose, l-rhamnose and others [77]. The semonosaccharides can be interconnected through 20 different link-ages [77] Within GalA monomers, the carboxylic groups or hydroxyls may be methyl-esterified) and/orO-acetyl-esterified respectively. O-acetyl-esterification occurs predominantly at the O-3 position and occasionally at the O-2 position. Native pectins in plants are composed of polysaccharidedomains: homogalacturonan (HGA), rhamnogalacturonan I (RG-I), rhamnogalacturonan II (RG-II), xylogalacturonan, and apiogalacturonan. The latter three have been classified as substituted HGA [74, 78, 79]. The deposition and the way in which these domains are joined to oneanother remains a matter of debate. Currently, it is thought that the polysaccharide domains are covalently linked, and to a greater or lesser extent, ionically cross-linked with other pectin strands toformpectic networks that branch throughout the primary cell walls [74, 80].

HGA is a linear homo polymer of -1,4-linked GalA and is known as the "smooth region." It is an abundant and wide spread domain of pectin, accounting for approximately 60–65% of total pectin amount [70] HGA appears to be synthesized in the Golgi apparatus and then transferred to the middle lamella and primary cell walls [73]. RG-I consists of up to 100 or more repeating units of the disaccharide -1,2-linked-l-rhamnose- 1,4-d-GalA. The rhamnose residues account for >30% of a pectin [81]. Within these rhamnose residues, 50–78% is estimated to be RGI [81]. The GalA residues of RG-I arenot methyl-esterified however, the GalA residues of RG-I could beO-acetyl-esterified [78]. RG-II is not structurally related to RG-I, despite their names running in sequence and the fact that it is also included in the hairy region. It is a branched pectic domain containing a HGA backbone. RG-II is a highly compact homopolymer of around nine -1,4-linked GalA units (of which some are methyl-esterified) with four structurally different polymeric side chains attached.

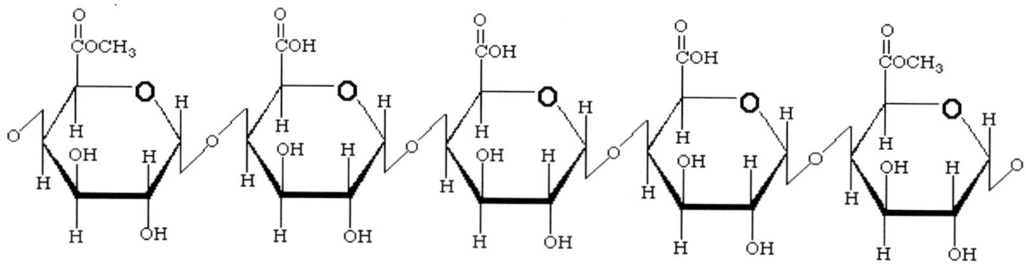

Figure 5.Structure of Pectin.

1.2.4. Chitosan

Chitosan is a linear semi crystalline polysaccharide, obtained by deacetylation of chitin and composed by combination of N-acetyl D-glucosamine and D-glucosamine units, linked through β (1→4) glycosidic bonds (Figure 6) [82,83]. Chitin or poly (β-(1→4)-N-acetyl-D-glucosamine) is synthesized by living organisms, structural component of the shells of crustaceous, cell walls of fungi and exoskeletons of insects [83-85]. Typically, shrimp and crab shell waste are the primary source for isolation and purification of this polysaccharide [83,87-89]. Deacetylation process, under alkaline conditions, part of the N-acetyl groups are lost and when at least 60% of the chitin units are D-glucosamine this polysaccharide is named CHI [90,91]. A soluble CHI is obtained after deacetylation, requiring 80–85% of free amino groups and acidic pH [85]. The free amino groups confer exceptional chemical and physical properties, among which are worth mentioning the positive charge in aqueous solutions at low pH [85]. At pH values above its pKa ≈ 6.5, the polymer loses the positive charge and precipitates, which makes CHI a pH responsive material [85]. Chitosan is one of the most commonly polymers in the scientific research dealing with a wide range of biopharmaceutical and biomedical applications including food science and technology. It has been strongly indicated as a suitable functional material in view of its properties that is biocompatibility, biodegradability and non-toxicity. Originally known as a marine polysaccharide from shrimps and crabs, chitosan was primarily thought to be an easily accessible substance from the wastes of food indusrty. The original samples of chitosan were often raw materials, with few exceptions of purification applied to chitosan in pharmaceutical and cosmetic uses. The main areas in its characterization were in terms of quality (either purity or harmfulness), intrinsic molecular properties and physical form. Independently for final human use, the first and most important issue was, and is still, related to the presence of impurities, metals, other inorganics, proteins, pyrogenic, endotoxic, cytotoxic agents, and bioburden, all of these have received over the years erratic attention [86].

Chitosan is the second most abundant polymer in nature after cellulose and and is usually available in the outer skeletons of crustaceans and many species of molluscs, insects and fungi. The most commonly obtained form of chitosan is the a-chitosan from crustacean chitin obtained from crab- and shrimp shell. Chitosan is prepared by deproteination and demineralization of ground shells by sequential treatment with alkali and acid, after which deacetylation by alkaline hydrolysis at high temperature of isolated chitin takes place to form chitosan. Production of chitosan from these sources is inexpensive and simple. It has also been suggested that another sources of chitin, e.g., β-chitin from squid pens, may be valuable in relation to the formation of chitosan. Chitosan is rare in nature, except in fungi. In recent years, chitosan is produced from fungi, fermentation method.

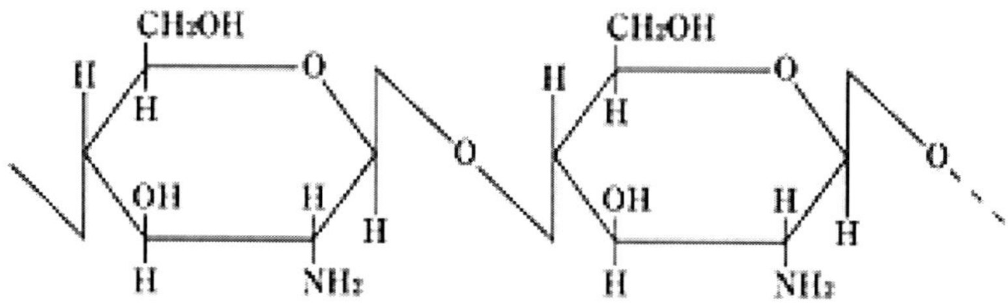

Figure 6.Structure of Chitosan.

Chitosan has been used by researchers to exploit several applications in very wide fields, in particular pharmaceutical industry (drug delivery, devices, and wound dressing), cosmetic, textile and food industries, as well as in agriculture and environment. A summary of antimicrobial properties of chitosan and its derivatives is given, keeping in mind the general assertion that chitosan is considered as a compound that is GRAS (Generally Recognized As Safe). Multiple constructs of chitosan, and its derivatives, have been evaluated to express the bactericidal potential in medical field (e.g., post-surgery wound healing, and bandages as antimicrobial dressing after burning, antimicrobial coating of prosthesis) [93-98]. In the field of nanobiopharmaceutics, antimicrobial chitosan nanoparticles have studied for their enhancing effect of surface-to-size value. Thus, a number of works report the use of chitosan in form of nanoparticles, formed in different conditions and blend, against certain microorganisms [99-103]. In food industry, maintaining the quality and extending the shelf life of food products. One of the most promising ways for effective preservation of food from alterations used bioactive films as packaging. The use of active and edible bio-film based on chitosan alone, enriched with different components such as plant extracts, other natural polysaccharides and antimicrobial peptides is emerging as answer to such difficulties [104-109]. Finally, the use of antibacterial agents to inhibit bacteria colonization is becoming a popular procedure in textile production, especially for goods employed in medical or services [110]. This antimicrobial supplementation is needed for natural fibers, which are more vulnerable to microbial attacks.

1.2.5. Guar Gum

Gums are plant exudates which can be dissolved in water to form viscous colloidal solution. Gums are naturally occurring substances mainly carbohydrates and are used for various purposes (e.g., as food and fodder) since from the beginning of civilization [111]. Guar gum also called Guaran, Clusterbean, Gum cyamposis, Cyamopsis gum, Calcutta lucern, Guarina, Glucotard and Guyan. Guar gum is a galactomannan. [112,113]. Guar gum is extracted from the seeds of the drought resistant plant *Cyamopsistetragonoloba* L. belonging to Leguminosae family [114-116]. Galactomannan is a naturally occurring polysaccharide [117]. Guar gum is native to north western parts of India and some parts of

Pakistan [118]. Guar seed endosperm is the major source of Guar gum. It is used as emulsifier, thickener and stabilizer in various food products and contributes to soluble dietary fiber (SDF) portion of seed total dietary fiber (TDF) [119]. As a food additive, it emulsifies, binds water, suppresses ice crystals in frozen products, thickens, stabilizes, moisturizes and suspends many liquid–solid systems. It is used in making of ice cream, cake mixes,sauces cheese spreads and fruit drinks generally in amount of <1% of the food weight [114,120]. Guar gum is a gel forming galactomannan which is extracted by grinding the endosperm portion of *Cyamopsistetragonolobus* L. [121].

Chemically, guar gum is a polysaccharide which is composed of the sugars galactose and mannose. It mainly consist of high molecular weight hydrocolloidal polysaccharide, composed of galactan and mannan units are combined through glycosidic linkages and shows degradation in the large intestine due to the presence of microbial enzymes [122,123]. The backbone is a linear chain of 1,4-linked mannose remaining part to which galactose remaining parts are 1,6-linked at every second mannose, forming short side-branches [124]. These galactose and mannose groups form the galactomannan portion of seed endosperm. It was believed that the side groups were added at regular intervals along with the mannan backbone [114]. Although, experiment using enzyme degradation of guar [125], spectroscopic method [126] and computer simulation method [127], point out arbitrary distribution of galactose side groups. One such model proposing a guar galactomannan in which the galactosyl units are randomly arranged mostly in pairs and triplets [128]. It has historically been reported that the ratio of mannose to galactose units is 2:1 [129]. Several research studies support ratios from the range of 1.6:1 to 1.8:1 [126-131]. Current data also suggest that galactomannans from different guar varieties have the same galactose and mannose arrangement [127]. The greater branches of guar are believed to be responsible for its easier hydration properties as well as its greater hydrogen bonding activity [132].

The biological properties of guar galactomannan and other such polysaccharides are It on their behavior in an aqueous medium. Guar gum swells when dissolves in water on dispersion and form strong hydrogen bonds. In non polar solvents, it forms weak hydrogen bonds. The rate of guar gum dissolution and viscosity formation generally increases with decreasing particle size, decreasing pH and increasing temperature. Hydration rates are decreased in presence of dissolved salts and other water binding agents e.g., sucrose [133]. The most important characteristic of guar gum is its ability to hydrate quickly in cold water systems to give highly viscous solutions. Guar gum forms a viscous colloidal solution when completely hydrated which is a thixotropic rheological system. As like the other gums, viscosity of guar gum also dependent on time, temperature, concentration, pH, ionic strength and on type of agitation. Thirteen different commercial samples of guar gum were analysed, and found great variation in the viscosity property, particle size and hydration rate [134]. Hydrogen bonding activity of guar gum is because of the presence of hydroxyl group in guar gum molecule. Guar gum display hydrogen bonding with cellulosic material and hydrated minerals. With slight addition of guar gum, the electrokinetic properties of a system changes markedly [135]. Substitution of hydroxyl groups in guar gum with hydroxypropyl causes steric hindrance which lowers the stability of hydrogen bonds [136].

Guar gum is hydrophilic in nature, economical, non-toxic, easily available and biodegradable in nature making it suitable for wide pharmaceutical applications and industrial applications [137]. Guar gum is used as a unique food additive in various food products for food stabilization and as fiber source [138]. Guar gum is also used in beverages for thickening and viscosity control because of its various inherent properties. Main purpose of using guar gum in frozen products is stabilization. Guar gum play important role in stabilization of ice cream because of its water binding properties [139]. Guar gum in ice cream improves the body, texture, chewiness and heat resistance. Partially hydrolyzed guar gum decreases syneresis and improves the textural and rheological properties of low fat yoghurt in comparison with full fat yoghurt [140].

Guar gum because its viscous nature it is effectively used as a binder and lubricant in making of sausage products and stuffed meat products. It performs specific functions in processed meat products such as syneresis control, viscosity control of liquid phase during processing and cooling and control of collection of the water in the can during storage. Guar gum increases enhances creaming stability and control rheology of emulsion prepared by egg yolk [141]. Guar gum in combination with starch is effective in prevention of dehydration, shrinking and cracking of frozen pie crust fillings [142]. Addition of guar gum in wheat bread dough results in significant increase in loaf volume on baking [143]. Guar gum also delay staling in chapati at room temperature as well as refrigerated temperature by controlling retrogradation of starch [144]. Guar gum has been found useful in thickening of pickle and relish sauces [145]. Guar gum increases the uniformity of tomato ketchup more chiefly than other hydrocolloids like carboxy methyl cellulose, Sodium alginate, gum acacia etc. By addition of guar gum serum loss and flow values of tomato ketchup decreases which makes it a unique thickener for tomato ketchup [146].

Various studies have shown that guar gum has both harmful and beneficial effects. Guar is completely degraded in the large intestine by *Clostridium butyricum* [147]. It has been shown that when high concentration of guar gum is given to animals it has harmful effects. High concentration leads to reduce the growth of animal because of decreased food intake, decreased protein efficacy and impaired digestion. The main cause of harmful effects is high viscosity of the intestinal tract contents which is resulting from the intake of guar gum at high concentration. Thus guargum is benefial when used in lower concentration. High viscosity of guar gum when used at a higher concentration will not only hinder the nutritional properties of the food but also hinder the physicochemical and sensory properties of the food product which is not accepted by the consumer. Partial hydrolysis of guar gum (PHGG) reduction of chain length and molecular weight of the polymer and ultimately the lower viscosity makes it a unique soluble fiber that resembles in basic chemical structure with native guar gum and has several applications in clinical nutrition related with ingestion of dietary fiber. PHGG help in solving all the problems of high viscosity of guar gum. With hydrolyzed guar gum it become possible to increase the dietary fiber content of various food products like beverages without disturbing the nutritional and sensory properties of the food products. Supplementation of PHGG in diet also reduces the laxative requirement, occurrence of diarrhea and symptoms of irritable bowel syndrome [148,149]. Non gelling fibers are preferred for the treatment of irritable bowel syndrome. Because of its water solubility and non gelling behavior partially hydrolyzed guar gum is helpful in decreasing the symptoms of both form of irritable bowel syndrome i.e., constipation predominant and diarrhea predominant [150]. Guar gum shows cholesterol and glucose lowering effects because of its viscosity. It also helps in weight loss and fat prevention. Guar gum supplemented diet help in lowering the appetite, hunger and desire for eating [151].

Conclusion

Biodegradable polymers could be either natural or synthetic polymers and they could be derived from either renewable or nonrenewable resources. In the development of "green" materials and processing techniques, developing biodegradable polymers from renewable resources appears to be the most excellent scenario. Naturally occurring biodegradable polymers such as starch, pectin, chitosan, cellulose etc are currently flourishing in the market at very high rate. The success of natural biodegradable polymers lies in their ability to design or modify existing biomaterials to achieve appropriate biocompatibility and physical properties to extract favorable biological responses.

References

[1] Ingrao, C., Tricase, C., Cholewa-Wójcik, A., Kawecka, A., Rana, R. and Siracusa, V. 2015. Polylactic acid trays for fresh-food packaging: A Carbon Footprint assessment.*Science of the Total Environment*,*537*, pp.385-398.

[2] Vieira, M.G.A., da Silva, M.A., dos Santos, L.O. and Beppu, M.M. 2011. Natural-based plasticizers and biopolymer films: A review.*European Polymer Journal*,*47*(3), pp.254-263.

[3] Colwill, J.A., Wright, E.I., Rahimifard, S. and Clegg, A.J. 2012. Bio-plastics in the context of competing demands on agricultural land in 2050.*International Journal of Sustainable Engineering*,*5*(1), pp.3-16.

[4] Jiang, L. and Zhang, J. 2017. Biodegradable and Biobased Polymers. Applied Plastics Engineering Handbook (Second Edition) Processing, Materials, and Applications A volume in Plastics Design Library. pp. 127–143.

[5] Sukan, A., Roy, I. and Keshavarz, T. 2015. Dual production of biopolymers from bacteria.Carbohydrate polymers,126, pp.47-51.

[6] Mohanty, A.K., Misra, M. and Drzal, L.T. eds. 2005.Natural fibers, biopolymers, and biocomposites. CRC press.

[7] Katti, D.S., Lakshmi, S., Langer, R. and Laurencin, C.T. 2002. Toxicity, biodegradation and elimination of polyanhydrides.Advanced drug delivery reviews,54(7), pp.933-961.

[8] Nair, L.S. and Laurencin, C.T. 2007. Biodegradable polymers as biomaterials.Progress in polymer science,32(8), pp.762-798.

[9] Geigenberger, P. 2011. Regulation of starch biosynthesis in response to a fluctuating environment. *Plant Physiology.* 155: 1566-1577.

[10] Smith, A.M., Martin, C. 1993. Starch biosynthesis and the potential for its manipulation. In: Grierson D, editor. Biosynthesis and Manipulation of Plant Products. *Plant Biotechnology Series.* 3: 1-4.

[11] Tetlow, I.J., Morell, M.K., Emes, M.J. 2004. Recent developments in understanding the regulation of starch metabolism in higher plants. *Journal of Experimental Botany.* 55: 2131-2145.

[12] James, M.G., Denyer, K., Myers, A.M. 2003. Starch synthesis in the cereal endosperm. *Current Opinion in Plant Biology.* 6: 215-222.

[13] Evers. A.D., Greenwood, CT., Muir, D.D., Venables, C.C. 1974. Studies on biosynthesis of starch granules. 8. A comparison of the properties of the small and the large granules. *Starch/ Stärke.* 26: 42-46.

[14] Bechtel, D.B., Zayas, I., Kaleikau, L., Pomeranz, Y. 1990. Size distribution of wheat starch granules during endosperm development. *Cereal Chemistry.* 67: 59-63.

[15] Yoo, S., Jane, J. 2002. Structural and physical characteristics of waxy and other wheat starches. *Carbohydrate Polymers.* 49: 297-305.

[16] Van, H.P., Maeda, T., Morita, N. 2006. Waxy and high-amylose wheat starches and flours characteristics, functionality and application. *Trends in Food Science and Technology.* 17: 448–456.

[17] Lafiandra, D., Sestili, F., D'Ovidio, R., Janni, M., Botticella, E., Ferrazzano, G., Silvestri, M., Ranieri, R., DeAmbrogio, E. 2010. Approaches for the modification of starch composition in durum wheat. *Cereal Chemistry.* 87: 28 34.

[18] Blennow, A., Bay-Smidt, A. M., Olsen, C. E., Moller, B. L. 2000. The distribution of covalently bound phosphate in the starch granule in relation to starch crystallinity. *International Journal of Biological Macromolecules.* 27: 211–218.

[19] Madhusudan, B., Tharanathan, R.N. 1995. Legume and Cereal starches why differences in digestibility? Part II - Isolation and characterization of starch from rice (*O.sativa*) and ragi (Finger millet, *E. coracona*). *Carbohydrate Polymers*. 28: 153-158.

[20] Tetlow, I.J., Morell, M.K., Emes, M.J. 2004. Recent developments in understanding the regulation of starch metabolism in higher plants. *Journal of Experimental Botany*. 55: 2131-2145.

[21] Ohdan, T., Francisco, P.B.Jr, Sawada, T., Hirose, T., Terao, T., Satoh, H., Nakamura, Y. 2005. Expression profiling of genes involved in starch synthesis in sink and source organs of rice. *Journal of Experimental Botany*. 56: 3229- 3244.

[22] Kang, G.Z., Xu, W., Liu, G.Q., Peng, X.Q., Guo, T.C. 2013. Comprehensive analysis of the transcription of starch synthesis genes and the transcription factor RSR1 in wheat (*Triticum aestivum*) endosperm. *Genome*. 56: 115- 122.

[23] Albani, D., Hammond-Kosack, M.C.U., Smith, C., Conlan, S., Colot, V., Holdsworth, M., Bevan, MW. 1997. The wheat transcriptional activator SPA: a seed-specific bZIP protein that recognizes the GCN4-like motif in the bifactorial endosperm box of prolamin genes. *The Plant Cell*. 9: 171-184.

[24] Aoki, N., Whitfeld, P., Hoeren, F., Scofield, G., Newell, K., Patrick, J., Offler, C., Clarke, B., Rahman, Furbank. R.T. 2002. Three sucrose transporter genes are expressed in the developing grain of hexaploid wheat. *Plant Molecular Biology*. 50: 453-462.

[25] Jobling, S. 2004. Improving starch for food and industrial applications. *Current Opinion in Plant Biology*. 7: 210–218.

[26] Wurzburg, O.B. 1986. Introduction. In O. B. Wurzburg (Ed.), Modified starches: Properties and uses (pp. 4e15). Boca Raton, FL: CRC Press.

[27] Shogren, R.L., Lawton, J.W., Tiefenbacher, K.F. 2002. Baked starch foams: starch modifications and additives improve process parameters, structure and properties. *Industrial Crops and Products*. 16: 69–79.

[28] Russo, M.A.L., O'Sullivan, C., Rounsefell, B., Halley, P.J., Truss, R., Clarke, W.P. 2009. The anaerobic degradability of thermoplastic starch: polyvinyl alcohol blends: potential biodegradable food packaging materials. *Bioresource Technology*. 100: 1705–1710.

[29] Follain, N., Joly, C., Dole, P., Bliard, C. 2005. Properties of starch based blends. Part 2: Influence of poly vinyl alcohol addition and photo cross linking on starch based materials mechanical properties. *Carbohydrate Polymers*. 60: 185–192.

[30] Composto. 2000. Life cycle assessment of Mater-Bi and EPS loose fills. Study prepared by COMPOSTO for Novamont, Novara, Italy.

[31] Matsumura, S., Kurita, H., Shimokobe, H. 1993. Anaerobic biodegradability of polyvinyl-alcohol. *Biotechnology Letters*. 15: 749–754.

[32] Ferrari, F., Rossi, S., Martini, A., Muggetti, I., De Ponti, R., Caramella, C. 1997. Technological induction of mucoadhesive properties on waxy starches by grinding. *European Journal of Pharmaceutical Sciences.* 5: 277-285.

[33] Huber, G.W., Iborra, S., and Corma, A. 2006. Synthesis of transportation fuels frombiomass: Chemistry, catalysts, and engineering. *Chemical Reviews*, 106, 4044–4098.

[34] Carlson, T.R., Vispute, T.P., and Huber, G.W. 2008. Green gasoline by catalytic fastpyrolysis of solid biomass derived compounds. *Chem Sus Chem*, 1, 397–400.

[35] Ragauskas, A.J., Williams, C.K., Davison, B.H., Britovsek, G., Cirney, J., Eckert, C. A.,et al. 2006. The path forward for biofuels and biomaterials. *Science*, 311,484–489.

[36] Lu, B., Xu, A.R., and Wang, J.J. 2014. Cation does matter: How cationic structureaffects the dissolution of cellulose in ionic liquids. *Green Chemistry*, 16, 1326–1335.

[37] Andanson, J.M., Bordes, E., Devémy, J., Leroux, F., Páduaa, A. A. H., and Costa Gomes, M.F. 2014. Understanding the role of co-solvents in the dissolution ofcellulose in ionic liquids. *Green Chemistry*, 16, 2528–2538.

[38] Brandt, A., Hallett, J.P., Leak, D.J., Murphy, R. J., and Welton, T. 2010. The effect ofthe ionic liquid anion in the pretreatment of pine wood chips. *Green Chemistry*, 12, 672–679.

[39] Brandt, A., Gräsvik, J., Halletta, J.P., and Welton, T. 2013. Deconstruction oflignocellulosic biomass with ionic liquids. Green Chemistry, 15, 550–583.

[40] Cannon, R.E., Anderson, S.M., 1991. Biogenesis of bacterial cellulose. *Crit. Rev. Microbiol.* 17, 435–447.

[41] Hon, D.N.S. 1994. Cellulose a random walk along its historical path. *Cellulose.* 1: 1–25.

[42] Perez, Serge, and Mackie, W. 2001. Structure and Morphology of Cellulose, in http://www.cermav.cnrs.fr/glyco3d/lessons/cellulose/index.html.

[43] Saxena, I.M. and Brown, R.M. 2005. Cellulose biosynthesis current views and evolving concepts. Ann. Bot. 96: 9–21.

[44] Peng, B.L., Dhar, N., Liu, H.L. and Tam, K.C. 2011. Chemistry and applications of nanocrystalline cellulose and its derivatives: A nanotechnology perspective. *Canadian Journal of Chemical Engineering.* 89(5): 1191–1206.

[45] Food and Agriculture Organization of the United Nations – FAO. 2013. Statistics Division. http://faostat3.fao.org/browse/Q/*/E, Accessed 15.10.15.

[46] Chandrasekhar, S., Satyanarayana, K.G., Pramada, P. N., Raghavan, P., and Gupta, T. N. 2003. Processing, properties and applications of reactive silica from rice husk an overview. *Journal of Materials Science*, 38, 3159–3168.

[47] Sims, I.M., Munro, S.L., Currie, G., Craik, D., and Bacic, A. 1996. Structural characterization of xyloglucan secreted by suspension-cultured cells of *Nicotiana plumbaginifoli. Carbohydrate Research*, 293, 147–172.

[48] Katsuraya, K., Okuyamab, K., Hatanakab, K., Oshimab, R., Satoc, T., and Matsuzakic, K. 2003. Constitution of konjac glucomannan: Chemical analysis and 13C NMR spectroscopy. *Carbohydrate Polymer*, 53, 183–189.
[49] Ullah, H., Wahid, F., Santos, H.A., and Khan, T. 2016. Advances in biomedical and pharmaceutical applications of functional bacterial cellulose-based nanocomposites. *Carbohydrate Polymers*, 150, 330–352.
[50] Iguchi, M., Yamanaka, S., and Budhiono, A. 2000. Bacterial cellulose-a masterpiece of nature's arts. *Journal of Materials Science*, 35, 261–270.
[51] Klemm, D., Heublein, B., Fink, H. and Bohn, A. 2004. Cellulose: Fascinating Biopolymer / Sustainable Raw Material. *Ang. Chemie* (Intl. Edn.). 44: 3358.
[52] Steinbuchel, A. 2003. Biopolymers: General Aspects and Special Applications. *Wiley-VCH: Weinheim* (Germany), 10: 516.
[53] Steinbuchel, A. and Doi, Y. 2005. Biotechnology of polymers. Vol.1, Wiley-VCH: Weinheim (Germany).
[54] Brown Jr., R.M. 1985. Cellulose microfibril assembly and orientation: recent developments. *J. Cell Sci.* Suppl. 2: 13-32.
[55] Englehart, J. 1995. Sources, industrial derivatives and commercial applications of cellulose *Carbohyrate Eur*. 5–14,
[56] Pérez, J., Muñoz-Dorado, J., De la Rubia, T. and Martínez, J. 2002. Biodegradation and biological treatments of cellulose, hemicellulose and lignin: an overview. *International Microbiology*. 5(2): 53–63.
[57] Siqueira, G., Bras, J. and Dufresne, A. 2010. Cellulosic bionanocomposites a review of preparation, properties and applications. *Polymer*. 2: 728–765.
[58] Fu, L., Zhang, J. and Yang, G. 2013. Present status and applications of bacterial cellulosebased materials for skin tissue repair. *Carbohydrate Polymer*. 92: 1432–42.
[59] Fukaya, Y., Hayashi, K., Wada, M., and Ohno, H. 2008. Cellulose dissolution withpolar ionic liquids under mild conditions: Required factors for anions. *Green Chemistry*, 10, 44–46.
[60] Sun, N., Rahman, M., Rogers, R.D., Qin, Y., and Maxim, M.L. 2009. Complete dissolution and partial delignification of wod in the ionic liquid1-ethyl-3-methylimidazolium acetate. *Green Chemistry*, 11, 645–655.
[61] Bobleter, O. 1994. Hydrothermal degradation of polymers derived from plants. *Progress in Polymer Science*, 19, 797–841.
[62] Kim, Y.H., Park, S., Won, K., Kim, H.J., and Lee, S.H. 2013. Bacterial cellulose-carbon nanotube composite as a biocompatible electrode for the direct electron transfer of glucose oxidase. *Journal of Chemical Technology and Biotechnology*, 88, 1067–1070.
[63] Shah, N., Ul-Islam, M., Khattak, W.A., and Park, J.K. 2013. Overview of bacterial cellulose composites: A multipurpose advanced material. *Carbohydrate Polymers*, 98, 1585–1598.

[64] Atmodjo, M.A., Hao, Z.Y., and Mohnen, D. 2013. Evolving views of pectin biosynthesis. *Annual Review of Plant Biology*. 64, 747–779.

[65] Somerville, C., Bauer, S., Brininstool, G., Facette, M., Hamann, T., and Milne, J. 2004. Toward a systems approach to understanding plant-cell walls. *Science*, 306(5705), 2206–2211.

[66] Lin, D., Lopez-Sanchez, P., and Gidley, M.J. 2016. Interactions of pectins with cellulose during its synthesis in the absence of calcium. *Food Hydrocolloids*. 52:57–68.

[67] Zykwinska, A., Gaillard, C., Buleon, A., Pontoire, B., Garnier, C., Thibault, J.F. 2007. Assessment of in vitro binding of isolated pectic domains to celluloseby adsorption isotherms, electron microscopy, and X-ray diffraction methods. *Biomacromolecules*, 8(1), 223–232.

[68] Vauquelin, M. 1790. Analyse du tamarin. *Annales de Chimie*, 5, 92–106.

[69] Braconnot, H. 1825. Nouvelles observations surl'acidpectique. *Annales De Chimie Et De Physique*, 30, 96–102.

[70] Voragen, A. J., Coenen, G.-J., Verhoef, R., and Schols, H. 2009. Pectin, a versatile polysaccharide present in plant cell walls. *Structural Chemistry*. 20(2), 263–275.

[71] Parre, E., and Geitmann, A. 2005. Pectin and the role of the physical properties of thecell wall in pollen tube growth of Solanum chacoense. *Planta*. 220(4), 582–592.

[72] Suárez, C., Zienkiewicz, A., Castro, A.J., Zienkiewicz, K., Majewska-Sawka, A., and Rodríguez-García, M.I. 2013. Cellular localization and levels of pectins and arabinogalactan proteins in olive (*Olea europaea* L.) pistil tissues during development: Implications for pollen–pistil interaction. *Planta*. 237(1), 305–319.

[73] Willats, W. T., McCartney, L., Mackie, W., and Knox, J.P. 2001. Pectin: cell biologyand prospects for functional analysis. In N. C. Carpita, M. Campbell, and M.Tierney (Eds.), *Plant Cell Walls*. 9–27. Netherlands: Springer.

[74] Harholt, J., Suttangkakul, A., and Vibe Scheller, H. 2010. Biosynthesis of pectin. *Plant Physiology*. 153(2): 384–395.

[75] Benedetti, M., Pontiggia, D., Raggi, S., Cheng, Z., Scaloni, F., Ferrari, S., and De Lorenzo, G. 2015. Plant immunity triggered by engineered in vivo release of oligogalacturonides, damage-associated molecular patterns. *Proceedings of the National Academy of Sciences*. 112(17): 5533–5538.

[76] Hahn, M. G., Darvill, A. G., and Albersheim, P. 1981. Host-pathogen Interactions: XIX. The endogenous elicitor, a fragment of a plant cell wall polysaccharide that elicits phytoalexin accumulation in soybeans. *Plant Physiology*. 68(5):1161–1169.

[77] Kaya, M., Sousa, A.G., Crépeau, M.J., Sorensen, S.O., and Ralet, M.C. 2014. Characterization of citrus pectin samples extracted under different conditions: Influence of acid type and pH of extraction. *Annals of Botany*. 114(6):1319–1326.

[78] Caffall, K.H., and Mohnen, D. 2009. The structure, function, and biosynthesis of plant cell wall pectic polysaccharides. *Carbohydrate Research*. 344(14): 1879–1900.

[79] Yapo, B.M. 2011. Pectic substances: From simple pectic polysaccharides to complex pectins-A new hypothetical model. *Carbohydrate Polymers*. 86(2):373–385.

[80] Willats, W.T., McCartney, L., and Knox, J.P. 2003. Pectin cell biology: complexity in context. *Advances in Pectin and Pectinase Research*. (147–157). Netherlands: Springer.

[81] Renard, C.M.G.C., Crépeau, M.J., and Thibault, J.F. 1995. Structure of therepeating units in the rhamnogalacturonic backbone of apple, beet and citruspectins. *Carbohydrate Research*. 275(1):155–165.

[82] Alves, N.M., Mano, F. 2008. Chitosan derivatives obtained by chemical modifications for biomedical and environmental applications. *International journal Biological Macromolecules*. doi: 10.1016/j.ijbiomac.09.007.

[83] Rinaudo, M . 2006. Chitin and chitosan: properties and applications. *Progress in Polymer Science*. doi: 10.1016/j.progpolymsci.

[84] Kumar, R. 2000. A review of chitin and chitosan applications. *Reactive and Functional Polymers*. doi: 10.1016/S1381-5148(00)00038-9.

[85] Mano, J. F. 2008. Stimuli-Responsive polymeric systems for biomedical applications. *Advanced Engineering Materials*. doi: 10.1002/(ISSN)1527-2648.

[86] Barbara, B., Agostino, I.D., Semeraro, S., Gamini, A., and Cesàro, A. 2016. The Good, the Bad and the Ugly" of Chitosans. Marine drugs.

[87] Croisier, F., Jérôme, C. 2013. Chitosan-based biomaterials for tissue engineering. *European Polymer Journal*. 49:780–792. doi: 10.1016/j.eurpolymj.2012.12.009.

[88] Riva, R., Ragelle, H., Rieux, A., Duhem, N., Jérôme, C., Préat, V. 2011.Chitosan and chitosan derivatives in drug delivery and tissue engineering.Berlin: Springer.

[89] Shahidi, F., Arachchi, J.K.V., Jeon, Y.J. 1999. Food applications of chitin and chitosans. *Trends in Food Science and Technology*. 10:37–51. doi: 10.1016/S0924-2244(99)00017-5.

[90] Acosta, N., Jiménez, C., Borau, V, et al.1993. Extraction and characterization of chitin from crustaceans. *Biomass and Bioenergy*. 5:145–153. doi: 10.1016/0961-9534(93)90096-M.

[91] Madihally, S.V., Matthe, H.W.T. 1999. Porous chitosan scaffolds for tissue engineering. *Biomaterials*. 20:1133–1142.doi:10.1016/S0142-9612(99)00011-3.

[92] Kong, M., Chen, X.G., Xing, K., Park, H.J.2010. Antimicrobial properties of chitosan and mode of action: A state of the art review. *International Journal of Food Microbiology*. 144, 51–63.

[93] Burkatovskaya, M., Castano, A.P., Demidova-Rice, T.N., Tegos, G.P. Hamblin, M.R. 2008. Effect of chitosan acetate bandage on wound healing in infected and noninfected wounds in mice. *Wound Repair and Regeneration*. 16, 425–431.

[94] Lee, S.J., Heo, D.N., Moon, J.H., Park, H.N., Ko, W.K., Bae, M.S., Lee, J.B., Park, S.W., Kim, E.C., Lee, C.H., et al. 2014 Chitosan/Polyurethane Blended Fiber Sheets Containing Silver Sulfadiazine for Use as an Antimicrobial Wound Dressing. *Journal of Nanoscience and Nanotechnology*. 14, 7488–7494.

[95] Tan, H.B., Wang, F.Y., Ding, W., Zhang, Y., Ding, J., Cai, D.X., Yu, K.F., Yang, J., Yang, L., Xu, Y.Q. 2015. Fabrication and Evaluation of Porous Keratin/chitosan (KCS) Scaffolds for Effectively Accelerating Wound Healing. *Biomedical and Environmental Sciences*. 28, 178–189.

[96] Miguel, S.P., Ribeiro, M.P., Brancal, H., Coutinho, P., Correia, I.J. 2014. Thermoresponsive chitosan-agarose hydrogel for skin regeneration. *Carbohydrate Polymers*. 111, 366–373.

[97] Anisha, B.S., Biswas, R., Chennazhi, K.P., Jayakumar, R. 2013. Chitosan-hyaluronic acid/nano silver composite sponges for drug resistant bacteria infected diabetic wounds. *International Journal of Biological Macromolecules*. 62, 310–320.

[98] Aziz, M.A., Cabral, J.D., Brooks, H.J.L., Moratti, S.C., Hanton, L.R. 2012. Antimicrobial properties of a chitosan dextran-based hydrogel for surgical use. *Antimicrobial. Agents Chemotherapy*. 56, 280–287.

[99] Du, W.L., Niu, S.S., Xu, Y.L., Xu, Z.R., Fan, C.L. 2009. Antibacterial activity of chitosan tripolyphosphate nanoparticles loaded with various metal ions. *Carbohydrate Polymers*. 75, 385–389.

[100] Qi, L., Xu, Z., Jiang, X., Hu, C., Zou, X. 2004. Preparation and antibacterial activity of chitosan nanoparticles. *Carbohydrate Research*. 339, 2693–2700.

[101] Shi, Z., Neoh, K.G., Kang, E.T., Wang, W. 2006. Antibacterial and mechanical properties of bone cement impregnated with chitosan nanoparticles. *Biomaterials* 27, 2440–2449.

[102] Alt, V., Bechert, T., Steinrücke, P., Wagener, M., Seidel, P., Dingeldein, E., Domann, E., Schnettler, R. 2004. An in vitro assessment of the antibacterial properties and cytotoxicity of nanoparticulate silver bone cement. *Biomaterials* 25, 4383–4891.

[103] Barreras, US., Méndez, F.T., Martínez, R.E.M., Valencia, C.S., Rodríguez, P.R.M., Rodríguez, J.P.L.2016. Chitosan nanoparticles enhance the antibacterial activity of chlorhexidine in collagen membranes used for periapical guided tissue regeneration. *Materials Science and Engineering. C, Materials for Biological Applications*. 58, 1182–1187.

[104] Aider, M.2010. Chitosan application for active bio-based films production and potential in the food industry Review. *LWT-Food Science and Technology*. 43, 837–842.

[105] Agulló, E., Rodríguez, M.S., Ramos, V., Albertengo, L. 2003. Present and Future Role of Chitin and Chitosan in Food. *Macromolecular Bioscience*. 3, 521–530.

[106] Tahiri, I., Desbiens, M., Lacroix, C., Kheadr, E., Fliss, I. 2009. Growth of Carnobacteriumdivergens M35 and production of Divergicin M35 in snow crab by-product, a natural-grade medium. *LWT-Food Science and Technology*. 42, 624–632.
[107] Devlieghere, F., Vermeulen, A., Debevere, J. 2004. Chitosan: Antimicrobial activity, interactions with food components and applicability as a coating on fruit and vegetables. *Food Microbiology*. 21, 703–714.
[108] Dutta, P.K., Tripathi, S., Mehrotra, G.K., Dutta, J. 2009. Perspectives for chitosan based antimicrobial films in food applications. *Food Chemistry*.114, 1173–1182.
[109] Elsabee, M.Z., Abdou, E.S. 2013. Chitosan based edible films and coatings: A review. *Materials Science and Engineering. C, Materials for Biological Applications*. 33, 1819–1841.
[110] Kenawy, E.R., Worley, S.D., Broughton, R. 2007.The chemistry and applications of antimicrobial polymers: A state-of-the-art review. *Biomacromolecules* 8, 1359–1384.

In: Modified Biopolymers: Challenges and Opportunities ISBN: 978-1-53612-116-2
Editors: D. Pathania, G. Sharma and A. Kumar

Chapter 7

PROGRESS FROM COMPOSITE MATERIALS TO BIOCOMPOSITE MATERIALS AND THEIR APPLICATIONS

Mu. Naushad[1,*], Sulaiman M. Alfadul[2], Ala'a H. Al-Muhtaseb[3], Gaurav Sharma[4], Senthil K. Ponnusamy[5], Zeid A. ALOthman[1] and Rani Bushra[6]

[1]College of Science, King Saud University, Riyadh, Saudi Arabia
[2]King Abdulaziz City for Science and Technology, Riyadh, Saudi Arabia
[3]Department of Petroleum and Chemical Engineering, Faculty of Engineering, Sultan Qaboos University, Muscat, Oman
[4]School of Chemistry, Shoolini University, Solan, India
[5]Department of Chemical Engineering, SSN College of Engineering, Chennai, India
[6]Department of Chemistry, Aligarh Muslim University, Aligarh, India

ABSTRACT

Biocomposite materials are made up of two or more constituent materials which have different physical or chemical properties out of which one is biogeneric in origin. There are two main categories of constituent materials: matrix and reinforcement. The matrix material surrounds & supports the reinforcement materials by maintaining their relative positions. In biocomposite materials, metals forms the matrix. The reinforcements materials could be metal oxides, cermicsor nanoparticles and they impart their special mechanical and physical properties to enhance the matrix properties. Biocomposite

* Corresponding author Tel: 0096614674198; Email: shad123@gmail.com.

materials are one of the most important classes of the engineered materials and these materials have been successfully used for various applications in the field of environmental remediation and medical science which have been described in the present chapter.

1. Introduction

The development of manhood today is directly or indirectly dependent on advanced technology materials that perform better and open new scopes in research and development. Among the major dependents in materials in recent years are biocomposite materials. The term "biocomposite materials" may possibly be simply defined on the classical definition of a composite material as given in Longman's dictionary "something combining the typical or essential characteristics of individuals making up a group". Thus, biocomposite materials are those materials which are formed by the combination of one or more materials with a biopolymer matrix to produce a material with the combination of desirable properties from individual components. Biocomposite materials are made up of two or more constituent materials with significantly different physical or chemical properties, that when combined; make a material with different characteristics from the individual components. The individual components remain separate and distinct within the finished structure. People have been making composite materials for many thousands of years. One early example of the composite materials is mud bricks. Wattle and daub is one of the oldest men who made composite materials at over 6000 years old [1]. Mud and straw are mixed together to make the bricks that are resistant to both squeezing and tearing and make excellent building blocks. Another early composite is concrete which is a mixture of aggregate (small gravel or stone), cement and sand. There are several natural composites like wood which is made up of long cellulose fibers (a polymer) held together by a much weaker substance called lignin. Cellulose is also found in cotton, which is also weak but two weak substances; lignin and cellulose together form a much stronger one and is best example of biopolymer based biocomposite material. Composites are used not only for their structural properties, but also for electrical, thermal and environmental applications. The first modern composite material was fiber glass which is widely used for boat hulls, sports equipment, building panels and many car bodies. The plastic matrix holds the glass fibers together and also protects them from damage by sharing out the forces acting on them. Some advanced composites are now made using carbon fibers instead of glass. These materials are lighter and stronger than fiber glass but more expensive to produce. They are used in aircraft structures and expensive sports equipment such as golf clubs. In planning of the composite materials, scientists and engineers have creatively combined various metals, ceramics, and biopolymers to produce a new generation of extraordinary materials. Most composite materials have been created to improve combinations of mechanical characteristics such as toughness, stiffness and high temperature strength. Many composite materials are composed of just two phases; matrix and dispersed phase. Composite materials have a bulk phase, which is continuous, called the matrix, and one or more dispersed, non-continuous phases, called the reinforcement, which usually has the superior mechanical or thermal properties to the matrix. The region between these two phases is called the interphase. The biggest advantage of modern composite materials is that they are light as well as strong. By choosing an appropriate combination of matrix and reinforcement material, a new material can be made that exactly meets the requirements of a particular application. The diverse phases of composite materials can be given as:

Matrix	Polymer	Thermosets (epoxy, polyester); thermoplastics (polystyrene, nylons) biopolymers
	Ceramic	Glass, cements
	Metal	Alloys (steels, aluminums)
	Carbon and Graphite	-
Reinforcements	Fibres	Long, short, organic, inorganic, threads
	Whiskers	Single crystal
	Laminar	-
	Flakes	-
	Filled	-
	Particulates	-
	Microspheres	-

Several common materials (metal alloys, doped ceramics and polymers mixed with additives) also have small amount of dispersed phases in their structure. However, they are not considered as composite materials since their properties are similar to those of their base constituents (physical properties of steel are similar to those of pure iron).

2. Classification of Composite Materials

Composite materials are commonly classified on the basis of two distinct levels: The first level of the classification is usually made with respect to the matrix constituent which is classified as:

2.1. Organic Matrix Composites (OMCs)

The term organic matrix composite is generally assumed to include two classes of composites, namely:

2.1.(a) Polymer matrix composites (PMCs)
2.1.(b) Carbon matrix composites commonly referred to as carbon-carbon composites

2.2. Metal matrix composites (MMCs)
2.3. Ceramic matrix composites (CMCs)
The second level of classification refers to the reinforcement form which is given as:
2.3.(a) Fibre reinforced composites
2.3.(b) Laminar composites
2.3.(c) Particulate composites

2.1.1. Polymer Matrix Composites (PMCs)

It consists of polymer resin as matrix with fibre as reinforcement medium. The function of the matrix is to bind the fibers together and to transfer loads between them. Polymer matrix composites are often divided into two categories: reinforced plastics, and advanced composites. The main advantage of PMCs is their light weight coupled with high stiffness and strength along the direction of the reinforcement. This combination is the basis of their usefulness in aircraft, automobiles and other moving structures. The main constituents of polymer matrix composites are matrix, thermosets, thermoplastic, reinforcement and interphase. The matrix properties determine the resistance of the PMC to most of the degradative processes that eventually cause failure of the structure. Sometimes, thermoplastic resins are called engineering plastics which include polyetherimide, polyesters, polyphenylene sulfide, polyether-etherketone (PEEK), and liquid crystal polymers. Thermosetting resins include epoxies, polyesters, vinyl esters and polyamides.

2.1.2. Carbon Carbon Composites

One of the most advanced and promising engineering material is the carbon fiber reinforced carbon-matrix composite which are frequently called carbon-carbon composite. As the name indicates, both reinforcement and matrix are carbon. They are costly and are not used comprehensively. Their desirable properties include high tensile moduli, tensile strength which is retained even at 2000°C. The two general categories of matrix precursors used for carbon-carbon densification are thermosetting resins, such as phenolic and furfurals, and pitch based on coal tar and petroleum. Carbon-carbon composites are attractive for space structure because of their relatively low density. The biggest shortcoming of these materials is the high temperature oxidation.

2.2. Metal Matrix Composites (MMCs)

These materials may be used at higher temperatures than their base metal counterparts. The reinforcement may improve specific strength, specific stiffness, abrasion resistance and stability. Metal matrix composites (MMCs) usually consist of a low-density metal, such as aluminum or magnesium, reinforced with particulate or fibers of a ceramic material, such as silicon carbide or graphite. Compared with unreinforced metals, MMCs offer higher specific strength and stiffness, higher operating temperature, and greater wear resistance, as well as the opportunity to tailor these properties for a particular application. Most metals and alloys make good matrix. However, practically, the choices for low temperature applications are limited. Titanium, aluminium and magnesium are the popular matrix metals currently in craze, which are mainly useful for aircraft applications. If metallic matrix materials have to offer high strength, they require high modulus reinforcements. The strength-to-weight ratios of resulting composites can be higher than most alloys. Current markets for MMCs are primarily in military and aerospace applications. The most important commercial application to date is the MMC diesel engine piston made by Toyota. A major drawback of MMCs compared to most other structural materials is that they are generally more costly. Cost/benefit ratios of most MMCs dictate that they be used only in high-performance applications.

2.3. Ceramic Matrix Composites (CMCs)

Ceramics can be explained as solid materials which show very strong ionic bonding in general and in few cases covalent bonding. Ceramic matrix composites (CMC) are developed from ceramic fibers embedded in a ceramic matrix. Various ceramic materials, oxide or non-oxide, are used for the fibers and the matrix. CMC are used in many high temperature processes. They have a very high thermal shock and creep resistance, which allow designs with large mechanical and thermal loads. Naturally, ceramic matrices are the obvious choice for high temperature applications.

2.3.1. Fibre Reinforced Composites (FRCs)

Fibre reinforced composites are those composite in which dispersed phase is in the form of fibre. FRC is high-performance fiber composite achieved and synthesized possible by cross-linking cellulosic fiber molecules with resins in the FRC material matrix through a proprietary molecular re-engineering process, yielding a product of exceptional structural properties. A fiber-reinforced composite is a composite building material that consists of three components: (i) the fibers as the discontinuous or dispersed phase, (ii) the matrix as the continuous phase, and (iii) the fine interphase region [2, 3]. This is a type of advanced composite group where the ingredients are rice hull, rice husk and plastic. This technology involves a method of refining, blending, and compounding natural fibers from cellulosic waste streams to form a high-strength fiber composite material in a polymer matrix. The selected waste or base raw materials used in this instance are those of waste thermoplastics and various categories of cellulosic waste including rice husk and saw dust. On the basis of diameter and character, fibers are classified into three different groups:

- Whiskers (these are thin single crystals that have extremely large length to diameter ratios.
- Fibers (these are either polycrystalline or amorphous and have small diameters)
- Wires (these are relatively large diameters)

2.3.2. Laminar Composites

This composite material is consisted of layers of various materials. Laminar composites are found in as many combinations as the number of materials. They can be described as materials comprising of layers of materials bonded together. These may be of several layers of two or more metal materials occurring alternately or in a determined order more than once, and in as many numbers as required for a specific purpose.

2.3.3. Particulate Composites

There are many different forms of particulate composites. The particulates can be very small particles (< 0.25 microns), platelets, chopped fibers (such as glass), hollow spheres, or new materials such as Bucky balls or carbon nano-tubes. In each case, the particulates provide desirable material properties and the matrix acts as binding medium necessary for structural applications. Particulate composites offer several advantages. They provide reinforcement to the matrix material thereby strengthening the material. The combination of reinforcement and matrix can provide for very specific material properties. The most common particulate composite materials are reinforced plastics which are used in a variety of industries.

3. Biopolymer Based Composites

3.1. Starch Based Biocomposites

Starch is a carbohydrate consisting of a large number of glucose units joined through glycosidic linkages. It contains two types of branching units: amylose and amylopectin. The concentration of amylose is about 150 times more than amylopectin. It is mainly found in large amount in maize, potatoes, wheat, bananas, pancakes, pasta and rice. Starch can be used as extenders, stabilizers, thickeners for processed meats. At high temperature, it is soluble in water. Starch based nanocomposites have been used for food packaging applications due to high ability of oil resistance and abundant availability. A short summary of starch based biocomposite materials with their applications has been given in Table 1.

Table 1. Starch based biocomposite materials with their applications

Starch based bionanocomposite materials	Applications	References
Corn-starch collagen composite films	Enhancement of several properties like tensile strength and thermal stability	[4]
Myofibrillar protein/cassava starch (MP/CS) composites	The hardness, adhesiveness, gumminess and chewiness of the gel decreased significantly	[5]
Gelatin/potato starch/glycerol edible biocomposite films	Mechanical properties (such as strength and elasticity),	[6]
Starch/polyvinyl alcohol (PVA) blend films	Wide application potentials as packaging materials.	[7]
Pectin/starch hydrogel	Encapsulated cells in pectin/starch hydrogels were resistant against adverse conditions of the gastro–intestinal tract and bile salt solution	[8]
Starch/poly-vinyl alcohol (PVA) film	Antimicrobial activity for food packaging applications	[9]
Starch-halloysite nanocomposites	Food preservation	[10]
Modified polylactic acid/starch/poly _-caprolactone nanocomposite	Antibacterial properties and medical applications	[11]
Starch/ethylene vinyl alcohol/forsterite nanocomposite	Bone tissue engineering	[12]
Starch/lysine and ZnO nanoparticles	Food packing applications	[13]
Starch/SnO_2 nanocomposites	Heavy metal ion removal	[14]

3.2. Pectin Based Biocomposites

Pectin is a non- toxic and one of the most abundant polysaccharide. It is mainly concentrated in citrus peels, oranges, apples, apricots, guavas and grapes. Its concentration varies from plant to plant and in various parts of the plant. It is commercially available as a brown powder. It is estimated that around 5g pectin is engulfed per day from fruits and vegetables. Pectin is mainly used a gelling and thickening agent in foods. It can provide viscosity and volume to stool so to overcome the issue of constipation. It helps in preparation of wound healing materials where it acts as a stabilizer. Certain pectin based biocomposites can be used food packaging application, controlled drug delivery, cosmetic applications, organic pollutant degradation, etc. as summarized in Table 2.

Table 2. Pectin based biocomposites with their applications

Pectin based biocomposite materials	Application	References
Pectin–cadmium sulfide nanocomposite	Photochemical degradation of methylene blue	[15]
Pectin/Starch hydrogel	Provide protection to cells against harsh conditions	[16]
Oxidised pectin/chitosan	Biomedical and cosmetic applications	[17]
Pectin–thorium (IV) tungstomolybdate nanocomposite	Photocatalytic degradation of methylene blue	[18]
Pectin-polyvinylpyrrolidone	Packaging Materials	[19]
Pectin-sodium alginate based films	As antimicrobial agent	[20]
Chitosan-zinc-pectinate-polyethylene glycol (PEG)	Colon-targeted delivery of resveratrol	[21]
Pectin thorium(IV) tungstomolybdate	Heavy metal separation and photoremediation of malachite green	[22]
Pectin/AgNPs nanocomposite film	Antimicrobial abilities	[23]
Pectin cerium(IV) iodate	Antimicrobial activity for Zn^{2+}	[24]
Amidated pectin/sodium carboxymethylcellulose microspheres	Drug delivery	[25]
Pectin-stabilized magnetic graphene oxide Prussian blue nanocomposites	Cesium removal from aqueous solution	[26]
Silver phosphate-pectin microspheres	Control release of levofloxacin	[27]
Pectin @ zirconium (IV) silicophosphate nanocomposite	Heavy metal separation	[28]
Pectin zirconium(IV) selenotungstophosphate	Toxic metal ions removal	[29]

3.3. Cellulose Based Biocomposites

Cellulose is a polysaccharide consisting of a number of β-linked D- glucose units. Different types of cellulose include cellulose acetate which is primarily used in textile and optical industries. Cellulose esters are produced by esterification of cellulose. Cellulose based nanocomposites are quite promising due to their relatively abundant, non- toxic, high mechanical strength, high barrier properties and degradation nature. These properties make them able to be used for packaging applications. Cellulose content in cotton and wood is about 90% and 50% respectively. Table 3 summarizes the applications of different cellulose based biocomposite materials.

Table 3. Cellulose based biocomposite materials with their applications

Cellulose Based biocomposite materials	Applications	References
Linseed gum/cellulose composite	Water conservation	[30]
Poly (lactic acid) (PLA) and poly-(3-hydroxybutyrate-co-3-hydroxyvalerate) (PHBV) reinforced nanocrystalline cellulose bionanocomposites	Packaging material	[31]
Cellulose-pectin composite	-	[32]
Cellulose nanocomposite films within situ generated copper nanoparticles	-	[33]
Alginate/bacterial cellulose nanocomposite	Application in lipaseimmobilization	[34]
Poly(carboxymethyl)-cellulose	Used for enhanced La(III) sorption	[35]
Potassium copper hexacyanoferrate-cellulose hydrogel	Selective and rapid cesium adsorption	[36]
Polyimide/cellulose/TiO_2 bionanocomposites	-	[37]
Cellulose nanofibrils/silver nanowires suspension	Used for the production of transparent and conductive electrodes by screen printing	[38]
Nano-tio2/cellulose composites	Used for rapid adsorption of Pb^{2+}	[39]
Cellulose/nano-hydroxyapatite nanofibers	Bone tissue engineering	[40]
cellulose nanocrystals (CNC) reinforced polyacrylonitrile (PAN) nanocomposite fibers	-	[41]
cellulose/magnetite/titanium dioxide(NFC@Fe_3O_4@TNP) nanocomposites	photocatalytic hydrogen generationXingye	[42]
polypyrrole/copper sulfide/bacterial cellulose nanofibrous composite membranes	supercapacitor electrodes	[43]
cellulose/polypyrrole composite films	-	[44]
Cellulose and gelatin based composite hydrogels	Removal of Cu (II) ion from water	[45]
curcumin-loaded cellulose acetate/polyvinylpyrrolidone fibrous materials	antibacterial activity	[46]
cellulose/graphene oxide nanocomposite	novel drug delivery system	[47]
Carbon-doped ZnO aided by carboxymethyl cellulose	photocatalytic applications	[48]
Laccase-immobilized bacterial cellulose/TiO2 functionalized composite membranes	Evaluation for photo- and bio-catalytic dye degradation	[49]
silver nanodendrite/cellulose acetate thin film	electrodes for flexible supercapacitors	[50]
ethyl cellulose-magnesium hydrogen phosphate composite	Sustained drug release	[51]
bacterial cellulose/polyaniline/single-walled carbon nano tubes membrane	potential application as biosensor	[52]

3.4. Chitosan Based Biocomposites

Chitosan is a linear polysaccharide composed of glucosamine units with some proportion of N-acetylglucosamine units. It is mainly synthesized by the deacetylation of chitin. The amino groups present in the chitosan structure helps in protonation which depends upon the pH value. Chitosan helps in the transport across the cell walls and is also biocompatible. Chitosan nanoparticles have attained more attention as drug delivery carriers due to their low toxicity, high stability, simple synthesis method and providing adjustable routes of management. Chitosan based bionanocomposites with their applications have been summarized in Table 4.

Table 4. Chitosan based biocomposites with their applications

Chitosan based biocomposite materials	Application	References
Nano-MgO reinforced chitosan nanocomposites	Packaging applications	[53]
Chitosan-coated Fe_3O_4	Magnetic adsorbent for patulin decontamination	[54]
Chitosan–xanthan gum microspheres	Selective antibacterial effects against the growth of *P. gingivalis*	[55]
Montmorillonite/chitosan nanocomposite	Improve paper smoothness	[56]
CFNP-chitosan-ampicillin	Controlled drug delivery	[57]
Calcium phosphate anhydrous/chitosan composites	Bone regeneration	[58]
Chitosan-hydroxyethyl methacrylate (HEMA) copolymer nanocomposites	In medicinal industry	[59]
Chitosan-g-poly(acrylamide)/Zn nanocomposite	Controlled drug delivery of ofloxacin	[60]
Chitosan/banana peel powder nanocomposites	Wound dressing	[61]
Chitosan/poly (ethylene oxide) nanofibrous membrane	Adsorption of divalent heavy metal ions	[62]
TLAC/Chitosan composite	Effective adsorbent for the removal of CV dye from aqueous solutions	[63]
Fe_3O_4/GO/Chitosan microspheres	Controlled drug delivery	[64]
K-carrageenan/chitosan complexes	Controlled drug release	[65]
Chitosan-crosslinked-poly(alginic acid) nanohydrogel	Chromium ion removal	[66]

3.5. Guargum Based Biocomposites

Guargum is a polysaccharide composed mainly of mannose and galactose units. Commonly, it is known as cluster beam. It has the ability of hydrating very fast in cold water even during low concentrations. It helps in the improvement of texture and water binding due to its ability to act as a stabilizer and emulsifier. It helps in improving the hydrogen bonding. Guargum based nanocomposites can be used in a number of industries such as cosmetics, textiles, pharmaceuticals, health care, packaging and oil well drilling. Applications of some Guargum based biocomposites have been given in Table 5.

Table 5. Guar gum based biocomposite materials with their potential application areas

Guargum Based biocomposite materials	Applications	References
Guargum oleate-graft-poly(methacrylic acid)	Colon-specific controlled drug delivery carrierD	[67]
Guargum embedded with silica nanoparticles	Adsorbent for removal of toxic cationic dyes and metal ions	[68]
Guargum/Montmorillonite Bionanocomposites	Controlled release of Ibuprofen	[69]
Guar gum-g-poly(acrylate) porous	Superabsorbent	[70]
Guargum succinate-sodium alginate	Carrier for colon-specific drug deliveryD	[71]
Guargum basedcomposite hydrogels	Enhanced adsorption of Cr(VI) from water	[72]
Collagen–poly(dialdehyde) guar gum based porous 3D scaffolds immobilized	Tissue engineering application	[73]
Platinum nanoparticles stabilised by guar gumSadanand	Catalytic reduction of p-nitrophenol	[74]
Silver/guargum/poly(acrylic acid)nanocomposite	-	[75]
Magnetic nanoparticle – Aminated guar gumhydrogel	Sustained release of doxorubicin hydrochloride	[76]
Carboxymethyl guar gum–alginate gel microspheres	Encapsulation of Congo Red	[77]
Guargum grafted sodiumacrylate/cloisite nanocomposites	Superabsorbent	[78]
Chitosan–guargum–silver nanoparticles hybrid matrix with immobilized enzymes	Used for fabrication of beta-glucan and glucose sensing photometric flow injection system	[79]
Guargum/Al_2O_3	Photocatalyst for the degradation of malachite green dye	[80]

4. Applications of Biocomposite Materials

Biocomposite materials have found increasingly wider utilities in the general areas of chemical sensors [81-87], toxic metal ion separation [88], chromatography, fabrication of selective materials [81-87] and electrical and optical applications [89, 90]. The disciplines like analytical chemistry, electro-analytical, medical, agriculture, potable water, power generation, textile and environmental have been using these materials. Several applications are as follows:

4.1. Environmental Protection

Environmental pollution is a universal hazard to public health and it has motivated the researchers for new initiatives for environmental protection. A number of pollutants constitute the industrial wastes such as dyes, acetic acid, heavy metal ions, naphthol and soaps. These eventually get added into the water bodies deteriorating the aquatic life. Their removal from water system is a priority for a nontoxic and clean environment. Different biocomposites have been prepared till date that can act as an adsorbent and exchanger. These materials have immense applications due to their low cost, abundant availability and biodegradable nature.

Organic pollutant such as malachite green is a cationic dye that mainly acts as a coloring agent. In high concentrations, it is carcinogenic and mutagenic in nature. So, its removal from the water is a very important. Guar gum/copper oxide biocomposite was synthesized by sol-gel method. It has been used as an efficient photocatalyst for the degradation of malachite green from the aqueous solution.

4.2. Optical Applications

- Absorbers, reflectors and transmitters

4.3. Magnetic Applications

- Magnetic field sensors
- Transformers
- Magnetic computer memories
- Magnetic recording (data storage)
- Magnetic shielding

- Robotics
- Micro machines

4.4. Biomedical Applications

Composite materials have been widely promoted as possible orthopedic biomaterials [91]. The bones and joints in the human body are made up of natural composite materials which are fractured due to excessive loads and impact stress. The various types of bone fractures take place in human body which depends upon the crack size orientation, morphology and its location. In general, the mean load on the hip joint is expected up to three to five times of the body weight during jumping, jogging, etc. [90]. Recently, significant interest has arisen in the development of biomaterials, capable of repairing bone defects due to a large increase in aging populations and patients suffering from trauma and bone cancer [92-97]. The metallic materials such as stainless steel, Co–Cr and Ti-6Al-4V alloys are typically used for the bone replacements in order to restore their lost functions [98]. Nevertheless, these materials mostly suffer from the wear and stress-shielding effect upon incorporation into the human body. These shortcomings and concerns have driven materials scientists to develop polymer composites with tailored mechanical stiffness and strength for making orthopedic load-bearing implants with excellent biocompatibility. Like other natural composite materials, bone consists of an organic part that forms the matrix and an inorganic part representing the filler. This natural nanocomposite material has superior strength and toughness than its individual components. Composite materials for medical application are developed for pathologies of osseous tissues from different parts of the body, like long bones, vertebrae, cartilage and teeth [99]. The existing methods and techniques for treatment of large bone defects, as a result of trauma or tumor, do not satisfactory restore the bone tissue. A series of collagen-calcium phosphate composite materials are tailored and used as temporary scaffolds in the studies on animals and humans for tissue regeneration. The addition of collagen to a ceramic material provides many advantages for medical applications such as shape control, spatial adaptation and ability for clot formation [100]. Collagen could also serve as efficient bonding agent for ceramic particles. Conversely, the addition of calcium phosphates to collagen scaffolds improves the osteoconductive properties of the material [98, 99]. The main advantage of a composite material tailored from collagen and calcium phosphate is the excellent biocompatibility property, due to collagen and its ability to allow bone cell attachment and differentiation. Synthetic materials are progressively becoming more common. Hydroxy apatite is prepared easily, but it is ceramic, which is too brittle to be used on its own for large scale applications. Composite materials of a hydroxy apatite with degradable polymers can also be used which allows the bone to regrow and fill the same. Ceramics are known for their good bio compatibility, corrosion resistance but main drawback is brittleness, low fracture strength, difficult to fabricate and low mechanical properties and high density. But polymer composite bio-materials provide better alternative choice for replacing because of bio-compatibility, corrosion resistant and easy to fabricate, etc. [101]. Composite materials are having the advantages of high specific modulus and strength to weight ratio besides, they have superior toughness to prevent crack propagation. The biomimetic synthesis of natural bone has awakened much research interest. Natural bone is in fact a fiber reinforced hybrid material composed of type-I collagen fibers and hydroxyapatite (HAp) minerals. The HAp minerals have outstanding biocompatibility, osteoconductivity and bioactivity. However, the mechanical properties of pure HAp are poor, hence limiting its medical application to small unloaded implants. Due to good electric conductivity as well as excellent biocompatibility, strong mechanical strength, light density and high stability, carbon nanomaterials in the form of graphite nano-sheets and nanotubes (CNTs) have emerged as very promising materials for bone tissue engineering applications [102, 103]. Calcium sulphate (plaster of Paris) acts as an osteoconductive scaffold and does not need to be removed after implantation due to its biodegradability [104]. Plaster of Paris was introduced in the 19th century to treat bone defects in humans by Dreesmann who filled bone with a slurry of plaster with 5% of phenol. The broad work of Peltier [105] on calcium sulphate led to the development of a medical grade alpha-calcium sulphate hemihydrate cylindrical pellet for the treatment of bony defects. Experimental and clinical studies have shown its positive effect in the filling of dead bony spaces [102-105]. Nowadays the most used ceramic material in biomedical applications is alumina which possesses a hexagonal (corundum-type) crystallographic structure (α-Al_2O_3 phase). Its strong ionic bond and oxygen-rich stoichiometry make alumina chemically bio-inert and, thus, exceptionally stable in the human body, leaving its structure conspicuously unaffected by corrosive processes [106]. Iontophoresis therapy is an extensively used technology in the drug delivery system (DDS) in which ionic or non-ionic drugs are imported *in vivo* by electric force [107, 108]. In the iontophoresis process, the drugs are imported into the body, while the other ions in the system also migrate. Ions with the same electrical charge as the drug ions will go the same direction as the drug ions while those with different electrical charges will go the opposite direction. To improve the ionic drug's iontophoresis efficiency, some researchers have purposefully tried to use ion-exchange materials in the process of iontophoresis, such as ion-exchange membrane. In 1934, Adams and Holmes synthesized phenol-formaldehyde resin and showed that this resin can be used as a substitute for zeolites for the medical applications. In 1939, the Resins Products and Chemical Company began investigations into the synthesis and production of ion-exchange resins (IER) with the help of the original Adams and Holmes patent. The suitability of the IER approach to drug delivery depends on the route and target of the delivery. Several scientists have succeeded in using microencapsulated or coated resonates [109-112] and most of the patented and marketed formulations belong to this category. Microencapsulation of resinates affords better control over the drug release because of the presence of a rate-controlling membrane. Some ion exchange resins, especially anoin exchange resin, such as cholestyramine, possess bio/mucoadhesive properties, which might be caused by their electrostatic interaction with the mucin and epithelial cell surface. The use of such bioadhesive IER is another attractive approach in the development of targeted formulations for the gastrointestinal tract. Polystyrene matrix cation exchange resins have been utilized to mask the bitter taste of chlorpheniramine maleate, ephedrine hydrochloride and diphenhydramine hydrochloride [112]. The ionic bonding of the drugs to polymeric materials such as carbopol is developing as an important mechanism of taste masking. One of the still open problems in the polymer resins for dental applications is the micro leakage between the sealant and tooth surface causing bacterial invasion and secondary caries. Several types of resin, both filled and unfilled, have been used as a pit and fissure sealants [113]. Composite ion exchange materials have also been used in the membrane technology which is an emerging field. Membrane technology is of major importance in medical applications, in particular in a number of lifesaving treatment methods. Membranes are used in tissue regeneration, drug delivery, artificial organs, diagnostic devices, as coatings for medical devices, bio-separations, etc. [114]. The biggest part of the medical market involves membranes in drug delivery, hemodialysis, other artificial organs (oxygenators, pancreas, etc.) and tissue engineering. Healthy kidneys are the essential part of metabolic processes of the body but failure of the kidney results in building up of harmful wastes and excess fluids in the body. Kidney diseases can be due to infections, high blood pressure (hypertension), diabetes and/orextensive use of medication [115-119]. Chronic kidney failure requires the treatment using an artificial kidney called dialysis. Blood is taken out of the body and passes through a special membrane that removes waste and extra fluids. The membranes consisting of composite ion exchange materials contains pores that allow small molecules such as urea, water, creatinine, and glucose to pass through the membrane readily, but the red cells, white cells, platelets and the most plasma proteins are retained. Theses membranes are also used in blood purification. Composite ion exchange materials based membranes have also played a big role in the field of tissue engineering.

CONCLUSION

The present book chapter describes the applications of composite ion exchange materials especially in the field of medical science. The composites materials offer several outstanding properties as compared to conventional materials. These materials are made up of two or more constituent materials with significantly different physical or chemical properties that when combined; make a material with different characteristics from the individual components. Composite materials have been widely used as dental materials, bone replacement materials, drug delivery applications, hemodialysis, other artificial organs (oxygenators, pancreas, etc.) and tissue engineering.

ACKNOWLEDGMENT

This work was supported by the Deanship of Scientific Research, King Saud University for funding through Vice Deanship of Scientific Research Chairs. We would also thank to Sultan Qaboos University, Oman for the help.

REFERENCES

[1] Shaffer, G. D. (1993). An Archaeomagnetic Study of a Wattle and Daub Building Collapse. *J. Field Arch.*, *20*, 59-75.

[2] Cantwell, W. J. & Morton, J. (1991). The impact resistance of composite materials - A review. *Composites*, *22*, 347–62.

[3] Kalpakjian, S. & Schmid, S. R. (2001). Manufacturing Engineering and Technology". International edition.4th Ed. Prentice Hall, Inc.

[4] Wang, K., Wang, W., Ye, R., Liu, A., Xiao, J., Liu, Y. & Zhao, Y. (2017). Mechanical properties and solubility in water of corn starch-collagen composite films : Effect of starch type and concentrations. *Food Chem.*, *216*, 209–216.

[5] Fan, M., Hu, T., Zhao, S., Xiong, S., Xie, J. & Huang, Q. (2017). Gel characteristics and microstructure of fish myofibrillar protein/cassava starch composites. *Food Chem.*, *218*, 221-230.

[6] Podshivalov, A., Zakharova, M., Glazacheva, E., & Uspenskaya, M. (2017). Gelatin/potato starch edible biocomposite films: Correlation between morphology and physical properties. *Carbohy. Polym.*, *157*, 1162-1172.

[7] Tian, H., Yan, J., Rajulu, A. V., Xiang, A. & Luo, X. (2017). Fabrication and properties of polyvinyl alcohol/starch blend films: Effect of composition and humidity. *Inter. J Bio. Macromol.*, *96*, 518-523.

[8] Dafe, A., Etemadi, H., Dilmaghani, A. & Mahdavinia, G. R. (2017). Investigation of pectin/starch hydrogel as a carrier for oral delivery of probiotic bacteria. *Inter. J Bio Macromol. 97*, 536-543.

[9] Liu, B., Xu, H., Zhao, H., Liu, W., Zhao, L. & Li, Y. (2017). Preparation and characterization of intelligent starch/PVA films for simultaneous colorimetric indication and antimicrobial activity for food packaging applications. *Carbohyd. Polym.*, *157*, 842-849.

[10] Meira, S. M. M., Zehetmeyer, G., Werner, J. O. & Brandelli, A. (2017). A novel active packaging material based on starch-halloysite nanocomposites incorporating antimicrobial peptides. *Food Hydrocolloids*, *63*, 561-570.

[11] Davachi, S. M., Heidari, B. S., Hejazi, I., Seyfi, J., Oliaei, E., Farzaneh, A. & Rashedi, H. (2017). Interface modified polylactic acid/starch/poly ε-caprolactone antibacterial nanocomposite blends for medical applications. *Carbohy. Polym.*, *155*, 336-344.

[12] Mahdieh, Z., Bagheri, R., Eslami, M., Amiri, M., Shokrgozar, M. A. & Mehrjoo, M. (2016). Thermoplastic starch/ethylene vinyl alcohol/forsterite nanocomposite as a candidate material for bone tissue engineering. *Materials Sci. Eng*: *C*, *69*, 301-310.

[13] Kotharangannagari, V. K. & Krishnan, K. (2016). Biodegradable hybrid nanocomposites of starch/lysine and ZnO nanoparticles with shape memory properties. *Materials & Design*, *109*, 590-595.

[14] Naushad, M., Ahamad, T., Sharma, G., Al-Muhtaseb, A. H., Albadarin, A. B., Alam, M. M., ALOthman, Z. A., Alshehri, S. M. & Ghfar, A. A. (2016). Synthesis and characterization of a new starch/SnO_2 nanocomposite for efficient adsorption of toxic Hg^{2+} metal ion. *Chem. Eng. J*, *300* (2016) 306-316.

[15] Gupta, V. K., Pathania, D., Asif, M. & Sharma, G. (2014). Liquid phase synthesis of pectin–cadmium sulfide nanocomposite and its photocatalytic and antibacterial activity. *J Molecular Liquids*, *196*, 107-112.

[16] Dafe, A., Etemadi, H., Dilmaghani, A. & Mahdavinia, G. R. (2017). Investigation of pectin/starch hydrogel as a carrier for oral delivery of probiotic bacteria. *Inter. J Bio. Macromol. 97*, 536-543.

[17] Chetouani, A., Follain, N., Marais, S., Rihouey, C., Elkolli, M., Bounekhel, M., Benachour, M. & Le Cerf, D. (2017). Physicochemical properties and biological activities of novel blend films using oxidized pectin/chitosan. *Inter. J Bio. Macromol.*, *97*, 348-356

[18] Gupta, V. K., Agarwal, S., Pathania, D., Kothiyal, N. C. & Sharma, G. (2013). Use of pectin–thorium (IV) tungstomolybdate nanocomposite for photocatalytic degradation of methylene blue. *Carbohy. polym*, *96*(1), 277-283.

[19] Nešić, A., Ružić, J., Gordić, M., Ostojić, S., Micić, D. & Onjia, A. (2017). Pectin-polyvinylpyrrolidone films: A sustainable approach to the development of biobased packaging materials. *Composites Part B: Engineering*, *110*, 56-61.

[20] Nešić, A., Onjia, A., Davidović, S., Dimitrijević, S., Errico, M. E., Santagata, G. & Malinconico, M. (2017). Design of pectin-sodium alginate based films for potential healthcare application: Study of chemico-physical interactions between the components of films and assessment of their antimicrobial activity. *Carbohyd. Polym.*, *157*, 981-990.

[21] Andishmand, H., Tabibiazar, M., Mohammadifar, M. A. & Hamishehkar, H. (2017). Pectin-zinc-chitosan-polyethylene glycol colloidal nano-suspension as a food grade carrier for colon targeted delivery of resveratrol. *Inter. J Bio. Macromol.*, *97*, 16-22

[22] Sharma, G., Naushad, M., Pathania, D. & Kumar, A. (2016). A multifunctional nanocomposite pectin thorium (IV) tungstomolybdate for heavy metal separation and photoremediation of malachite green. *Desalination and Water Treatment*, *57*(41), 19443-19455.

[23] Shankar, S., Tanomrod, N., Rawdkuen, S. & Rhim, J. W. (2016). Preparation of pectin/silver nanoparticles composite films with UV-light barrier and properties. *Inter. J Bio. Macromol*, *92*, 842-849.

[24] Pathania, D., Sharma, G., Naushad, M. & Priya, V. (2016). A biopolymer-based hybrid cation exchanger pectin cerium (IV) iodate: synthesis, characterization, and analytical applications. *Desalination and Water Treatment*, *57*(1), 468-475.

[25] Gadalla, H. H., El-Gibaly, I., Soliman, G. M., Mohamed, F. A. & El-Sayed, A. M. (2016). Amidated pectin/sodium carboxymethylcellulose microspheres as a new carrier for colonic drug targeting: Development and optimization by factorial design. *Carbohy. Polym.*, *153*, 526-534.

[26] Kadam, A. A., Jang, J. & Lee, D. S. (2016). Facile synthesis of pectin-stabilized magnetic graphene oxide Prussian blue nanocomposites for selective cesium removal from aqueous solution. *Bioresource technology*, *216*, 391-398.

[27] Bayón, B., Bucalá, V. & Castro, G. R. (2016). Development of antimicrobial hybrid mesoporous silver phosphate–pectin microspheres for control release of levofloxacin. *Microporous and Mesoporous Materials*, *226*, 71-78.

[28] Pathania, D., Sharma, G. & Thakur, R. (2015). Pectin@ zirconium (IV) silicophosphate nanocomposite ion exchanger: photo catalysis, heavy metal separation and antibacterial activity. *Chem. Eng. J*, *267*, 235-244.

[29] Sharma, G., Pathania, D. & Naushad, M. (2014). Preparation, characterization and antimicrobial activity of biopolymer based nanocomposite ion exchanger pectin zirconium (IV) selenotungstophosphate: application for removal of toxic metals. *J Industrial and Eng Chem.*, *20*(6), 4482-4490.

[30] Zhang, H., Luan, Q., Huang, Q., Tang, H., Huang, F., Li, W. & Zhou, Q. (2017). A facile and efficient strategy for the fabrication of porous linseed gum/cellulose superabsorbent hydrogels for water conservation. *Carbohy Polym.*, *157*, 1830-1836.

[31] Dasan, Y. K., Bhat, A. H. & Ahmad, F. (2017). Polymer blend of PLA/PHBV based bionanocomposites reinforced with nanocrystalline cellulose for potential application as packaging material. *Carbohy Polym*, *157*, 1323-1332.

[32] Lopez-Sanchez, P., Martinez-Sanz, M., Bonilla, M. R., Wang, D., Gilbert, E. P., Stokes, J. R. & Gidley, M. J. (2017). Cellulose-pectin composite hydrogels: Intermolecular interactions and material properties depend on order of assembly. *Carbohy Polym.*, *162*, 71-81.

[33] Muthulakshmi, L., Rajini, N., Nellaiah, H., Kathiresan, T., Jawaid, M. & Rajulu, A. V. (2017). Preparation and properties of cellulose nanocomposite films with in situ generated copper nanoparticles using Terminalia catappa leaf extract. *Inter. J Bio. Macromol*, *95*, 1064-1071.

[34] Kim, J. H., Park, S., Kim, H., Kim, H. J., Yang, Y. H., Kim, Y. H. & Lee, S. H. (2017). Alginate/bacterial cellulose nanocomposite beads prepared using Gluconacetobacter xylinus and their application in lipase immobilization. *Carbo Polym.*, *157*, 137-145.

[35] Tolba, A. A., Mohamady, S. I., Hussin, S. S., Akashi, T., Sakai, Y., Galhoum, A. A. & Guibal, E. (2017). Synthesis and characterization of poly (carboxymethyl)-cellulose for enhanced La (III) sorption. *Carbohyd Polym. 157*, 1809-1820.

[36] Kim, Y., Kim, Y. K., Kim, S., Harbottle, D. & Lee, J. W. (2016). Nanostructured potassium copper hexacyanoferrate-cellulose hydrogel for selective and rapid cesium adsorption. *Chem. Eng J.*, *42*, 56-67.

[37] Ahmadizadegan, H. (2017). Surface modification of TiO2 nanoparticles with biodegradable nanocellolose and synthesis of novel polyimide/cellulose/TiO2 membrane. *J Colloid and Interface Science*, *491*, 390-400.

[38] Hoeng, F., Denneulin, A., Reverdy-Bruas, N., Krosnicki, G. & Bras, J. (2017). Rheology of cellulose nanofibrils/silver nanowires suspension for the production of transparent and conductive electrodes by screen printing. *Applied Surface Science*, *394*, 160-168.

[39] Zhang, J., Li, L., Li, Y. & Yang, C. (2016). Microwave-assisted synthesis of hierarchical mesoporous nano-TiO2/cellulose composites for rapid adsorption of Pb 2+. *Chem. Eng J.*

[40] Ao, C., Niu, Y., Zhang, X., He, X., Zhang, W. & Lu, C. (2017). Fabrication and characterization of electrospun cellulose/nano-hydroxyapatite nanofibers for bone tissue engineering. *Inter. J Bio. Macromol.*, *132*, 887-890.

[41] Chang, H., Luo, J., Liu, H. C., Davijani, A. A. B., Wang, P. H. & Kumar, S. (2017). Orientation and interfacial stress transfer of cellulose nanocrystal nanocomposite fibers. *Polymer.*, *110*, 228-234.

[42] An, X., Cheng, D., Dai, L., Wang, B., Ocampo, H. J., Nasrallah, J. & Ni, Y. (2017). Synthesis of nano-fibrillated cellulose/magnetite/titanium dioxide (NFC@ Fe3O 4@ TNP) nanocomposites and their application in the photocatalytic hydrogen generation. *Applied Catalysis B: Environmental.*, *206*, 53-64

[43] Peng, S., Fan, L., Wei, C., Liu, X., Zhang, H., Xu, W. & Xu, J. (2017). Flexible polypyrrole/copper sulfide/bacterial cellulose nanofibrous composite membranes as supercapacitor electrodes. *Carbohy Polym.*, *157*, 344-352.

[44] Raghunathan, S. P., Narayanan, S., Poulose, A. C. & Joseph, R. (2017). Flexible regenerated cellulose/polypyrrole composite films with enhanced dielectric properties. *Carbohy Polym.*, *157*, 1024-1032.

[45] Maity, J. & Ray, S. K. (2017). Removal of Cu (II) ion from water using sugar cane bagasse cellulose and gelatin based composite hydrogels. *Inter. J Bio. Macromol.*

[46] Tsekova, P. B., Spasova, M. G., Manolova, N. E., Markova, N. D. & Rashkov, I. B. (2017). Electrospun curcumin-loaded cellulose acetate/polyvinylpyrrolidone fibrous materials with complex architecture and antibacterial activity. *Materials Science and Engineering: C*, *73*, 206-214.

[47] Luo, H., Ao, H., Li, G., Li, W., Xiong, G., Zhu, Y. & Wan, Y. (2017). Bacterial cellulose/graphene oxide nanocomposite as a novel drug delivery system. *Current Applied Physics*, *17*(2), 249-254.

[48] Zhao, S. W., Zuo, H. F., Guo, Y. R. & Pan, Q. J. (2017). Carbon-doped ZnO aided by carboxymethyl cellulose: Fabrication, photoluminescence and photocatalytic applications. *Journal of Alloys and Compounds*, *695*, 1029-1037.

[49] Li, G., Nandgaonkar, A. G., Wang, Q., Zhang, J., Krause, W. E., Wei, Q. & Lucia, L. A. (2017). Laccase-immobilized bacterial cellulose/TiO2 functionalized composite membranes: Evaluation for photo-and bio-catalytic dye degradation. *Journal of Membrane Science*, *525*, 89-98.

[50] Devarayan, K., Park, J., Kim, H. Y. & Kim, B. S. (2017). Facile green synthesis of silver nanodendrite/cellulose acetate thin film electrodes for flexible supercapacitors. *Carbohydrate Polymers*, *163*, 153-161.

[51] Mohammad, F., Arfin, T. & Al-Lohedan, H. A. (2017). Sustained drug release and electrochemical performance of ethyl cellulose-magnesium hydrogen phosphate composite. *Materials Science and Engineering: C*, *71*, 735-743.

[52] Jasim, A., Ullah, M. W., Shi, Z., Lin, X. & Yang, G. (2017). Fabrication of bacterial cellulose/polyaniline/single-walled carbon nanotubes membrane for potential application as biosensor. *Carbohydrate Polymers*, *163*, 62-69.

[53] De Silva, R. T., Mantilaka, M. M. M. G. P. G., Ratnayake, S. P., Amaratunga, G. A. J. & de Silva, K. N. (2017). Nano-MgO reinforced chitosan nanocomposites for high performance packaging applications with improved mechanical, thermal and barrier properties. *Carbohydrate Polymers*, *157*, 739-747.

[54] Luo, Y., Zhou, Z. & Yue, T. (2017). Synthesis and characterization of nontoxic chitosan-coated Fe3O4 particles for patulin adsorption in a juice-pH simulation aqueous. *Food Chemistry*, *221*, 317-323.

[55] Kim, J., Hwang, J., Seo, Y., Jo, Y., Son, J. & Choi, J. (2017). Engineered chitosan–xanthan gum biopolymers effectively adhere to cells and readily release incorporated antiseptic molecules in a sustained manner. *Journal of Industrial and Engineering Chemistry*, *46*, 68-79.

[56] Wang, S. & Jing, Y. (2017). Effects of formation and penetration properties of biodegradable montmorillonite/chitosan nanocomposite film on the barrier of package paper. *Applied Clay Science*, *138*, 74-80.

[57] Bilas, R., Sriram, K., Maheswari, P. U. & Begum, K. M. S. (2017). Highly biocompatible chitosan with super paramagnetic calcium ferrite (CaFe2O4) nanoparticle for the release of ampicillin. *International Journal of Biological Macromolecules*.

[58] Park, K. H., Kim, S. J., Lee, W. Y., Song, H. J. & Park, Y. J. (2017). Hydrothermal fabrication and characterization of calcium phosphate anhydrous/chitosan composites. *Ceramics International*, *43*(2), 2786-2790.

[59] Bari, S. S. & Mishra, S. (2017). Effect of calcium sulphate nanorods on mechanical properties of chitosan-hydroxyethyl methacrylate (HEMA) copolymer nanocomposites. *Carbohydrate Polymers*, *157*, 409-418.

[60] Pathania, D., Gupta, D., Kothiyal, N. C., Eldesoky, G. E. & Naushad, M. (2016). Preparation of a novel chitosan-g-poly (acrylamide)/Zn nanocomposite hydrogel and its applications for controlled drug delivery of ofloxacin. *International journal of biological macromolecules*, *84*, 340-348.

[61] Kamel, N. A., El-messieh, S. L. A. & Saleh, N. M. (2017). Chitosan/banana peel powder nanocomposites for wound dressing application: Preparation and characterization. *Materials Science and Engineering: C*, *72*, 543-550.

[62] Shariful, M. I., Sharif, S. B., Lee, J. J. L., Habiba, U., Ang, B. C. & Amalina, M. A. (2017). Adsorption of divalent heavy metal ion by mesoporous-high surface area chitosan/poly (ethylene oxide) nanofibrous membrane. *Carbohydrate Polymers*, *157*, 57-64.

[63] Kumari, H. J., Krishnamoorthy, P., Arumugam, T. K., Radhakrishnan, S. & Vasudevan, D. (2017). An efficient removal of crystal violet dye from waste water by adsorption onto TLAC/Chitosan composite: A novel low cost adsorbent. *International Journal of Biological Macromolecules*, *96*, 324-333.

[64] Li, S., Xiao, L., Deng, H., Shi, X. & Cao, Q. (2016). Remote controlled drug release from multi-functional Fe3O4/GO/Chitosan microspheres fabricated by an electrospray method. *Colloids and Surfaces B: Biointerfaces.*, *151*, 354-362.

[65] Mahdavinia, G. R., Mosallanezhad, A., Soleymani, M. & Sabzi, M. (2017). Magnetic-and pH-responsive κ-carrageenan/chitosan complexes for controlled release of methotrexate anticancer drug. *International Journal of Biological Macromolecules.*, *97*, 209-217.

[66] Sharma, G., Naushad, M., Ala'a, H., Kumar, A., Khan, M. R., Kalia, Shweta, Bala, M. & Sharma, A. (2017). Fabrication and characterization of chitosan-crosslinked-poly (alginic acid) nanohydrogel for adsorptive removal of Cr (VI) metal ion from aqueous medium. *International Journal of Biological Macromolecules*, *95*, 484-493.

[67] Seeli, D. S. & Prabaharan, M. (2017). Guar gum oleate-graft-poly (methacrylic acid) hydrogel as a colon-specific controlled drug delivery carrier. *Carbohydrate Polymers*, *158*, 51-57.

[68] Patra, A. S., Ghorai, S., Sarkar, D., Das, R., Sarkar, S. & Pal, S. (2017). Anionically functionalized guar gum embedded with silica nanoparticles: An efficient nanocomposite adsorbent for rapid adsorptive removal of toxic cationic dyes and metal ions. *Bioresource Technology*, *225*, 367-376.

[69] Dziadkowiec, J., Mansa, R., Quintela, A., Rocha, F. & Detellier, C. (2017). Preparation, characterization and application in controlled release of Ibuprofen-loaded Guar Gum/Montmorillonite Bionanocomposites. *Applied Clay Science*, *135*, 52-63.

[70] Chandrika, K. P., Singh, A., Rathore, A. & Kumar, A. (2016). Novel cross linked guar gum-g-poly (acrylate) porous superabsorbent hydrogels: Characterization and swelling behaviour in different environments. *Carbohydrate polymers*, *149*, 175-185.

[71] Seeli, D. S., Dhivya, S., Selvamurugan, N. & Prabaharan, M. (2016). Guar gum succinate-sodium alginate beads as a pH-sensitive carrier for colon-specific drug delivery. *International journal of biological macromolecules*, *91*, 45-50.

[72] Maity, J. & Ray, S. K. (2016). Enhanced adsorption of Cr (VI) from water by guar gum based composite hydrogels. *International journal of biological macromolecules*, *89*, 246-255.

[73] Ragothaman, M., Palanisamy, T. & Kalirajan, C. (2014). Collagen–poly (dialdehyde) guar gum based porous 3D scaffolds immobilized with growth factor for tissue engineering applications. *Carbohydrate polymers*, *114*, 399-406.

[74] Pandey, S. & Mishra, S. B. (2014). Catalytic reduction of p-nitrophenol by using platinum nanoparticles stabilised by guar gum. *Carbohydrate polymers*, *113*, 525-531.

[75] Abdel-Halim, E. S. & Al-Deyab, S. S. (2014). Electrically conducting silver/guar gum/poly (acrylic acid) nanocomposite. *International journal of biological macromolecules*, *69*, 456-463.

[76] Murali, R., Vidhya, P. & Thanikaivelan, P. (2014). Thermoresponsive magnetic nanoparticle–aminated guar gum hydrogel system for sustained release of doxorubicin hydrochloride. *Carbohydrate polymers*, *110*, 440-445.

[77] Bosio, V. E., Basu, S., Abdullha, F., Villalba, M. E. C., Güida, J. A., Mukherjee, A. & Castro, G. R. (2014). Encapsulation of Congo Red in carboxymethyl guar gum–alginate gel microspheres. *Reactive and Functional Polymers*, *82*, 103-110.

[78] Likhitha, M., Sailaja, R. R. N., Priyambika, V. S. & Ravibabu, M. V. (2014). Microwave assisted synthesis of guar gum grafted sodium acrylate/cloisite superabsorbent nanocomposites: Reaction parameters and swelling characteristics. *International journal of biological macromolecules*, *65*, 500-508.

[79] Bagal-Kestwal, D. R., Kestwal, R. M., Hsieh, W. T. & Chiang, B. H. (2014). Chitosan–guar gum–silver nanoparticles hybrid matrix with immobilized enzymes for fabrication of beta-glucan and glucose sensing photometric flow injection system. *Journal of pharmaceutical and biomedical analysis*, *88*, 571-578.

[80] Pathania, D., Katwal, R., Sharma, G., Naushad, M., Khan, M. R. & Ala'a, H. (2016). Novel guar gum/Al2O3 nanocomposite as an effective photocatalyst for the degradation of malachite green dye. *International journal of biological macromolecules*, *87*, 366-374.

[81] Naushad, M., Inamuddin, Rangreez, T. A. & ALOthman, Z. A. (2014). A Mercury Ion Selective Electrode based on Poly-o-toluidine Zr(IV) Tungstate Composite Membrane, *J. ElectroanalyticalChem.*, *713*, 125-130

[82] Inamuddin, Khursheed, A. & Naushad, M. (2014). Optimization of glassy carbon electrode based graphene/ferritin/glucose oxidase bioanode for biofuel cell applications, *Int. J. Hydrogen Energy.*, *38*, 7417-7421

[83] Naushad, M. (2008). PVC based membrane of Ti(IV) iodovanadate for Pb(II) determination, *Sensors & Trans. J*, *95*, 86-96.

[84] Naushad, M. (2008). A new ion selective electrode based on aluminium tungstate for Fe(III) determination in rock sample, pharmaceutical sample and water sample, *Bull. Mater. Sci.*, *7*, 1–9.

[85] Naushad, M., (2014). Surfactant assisted nano-composite cation exchanger: Dev., charact. And applications for the removal of toxic Pb^{2+} from aqueous medium, *Chem. Eng. J.*, *235*, 100-108.

[86] Nabi, S. A., Naushad, M. & Bushra, R. (2009). Synthesis and characterization of a new organic-inorganic Pb^{2+} selective nano-composite cation exchanger acrylonitrile stannic(IV) tungstate and its analytical applications, *Chem. Eng. J.*, *152*, 80-87.

[87] Nabi, S. A., Naushad, M. & Bushra, R. (2009). A new hybrid EDTA-Zirconium phosphate cation exchanger, Synthesis, characterization and adsorption behaviour for environmental monitoring, *Adsor. Sci. Tech.*, *27*, 423-435.

[88] Khan, A. A., Alam, M. M. & Mohammad, F. (2003). *Electrochimica Acta*, *48*, 2463-2472.

[89] Evans, S. L. & Gregson, P. J. (1998). Composite technology in load-bearing orthopaedic implants, *Biomat.*, *19*, 1329-42.

[90] Haneef, M., Rahman, J. F., Yunus, M., Zameer, S., Patil, S. & Yezdani, T. (2013). International *Journal of Modern Engineering Research*, *3*, 970-979.

[91] Uhthoffm, H. K. & Finnegan, M. (1983). *J. Bone Joint Surg.*, *65*, 66–69.

[92] Sachlos, E., Wahl, D. A., Triffitt, J. T. & Czernuszka, J. T. (2008). *Acta Biomater.*, *4*, 1322–1331.

[93] Ngiam, M., Liao, S. S., Patil, A. J., Cheng, Z. Y., Chan, C. K. & Ramakrishna, s. (2009). *Bone*, *45*, 4–16.

[94] Huang, J., Disilvio, L., Wang, M., Tanner, K. E. & Bonfield, W. (1997). *J. Mater. Sci. Mater. Med.*, *8*, 775–779.

[95] Guild, F. J. & Bonfield, W. (1998). *J. Mater. Sci. Mater. Med.*, *9*, 497–502.

[96] Liao, C. Z., Li, K., Wong, H. M., Tong, W. Y., Yeung, K. W. K. & Tjong, S. C. (2013). Novel polypropylene biocomposites reinforced with carbon nanotubes and hydroxyapatite nanorods for bone replacements, *Materials Science and Engineering C*, *33*, 1380–1388.

[97] Craciunescu, O. & Moldovan, L. (2013). Designing Bio-Inspired Composite Materials for Medical Applications.

[98] Wang, X., Grogan, S. P., Rieser, F., Winkelmann, V., Maquet, V., Berge, M. L. & Mainil-Varlet, P. (2004). Tissue engineering of biphasic cartilage constructs using various biodegradable scaffolds: an *in vitro* study, *Biomaterials*, *25*, 3681-3688.

[99] Chen, Y. S., Kuo, S. M. & Yao, C. H. (2009). A review for gelatin used for artificial nerve and bone implants, *BioMedical Engineering-Applications Basis Communications*, *21*, 233-238.

[100] Scabbia, A. & Trombelli, L. (2004). A comparative study on the use of a HA/collagen/chondroitin sulphate biomaterial (Biostite) and a bovine-derived HA xenograft (Bio-Oss) in the treatment of deep intra-osseous defects, *Journal of Clinical Periodontology*, *31*, 348-355

[101] Takahashi, Y., Yamamoto, M. & Tabata, Y. (2005). Enhanced osteoinduction by controlled release of bone morphogenetic protein-2 from biodegradable sponge composed of gelatin and beta-tricalcium phosphate, *Biomaterials*, *26*, 4856–4865.

[102] Meiyu, Wu., Qiaoying, W., Xinqing, L. & Haiqing, L. (2013). Biomimetic synthesis and characterization of carbon nanofiber/hydroxyapatite composite scaffolds, *Carbon*, *51*, 335-345

[103] Zhao, B., Hu, H., Mandal, S. K. & Haddon, R. C. (2005). A bone mimic based on the self-assembly of hydroxyapatite on chemically functionalized single-walled carbon nanotubes. *Chem Mater*, *17*(12), 3235–41.

[104] White, A. A., Best, S. M. & Kinloch, I. A. (2007). Hydroxyapatite–carbon nanotube composites for biomedical applications: a review. *Int J Appl Ceram Technol*, *4*, 1–13.

[105] Peltier, L. F. (1961). The use of Plaster of Paris to fill defects in bone. *ClinOrthop*, *21*, 1–29.

[106] Alexander, D. I. & Manson, N. A. (2001). Mitchell MJ. Efficacy of calcium sulfate plus decompression bone in lumbar and lumbosacral spinal fusion: preliminary results in 40 patients. *Can J Surgery*, *44*, 262–6.

[107] Gitelis, S., Piasecki, P., Turner, T., Haggard, W., Charters, J. & Urban, R. (2001). Preliminary report: use of a calcium sulfate-based bone graft—substitute for benign bone lesions. *Orthopedics*, *24*, 162–6.

[108] Kelly, C. M., Wilkins, R. M., Gitelis, S., Hartjen, C., Watson, J. T. & Kim, P. T. (2001). The use of a surgical grade calcium sulfate as a bone graft substitute. *ClinOrthop*, *382*, 44–50.

[109] Mirzayan, R., Panossian, V., Avedian, R., Forrester, D. M. & Menendez, L. R. (2001). The use of calcium sulfate in the treatment of benign bone lesions. *J Bone JtSurg*, 83-A, 355–8.

[110] Pezzotti, G. & Yamamoto, K. (2014). Artificial hip joints: The bio-materials challenge, *journal of the mechanical behavior of biomedical materials*, *31* (2014), 3–20.

[111] Banga, A. K. (1998). Electrically Assisted Transdermal and Topical Drug Delivery. Taylor & Francis, London, Bristol, PA.

[112] Kasting, G. B. (1992). Theoretical-models for iontophoretic delivery. *Adv. Drug Deliv. Rev.*, *9*, 177–199.

[113] Irwin, W. J. & Belaid, K. A. (1987). Drug delivery by ion exchange Part I: Ester prodrugs of Propranolol. *Drug Dev. Ind. Pharm.*, *13*, 2017–2031.

[114] Amsel, L. P. (1984). Recent advances in sustained release technology using ion-exchange polymers. *Pharm. Technol.*, April, 28–48.

[115] Sayed, U. G. & Bajaj, A. N. (2000). Oral controlled release bromhexine ion-exchange resinate suspension formulation. *Indian Drugs*, *37*, 185–189.

[116] Motycka, S. & Nairn, J. G. (1979). Preparation and evaluation of microencapsulated ion-exchange resin beads. *J. Pharm. Sci.*, *68*, 211–215.

[117] Manek, S. P. & Kamat, V. S. (1981). Evaluation of Indion CRP 244 and CRP 254 as sustained release and taste masking agent. *Indian J. Pharm. Sci.*, *26*, 773–776.

[118] Anusavice, K. J. (2003). Phillips' science of dental materials. 11th ed. St. Louis, MO: Saunders.

In: Modified Biopolymers: Challenges and Opportunities ISBN: 978-1-53612-116-2
Editors: D. Pathania, G. Sharma and A. Kumar

Chapter 8

Biological Traits of Nanocomposites: Nanofertilizers, Nanopesticides, Anticancer and Antimicrobials

Adesh K. Saini[1,*], Himika Gupta[2], Ashu M. Poswal[1], Reena Kumari[1], Rakesh Kumar[1] and Reena V. Saini[1,†]

[1]Faculty of Applied Sciences and Biotechnology, Shoolini University, Solan, India
[2]Department of Animal, Plant and Soil Sciences, AgriBio, La Trobe University, Bundoora, Australia

Abstract

In the last few decades, nanocomposites have gained attention due to their wide applications. These include energy production, water purification, improving the catalytic as well as physical properties of compounds, automobile industry and many more. Importantly, nanocomposites are also been tested for their application in agriculture and medical sciences. In the present chapter, we reviewed the role of these composites to reduce the burden of fertilizer and pesticides and, for their use in antimicrobial and anticancer activities.

Keywords: nano, environment, cytotoxicity, cancer, microbes

* Corresponding Author Email: sainiade@gmail.com.
† Corresponding Author Email: reenavohra10@gmail.com.

INTRODUCTION

Nanocomposites and Their Antimicrobial Activity

Metals like silver (Ag), zinc (Zn), and copper (Cu) have been traditionally used as antimicrobial agents for many centuries. Metals and their oxides have been used in combination with other components to generate new materials called nanocomposites. The antimicrobial activity of nanocomposite is attributed to their increased surface area and the possibility to interact directly with the bacterial cell wall because of their small size. Metal, oxide, or salt compounds based on copper and silver are among the most widely applied antimicrobial agents.

Ag and Cu nanoparticles (NPs) have been proved effective against a wide range of bacterial strains including *Staphylococcus aureus*, *Streptococcus mutans*, *Pseudomonas aeruginosa*, *Escherichia coli*, *Enterococcus faecalis*, *Streptococcus sobrinus*, and *Staphylococcus epidermidis* [1, 2]. Against this, some NPs are more bacteria-specific depending on their mechanism of action [e.g., ZnMgO NPs show highly specific antibacterial activity toward Gram-positive bacteria [3].

Among all the metals, silver nanoparticle (AgNPs) is the most effective antimicrobial agent [4, 5] followed by TiO_2 [6]. Silver is well known for its strong toxicity to a wide range of microorganisms and have been widely used in a broad spectrum of applications including water purification, dentistry, wound care, textiles and clothing, food packaging, medical and cosmetic ingredients, and household usage [7, 8, 9]. In a study, Besinis et al. showed that AgNPs are more antibacterial against cariogenic bacterial species as compared to other metal NPs. The minimum bactericidal concentration (MBC) of Ag, ZnO, and CuO against *P. aeruginosa* to be 100, >5000, and 5000 μg mL^{-1}, respectively. The corresponding MBC values against *S. aureus* were 100, 2500, and 2500 $\mu g mL^{-1}$; confirming the effectiveness of AgNPs [10, 11].

Silver ions and silver nanoparticles can cause direct damage to the bacterial cell membrane. The main antibacterial effect of silver is due to its partial oxidation and release of silver ions (Ag^+). The released Ag ions increase the cell permeability and penetrate the cells, bind to microbial DNA, generate reactive oxygen species, inhibit bacterial replication, and bind to the thiol groups of the metabolic enzymes of the bacterial electron transport chain, causing their alteration and inactivation [12, 13, 14, 15] (Figure 1).

However, bare AgNPs aggregate when they come in contact with bacteria; thus, they lose their active surface area and show weaker antibacterial activity. To overcome this problem, many technologies and strategies have been developed to protect AgNPs from aggregation and so that Ag ions can be slowly released to the solution system. To improve its stability Wang et al. (2016), synthesized hollow multiple-Ag-nanoclustes-C-shell nanocomposites by using silane coupling agent to graft carbon dots (CDs) with silver nanoparticles (AgNPs). CDs act as coating and stabilizing agent that prevent AgNPs aggregation and oxidation. The disk diffusion test results indicated that multiple-Ag-nanoclustes-C-shell nanocomposites showed stronger bactericidal effect on both gram-negative and gram-positive bacteria [16].

Han et al. prepared Ag-palygorskite (Ag-Pal) nanocomposites by a convenient ultraviolet irradiation technique, using carbon dots derived from wool fiber as a reducing agent. The antibacterial properties of the Ag-Pal nanocomposite against Gram-negative (*E. coli*) and Gram-positive *S.* (*aureus*) bacteria was evaluated using the Kirby-Bauer disk diffusion method and minimum inhibitory concentration (MIC). The antibacterial activity tests revealed Ag-Pal nanocomposite allowed free diffusion and sustained release of silver ions into the cultivation medium thereby producing an excellent biocidal effect on both Gram-positive (*S. aureus*) and Gram-negative (*E. coli*) bacteria.

Chitosan/Ag Nanocomposites

Chitosan is a natural polymer and has been extensively studied due to its biological properties, such as antifungal activity, antibacterial activity and antitumor activity. Due to its excellent antimicrobial properties, chitosan nanoparticle can be used as a preservative in food packaging and in dentistry applications. The antimicrobial properties can be enhanced by combining chitosan with other metals such as copper and silver. Kaur et al. prepared Chitosan/Silver nanocomposites by embedding silver nanoparticles in Chitosan polymer. The results showed that chitosan/silver nanocomposites have greater inhibitory effect against *S. aureus*, *P. aeruginosa* and *S. entrica* in comparison to chitosan alone [17]. Ali et al. and Cao et al. also reported the similar antibacterial activity of silver/chitosan nanocomposites against *S. aureus* [18, 19].

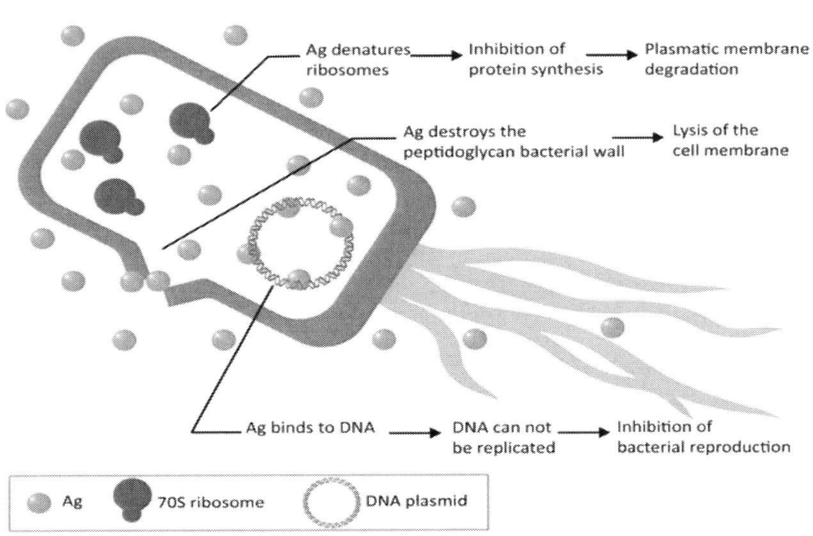

Figure 1. Antibacterial activity of silver.

Polyacrylic Acid/Silver Nanocomposite Hydrogels

Wei et al. Used fast and novel method to prepare nanocomposite hydrogels with uniformly distributed silver nanoparticles that show strong antimicrobial activity. Hydrogels are three-dimensional cross-linked hydrophilic polymers that can absorb considerable quantities of water. Because of their high swelling ratio (SR) and biocompatibility, hydrogels have several applications in biomedical engineering, agriculture, biosensors, drug delivery, and adsorption of heavy metals. This method used a solution of the Polyacrylic acid (PAAc) with carboxylic acid functional groups and silver nitrate to prepare nanocomposite hydrogels through ultraviolet (UV)-light. The antimicrobial test was conducted against four standard strains *Candida albicans, S. aureus, P. aeruginosa,* and *E. coli;* and five clinical isolates *S. aureus, S. epidermidis, P. aeruginosa, Acinetobacter baumannii* and *Klebsiella pneumoniae.* The silver nanocomposite hydrogels exhibited complete inhibition of bacteria and yeast proliferation, which can be useful in developing new superabsorbent antimicrobial pharmaceutical products [20].

Polyaniline/Polyvinyl Alcohol/Ag Nanocomposites

Polyvinyl alcohol (PVA) and polyaniline (PANI) are organic polymer which can be used for coating of Ag. Nanocomposites based on organic polymers have many advantages such as long-term stability, biocompatibility and biodegradability [21, 22, 23, 24]. A novel nanocomposite based on Ag nanoparticle coating by PANI and PVA polymers have been successfully synthesized. The Ag nanoparticles were produced from Ag^{+} ions using the chemical reduction method. The antibacterial activity of the nanocomposite was evaluated against Gram positive bacteria *S. aureus* and Gram negative *E.coli* using the paper disk diffusion method. The antibacterial study showed that the PANI/PVA composite did not have a very good antibacterial activity but PANI/PVA/Ag nanocomposites were found to be effective against both bacterial strains [25].

Copper-Polymer Nanocomposites

Copper nanoparticles have been shown to exhibit a broad spectrum of anti-microbial activity, inhibiting the growth of bacteria, fungi, viruses and algae [26, 27, 28]. In addition, copper is inexpensive as compared to other metals therefore, proving to be a cost-effective material. Antibacterial property of copper is associated with multiple toxic effects on cells, such as generation of reactive oxygen species (ROS), disruption of bacterial cell wall, lipid peroxidation, protein oxidation, DNA binding and disrupting biochemical processes [29] (Figure 2).

Large number of polymers are used as matrices to prepare copper-based nanocomposites and it has increased the apllications of copper nanoparticles (Table 1). The use of polymers not only provided a supporting function for nanoparticles, but also enhanced its antibacterial activity. The multitude applications of copper based nanocomposites and its costeffectiveness as compared to other metals like silver, make it an outstanding biocide. Although copper-based nanocomposites have diverse applications, some studies have reported its long-term cytotoxic effects, bioaccumulation and toxicity in non-target organisms exposed to copper nanoparticles (example: fish, plants, nematodes, algae, mammalian cell lines, etc.) [30, 31].

Table 1. Different types of copper-Polymer matrix and their applications

Polymer matrix	Micoorganism	Applications	Reference
Cellulose	*S. cerevisiae*	Food packaging	[32]
Polyethylene	*E. coli*	Medical devices	[31]
Hydroxypropyl methylcellulose	*Streptococcus spp., S. epidermis, B. cereus, E. faecalis, Salmonella spp., P. aeruginosa, S. aureus*	Food packaging	[33]
Cotton	*E. coli, S. aureus*	Medical usage	[34]
Cotton	*E. coli, S. aureus*	Medical apparel	[35]
Polylactic acid	*Pseudomonas spp.*	Food packaging	[36]
Agar	*L. monocytogenes, E. coli*	Food packaging	[37]
Bioglass	*E. coli*	Medical implant	[38]
Polyaniline	*S. aureus, E. coli, C. albicans*	Water disinfection	[39]
Cotton	*E. coli*	Dressing materials	[40]
Bamboo rayon	*S. aureus, E. coli*	Medical clothes	[41]
High density polyethylene	*E. coli*	Food packaging	[42]
Epoxy resin	*S. aureus, C. albicans, Chlorella sp.*	Implantable antimicrobial biomaterial	[43]
Cotton	*E. coli, S. aureus*	Wound dressing, bed lining and active bandages	[44]
Polymers based on acrylic	*Chlamydomonas spp., Synechocystis, Phaeodactylum tricornutum*	Marine antifouling coatings	[38]
Polypropylene	*S. aureus, P. aeruginosa*	Medical and health care applications	[45]

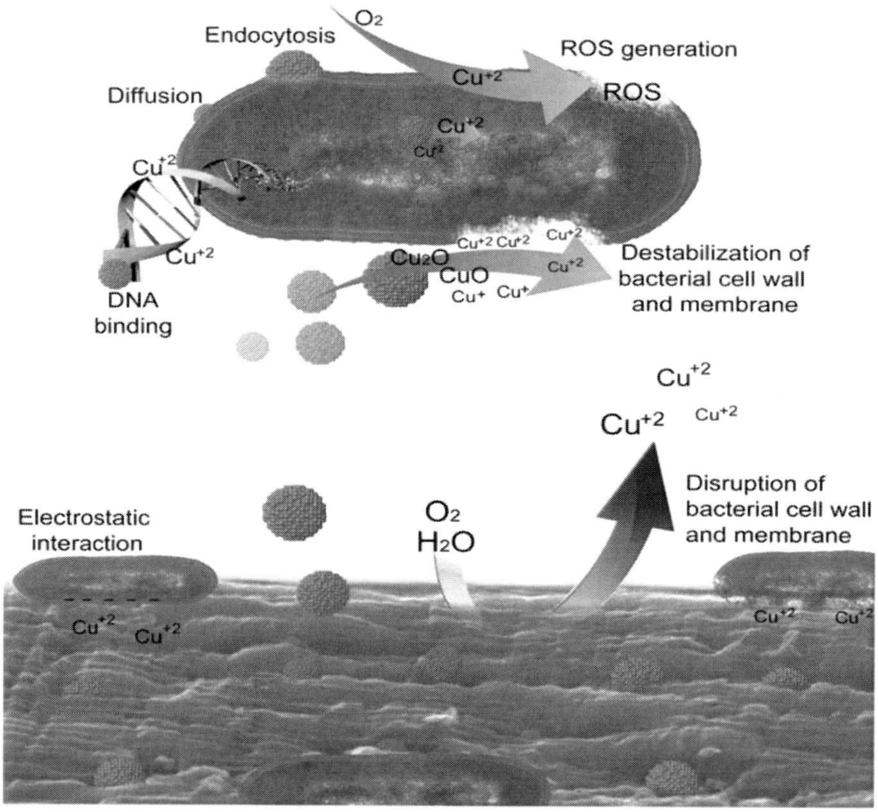

Figure 2. Antibacterial activity of copper.

Nanocomposite as a Potential Anticancer Agent

Nanocomposite as a Potential Anticancer Agent

Worldwide cancer is among the most common health concern. Chemotherapy and radiotherapy are the main treatments currently used for cancer. However, these therapies, have their own side effects and these are toxic to normal cells too. These side effects can be reduced by encapsulating the anti-cancer drug into a bio-nanocomposite (Table 2).

Recently, the chitosan nanocomposite coated with Mn-Zn ferrite nanoparticles and modified by hydroxyapatite nanoparticles (the main inorganic constituent of human bones and teeth) was synthesized and was used against colon cancer cells after loading with an anticancer drug, paclitaxel. The paclitaxel loaded nanocomposite showed encouraging results while they are non-toxic to healthy monocytes [46]. Another nano-formulation composed of sulfated polysaccharide (iota carrageenan) and a magnetic nanoparticle (maghemite) showed non-toxic effects to normal human cell lines HEK-293 (Human embryonic kidney cell lines) and L6 (Rat myoblast cell lines) but showed cytotoxicity against human colon cancer cell line (HCT116) by inducing apoptosis [47].

Table 2. Nanocomposites and their Cytotoxic studies

Nanocomposite	Composition	Synthesis method	Cell Type	Assay Performed	Reference
Bionanocomposite + Paclitaxel [bio-NCP + PTX)	Chitosan + hydroxyapatite [Ca10[PO4)6[OH)2 + Mn-Zn ferrite nanoparticles + Paclitaxel	Double emulsion method	HCT116 3LL	MTT assay	[42]
ι-carrageenan-γ-maghemite nanocomposite [ι-CGN-γ-Fe_2O_3)	γ - maghemite [Fe_2O_3) nanoparticles+ ι-carrageenan	Co-precipitation	HCT116	MTT assay	[47]
PGE-HRG-Ag nanocomposite	*Pulicaria glutinosa* extract + reduced graphene oxide+ silver	Green synthesis	MCF-7 A549 Hela DU-145 HepG2	MTT assay	[48]
TAE-rGO-Agnanocomposite	*Tilia amurensis* + reduced graphene oxide+silver	Green synthesis	A2780	WST-8 assay	[49]
GOGA nanocomposite	Graphene oxide + gallic acid				[50]
Withania somnifera extract/Gadolinium III oxide nanocomposite [WSGNC)	*Withania somnifera* extract+Gadolinum III oxide nano powder	Green synthesis	EAC	MTT assay	[51]
Hyaluronan/zinc oxide nanocomposite	*Sargassum muticum* + hyaluronan biopolymer	Green synthesis	PANC-1 CaOV-3 COLO205 HL-60	MTT assay	[52]
CNT-PMAA-Q	Nanotubes + Carbon polymethacrylic acid + quercetin	free radical grafting reaction	HeLa	WST-8	[53]
Magnetic fluid nanocomposite	Colloid dispersion of magnetic nanoparticles + Cisplastin		MCF-7	MTTassay	[55]
AgCl@PPCCore-Shell Nanocomposites	Pyrrole + chitosan + silver nitrate + 3-amino-2-phenyl-4[3H)-quinazolinone [I)	Oxidative polymerization	EAC	SRBassay	[56]
P22 viral capsid nanocomposite	P22 viral capsid + Catechol ligands + bortezomib		HepG2	MTT assay	[57]

Graphene and graphene-based nanocomposites have been used in various biological applications due to their extraordinary physico-chemical properties. Graphene oxide nanocomposites coated with silver and encapsulating *Pulicaria glutinosa* extract (PGE) as a reducing agent displayed promising anticancer activity against MCF-7 (breast), A549 (lung), Hela (cervical), DU-145 (prostate), and HepG2 (liver) cancer cell lines. These silver coated nanocomposites has highly reduced the graphene oxide nanocomposites and were found to be the most potent in inducing cell death by apoptosis against human lung cancer cells, A549 [48]. Similarly, reduced graphene oxide-silver nanocomposite developed by using *Tilia amurensis* plant extracts showed significant cytotoxicity against A2780 (ovarian cancer cells) [49]. In another study, graphene oxide was used as a nanocarrier for delivering gallic acid [an anticancer agent) to cancer cells [50].

Nanoscale-formulations have been synthesized by using medicinal plants extracts and metal oxides which can act as efficient anticancer agents. For example, *Withania somnifera* extract/Gadolinium III oxide nanocomposite showed significantly enhanced antitumor and radio-sensitization efficacy in mice bearing Ehrlich carcinoma [51]. Additionally, green biosynthesis of zinc oxide nanocomposite using the seaweed *Sargassum muticum*aqueous extract and hyaluronan biopolymer lead to apoptotic cell death induction in acute promyelocytic leukemia (HL-60) whereas hyaluronan/zinc oxide (HA/ZnO) nanocomposite treatment did not cause toxicity to the normal human lung fibroblast (MRC-5) cell line [52].

Quercetin, a well-known phyto-compound possessing anticancer properties, was loaded on to polymethacrylic acid and carbon nanotubes which resulted in its enhanced anticancer effects against HeLa cells [53]. Various anticancer drugs have been encapsulated in nanocomposites to enhance their anticancer potential. For example, cisplastin entrapped in colloidal dispersion of magnetic nanoparticles exhibited enhanced cytotoxicity towards MCF-7 cells than unconjugated cisplastin [54]. Similarly, doxorubicin loaded ontomagnetite/chitosan-L-glutamic acid (core/shell) nanocomposite with specific hypothermic effect have been synthesized for antitumor therapy [55].

3-amino-2-phenyl-(3H)-quinazolinone encapsulated Polypyrrole/chitosan silver chloride nanocomposite also showed cytotoxic effect against Erlich ascites carcinoma cells [56]. Junseon Min et al. synthesized a P22 viral capsid nanocomposite for the targeted delivery of bortezomib (BTZ), an anticancer drug. The interior surface of P22 viral capsid was chemically modified by attaching catechol and for the targeted delivery hepatocellular carcinoma (HCC)-targeting peptide (SP94, SFSIIHTPILPL) was conjugated to the exterior surface of the P22 viral capsid nanocomposites. This nanoformulation carrying BTZ displayed cytotoxicity towards HepG2 which was similar to the cells treated with free BTZ. Thus, this study indicates that P22 viral capsid-nanocomposite can be used as multifunctional nanoscale delivery platform [57].

From all these studies we can conclude that functionalization of nanocomposites can open new prospects for the growing field of drug delivery and cancer treatment.

Table 3. Comparison of nanostructured formulations and conventional method [60]

Properties	Nano-fertilizers and Nano-pesticides based technology	Conventional technology
Efficiency of nutrient uptake	Increase in uptake ratio of nutrients in crop production Nanoparticles may increase fertilizer and pesticide efficiency Save nutrient resource	Bulk composite is not available to roots It decreases efficiency
Solubility and dispersion of nutrients	Nanoparticle formulation improves solubility and dispersion of nutrient, enhance the availability to plants	Large particle size and less solubility results in less bioavailability
Controlled release mode	Precisely control release of particles due to microencapsulation and immobilization	Excess release of fertilizers and pesticides may cause toxicity and destroy ecological balance of the soil
Effective duration	Nanostructured formulation can increase effective duration of nutrient supply	Plants can use nutrients at the time of delivery, else will be converted into insoluble salts
Nutrient loss rate	Nanoparticles have low nutrient loss rate	Bulk composite have high loss rate due to leaching, rain off and drift

Nanofertilizers and Nanopesticides

Explosion of world population over the past few decades, has compelled the agricultural sector to increase crop production. To raise the crop production, farmers are using large quantities of chemical fertilizers and pesticides that interfere with the loss of biodiversity, fertility of soil and soil mineral balance. To overcome this problem, nanoparticles in the form of nano-pesticides and nanofertilizers are designed for agricultural purposes that work to solve concerns for the soil, environment and occupational health of workers (Table 3).

Nanotechnology has potential application in agri-biotechnology. Nanotechnology in the form of nano-fertilizers and nano-pesticides has tremendous role in agricultural system. Nano-fertilizers and nano-pesticides have application in detection of pest, pathogens, plant diseases; site targeted delivery; enhanced utilization of nutrients; controlled release of fertilizers, pesticides, and genetic material; build and protect the make-over of soil; reduction in toxicity of soil, water [58, 59, 60]. These trait of nanomaterials are due to its small size, stability, high mobility, high surface area to volume ratio, high solubility in water, act on specific target, low toxicity due to elimination of organic solvents [61].

Aerosol based foliar delivery of nanoparticles may help to uptake, translocate, accumulate nanoparticles and reduce environmental impact as compared to conventional fertilizers applied through soil route. A study has reported that gold nanoparticles taken by watermelon plant through direct or stomatal opening; transported to root tissues by phloem transport mechanism. Accumulation and translocation of nanoparticles depend on nanoparticle shape and nature of plant part [62].

Fertilizers and pesticides are attached with nanoparticles is done by adsorption on nanoparticle mediated ligand, microencapsulation in nanoparticulate polymeric shell, immobilisation of polymeric nanomaterial, absorption in nanoparticles [63]. Nanoparticles have potential application in pest management strategies. Nanoparticles offer a path to achieve better effects with lower pesticide use; control release and site targeted delivery [64]. Nanopesticides offer a novel route to control pests and pathogens even with low chemical dose. Silver nanoparticles with concentration of 100 ppm is effective to inhibit the growth of powdery mildews in pumpkins and cucurbits [65]. Silver nanoparticles have potential application in insect pest management. CdS, Nano-Ag and Nano-TiO_2 nanoparticles have been shown to be effective against *Spodoptera litura* and caused larval mortality of 93.79%, 56.89% and 73.79%, respectively. Therefore, these three nanoparticles are useful component of an integrated pest management strategy [66].

Antibacterial and antifungal activity of soda lime glass powder copper nanoparticles have been reported against *Escherichia coli*, *Micrococcus luteus* and *Issatchenkia orientalis* [67]. Silica silver nanoparticles showed efficient antifungal activity against various plant pathogenic fungus viz. *Blumeria* spp., *Sphaerotheca* spp., *Phytophthora* spp., *Rhizoctonia* spp., *Colletotrichum* spp., *Botrytis* spp., *Magnaporthe* spp. and *Pythium* spp. [68]. Zinc oxide nanoparticles are effective against postharvest pathogenic fungi (*Botrytis cinerea* and *Penicillium expansum*) by inhibiting the growth of cellular functions in fungal hyphae and preventing the development of conidiophores, respectively [69].

Goswami et al. [70] reported that SiO_2 nanoparticles showed approximately 100% mortality against insect pest. Treatment of silkworm with hydrophobic nano-aluminosilicate significantly reduced the grasserie disease. Burman et al. reported that chickpea sprayed with 1.5 ppm zinc oxide nanoparticles showed more biomass accumulation as compared to bulk treatment [71]. This may be due to lower reactive oxygen species (ROS) levels which resulted in less lipid peroxidation. Nano-fertilizers is a way to gradually release nutrients into the soil in a controlled manner and preventing pollution of water resources [72]. Zhao et al. 2014, observed the effect of ZnO and CeO_2 nanoparticles at 400 ppm on cucumber fruit quality and found that both the tested nanoparticles resulted in increased starch content. Fertilizers based on nanotechnology have the ability to surpass conventional fertilizers. In this context, experts also showed that multiwalled carbon nanotubes (MWCNT) have ability to increase the cell growth by 55-64% in tobacco as compared to the activated carbon resulted in enhancement of seed germination and plant growth [73]. Similarly, low concentration of oxidised MWCNT enhance the seed germination and plant growth in mustard seeds as compared to non-oxidised MWCNT or high concentration of oxidised MWCNT. The beneficial effects are due to enhancement in the moisture content of seeds and water absorption capacity of root tissues [74].

Comparision of nanoparticles made out of silicon, palladium, gold and copper for their effect on lettuce seeds germination by measuring the length of shoot and root of the plant, it was found that nanoparticles enhanced the seed germination [75]. Treatment of nanocrystalline powders of Fe, Co and Cu nanoparticles increase soybean seed germination by 25%, 25% and 10%, respectively as compared to control sample. Along with that, a noticeable increase has also been observed in chlorophyll index, number of nodules and crop yield [76]. Yang et al. 2007 [77], reported that treatment of nano-anatase TiO_2 increased the nitrogen fixation and significantly, improves the growth and yield of spinach leaves as compared to control. Nano-anatase TiO_2 also enhanced the fresh weight, dry weight, chlorophyll, total nitrogen content and protein of spinach as compared to conventional chemical fertilizers.

Conventional fertilizers are generally sprayed or spread on plants. However, very less concentration of fertilizers succeed in reaching their specific site due to leaching of chemicals, rain run-off, microbial degradation, etc. The findings of Ombaodi and Saigusa (2000) showed that more than 90% of the nitrogen content are uptake by the *Capsicum annuum* L. in controlled release fertilizers than conventional fertilizers [78].

We also have to understand that the use of nanoparticles need to be tested very cautiously because they may cause health hazard and could threat to ecosystem. In a study, cerium oxide nanoparticles were shown to be taken by tomato roots and then translocated to edible tissues in fruits. This study sheds light on the long term effect on human health by affecting the heart, brain and skin [79]. Similarly, the trophic level transfer of La_2O_3 had also been observed at higher level trophic organisms fed with plant leaves grown in nanoparticle amended soil as compared to control [58]. In conclusion, nanoparticles as fertilizers and pesticides can increase the efficiency of nutrient use and could decrease the cost of environmental pollution but these claims need to be tested in depth.

REFERENCES

[1] Ruparelia, J. P., Chatterjee, A. K., Duttagupta, S. P. and Mukherji S., (2008). "Strain specificity in antimicrobial activity of silver and copper nanoparticles." *Acta Biomaterialia.* 4, 707-16.

[2] Llorens, Amparo, Lloret, E., Picouet, P. A., Trbojevich R, and Fernandez, A., (2012). "Metallic-based micro and nanocomposites in food contact materials and active food packaging." *Trends in Food Science and Technology.* 24, 19-29.

[3] Vidic, Jasmina, Stankic, S., Haque F., Ciric D., Ronan Le Goffic, Vidy, A., Jupille J, and Delmas B (2013). "Selective antibacterial effects of mixed ZnMgO nanoparticles." *Journal of Nanoparticle Research.* 15, 5.

[4] Nam, Ki-Young, (2011). "*In vitro* antimicrobial effect of the tissue conditioner containing silver nanoparticles." *The Journal of Advanced Prosthodontics.* 3, 20.

[5] Travan, Andrea, Marsich, E., Donati, I., Benincasa, M., Giazzon, M., Felisari, L. and Paoletti, S., (2011). "Silver-polysaccharide nanocomposite antimicrobial coatings for methacrylic thermosets." *Acta Biomaterialia.* 7, 337-46.

[6] Elsaka, Shaymaa E., Ibrahim, M., Hamouda, and Swain, M. V., (2011). "Titanium dioxide nanoparticles addition to a conventional glass-ionomer restorative: Influence on physical and antibacterial properties." *Journal of Dentistry.* 39, 589-98.

[7] Honary, S., Ghajar, K., Khazaeli P., and Shalchian, P., (2011). "Preparation, Characterization and Antibacterial Properties of Silver-Chitosan Nanocomposites Using Different Molecular Weight Grades of Chitosan." *Tropical Journal of Pharmaceutical Research.* 10.

[8] Maryan, Sadeghian, A., Montazer, M. and Harifi, Tina, (2015). "Synthesis of nano silver on cellulosic denim fabric producing yellow colored garment with antibacterial properties." *Carbohydrate Polymers.* 115, 568-74.

[9] Calderon, V. S., Ferreri, I., Galindo, R. E., Henriques, M., Cavaleiro, A. and Carvalho, S., (2015). "Electrochemical vs antibacterial characterization of ZrCN-Ag coatings." *Surface and Coatings Technology.* 275, 357-62.

[10] Besinis, Alexandros, Peralta, T.D. and Richard, D.H., (2012). "The antibacterial effects of silver, titanium dioxide and silica dioxide nanoparticles compared to the dental disinfectant chlorhexidine on Streptococcus mutansusing a suite of bioassays." *Nanotoxicology.*1, 1-16.

[11] Ren, Guogang, Hu, D., Eileen, W.C., Cheng, M.A., Vargas-Reus, Paul R. and Robert P.A., (2009). "Characterisation of copper oxide nanoparticles for antimicrobial applications." *International Journal of Antimicrobial Agents.* 33, 587-590.

[12] Nawaz, M., Han, M.Y., Tschung-Il Kim, Manzoor, U. and Amin, M.T., (2012). "Silver disinfection of Pseudomonas aeruginosa and E. coli in rooftop harvested rainwater for potable purposes." *Science of the Total Environment* 431, 20-25.

[13] Marambio-Jones, Catalina, and Hoek, E.M.V., (2010). "A review of the antibacterial effects of silver nanomaterials and potential implications for human health and the environment." *Journal of Nanoparticle Research.* 12, 1531-551.

[14] Morones, Jose Ruben, Jose Luis Elechiguerra, Alejandra Camacho, Katherine Holt, Kouri, J.B., Jose Tapia Ramírez, and Miguel Jose Yacaman, (2015). "The bactericidal effect of silver nanoparticles." *Nanotechnology.* 16, 2346-353.

[15] Hwang, Taek, E., Lee, J., Chae, H.Y. J., Kim, Y.S., Byoung, C. K., Sang, B.I. and Gu, M.B., (2008). "Analysis of the Toxic Mode of Action of Silver Nanoparticles Using Stress-Specific Bioluminescent Bacteria." *Small.* 4, 746-750.

[16] Wang, Kunjie, Qingjuan, J., Hongxia L., Guan, F., Zhang, D., Feng, H. and Fan, H., (2017). "Synthesis and antibacterial activity of silver@carbon nanocomposites." *Journal of Inorganic Biochemistry*. 166, 64-67.

[17] Kaur, Pawan, Choudhary, A. and Thakur, R., (2013). "Synthesis of Chitosan-Silver Nanocomposites and their Antibacterial Activity," *International Journal of Scientific and Engineering Research*. 4, 869.

[18] Cao, X.L., Cheng, C., Ma, Y.L. and Zhao, C.S., (2010). "Preparation of silver nanoparticles with antimicrobial activities and the researches of their biocompatibilities." *Journal of Materials Science: Materials in Medicine*. 21, 2861-868.

[19] Ali, S., Wazed, Joshi, M. and Rajendran, S., (2010). "Modulation of Size, Shape and Surface Charge of Chitosan Nanoparticles with Reference to Antimicrobial Activity." *Advanced Science Letters*. 3, 452-60.

[20] Wei, Yi-Syuan, Chen, K.S. and Wu, L.T., (2016). "*In situ* synthesis of high swell ratio polyacrylic acid/silver nanocomposite hydrogels and their antimicrobial properties." *Journal of Inorganic Biochemistry*. 164, 17-25.

[21] Shin, Junhwa, Kim, Y., Lee, K., Lim, Y.M. and Chang, Y.N., (2008). "Significant effects of sodium acetate, an impurity present in poly[vinyl alcohol]solution on the radiolytic formation of silver nanoparticle." *Radiation Physics and Chemistry*. 77, 871-76.

[22] Khanna, P.k., Singh, N., Charan, S. and Viswanath, A.K., (2005). "Synthesis of Ag/polyaniline nanocomposite via an *in situ* photo-redox mechanism." *Materials Chemistry and Physics*. 92, 214-19.

[23] Barkade, Shrikant S., Jitendra B. Naik, and Sonawane, S.H., (2011). "Ultrasound assisted miniemulsion synthesis of polyaniline/Ag nanocomposite and its application for ethanol vapor sensing." *Colloids and Surfaces A: Physicochemical and Engineering Aspects*. 378, 94-98.

[24] Porramezan, Mohammad, and Eisazadeh, H., (2011). "Fabrication and characterization of polyaniline nanocomposite modified with Ag_2O nanoparticles." *Composites Part B: Engineering*. 42, 1980-986.

[25] Ghaffari-Moghaddam, Mansour, and Eslahi, H., (2013). "Synthesis, characterization and antibacterial properties of a novel nanocomposite based on polyaniline/polyvinyl alcohol/Ag." *Arabian Journal of Chemistry*. 7, 846-55.

[26] Cady, Nathaniel, C., Jason, L., Behnke, and Aaron D., Strickland, (2011). "Copper-Based Nanostructured Coatings on Natural Cellulose: Nanocomposites Exhibiting Rapid and Efficient Inhibition of a Multi-Drug Resistant Wound Pathogen, A. baumannii, and Mammalian Cell Biocompatibility *In Vitro*." *Advanced Functional Materials*. 21, 2506-514.

[27] Shionoiri, Nozomi, Tetsuya S., Yoshie F., Nakayama, T., Nemoto, M., Matsunaga, T. and Tanaka, T., "Investigation of the antiviral properties of copper iodide nanoparticles against feline calicivirus." *Journal of Bioscience and Bioengineering.* 113, 580-86.

[28] Anyaogu, Kelechi C., Andrei V., Fedorov, and Douglas C., (2008). "Synthesis, Characterization, and Antifouling Potential of Functionalized Copper Nanoparticles." *Langmuir.* 24, 4340-346.

[29] Chatterjee, Kumar, A., Chakraborty R., and Basu T., (2014). "Mechanism of antibacterial activity of copper nanoparticles." *Nanotechnology.* 25, 135101.

[30] Lee, Woo-Mi, Youn-Joo An, Hyeon Yoon, and Kweon, H.S., (2008). "Toxicity and Bioavailability of Copper Nanoparticles To The Terrestrial Plants Mung Bean (Phaseolus Radiatus) And Wheat (Triticum Aestivum): Plant Agar Test For Water-Insoluble Nanoparticles." *Environmental Toxicology and Chemistry* 27, 1915.

[31] Tamayo, L. A., Zapata, P. A., Rabagliati, F. M., Azócar, M. I., Muñoz, L. A., Zhou, X., Thompson G. E., and Páez, M. A., (2015). "Antibacterial and non-cytotoxic effect of nanocomposites based in polyethylene and copper nanoparticles." *Journal of Materials Science: Materials in Medicine.* 26.

[32] Llorens, Amparo, Lloret, E., Picouet, P. and Fernandez, A., (2012). "Study of the antifungal potential of novel cellulose/copper composites as absorbent materials for fruit juices." *International Journal of Food Microbiology* 158, 113-19.

[33] Ebrahimiasl, Saeideh, and Rajabpour, A., (2014). "Synthesis and characterization of novel bactericidal Cu/HPMC BNCs using chemical reduction method for food packaging." *Journal of Food Science and Technology.* 52, 5982-988.

[34] Berendjchi, Amirhosein, Khajavi, R. and Yazdanshenas, M., (2011). "Fabrication of superhydrophobic and antibacterial surface on cotton fabric by doped silica-based sols with nanoparticles of copper." *Nanoscale Research Letters.* 6, 594.

[35] Anita, S., Ramachandran, T., Rajendran, R., Koushik, C. and Mahalakshmi, M., (2011). "A study of the antimicrobial property of encapsulated copper oxide nanoparticles on cotton fabric." *Textile Research Journal.* 81, 1081-088.

[36] Longano, D., Ditaranto, N., Cioffi, Niso, N., Ancona, A., Conte, A., Nobile, M.A.D.L., Sabbatini, and Torsi, L., (2012). "Analytical characterization of laser-generated copper nanoparticles for antibacterial composite food packaging." *Analytical and Bioanalytical Chemistry.* 403, 1179-186.

[37] Shankar, S. and Jong-Whan Rhim, (2014). "Effect of copper salts and reducing agents on characteristics and antimicrobial activity of copper nanoparticles." *Materials Letters.* 132, 307-11.

[38] Goh, Yi-Fan, Ammar, Z., Alshemary, Muhammad, A., Mohammed R.A.K. and Hussain, R., (2014). "Bioactive Glass: An *In-Vitro* Comparative Study of Doping with Nanoscale Copper and Silver Particles." *International Journal of Applied Glass Science.* 5, 255-66.

[39] Bogdanović, Una, Vodnik, V., Miodrag, M., Dimitrijević, S., Srečo D., Škapin, Žunič, V., Budimir, M. and Stoiljković, M., (2015). "Nanomaterial with High Antimicrobial Efficacy - Copper/Polyaniline Nanocomposite." *ACS Applied Materials and Interfaces* 7, 1955-966.

[40] Mary, Grace, Bajpai, S. K. and Chand, N., (2009). "Copper alignate-cotton cellulose [CACC] fibers with excellent antibacterial properties," *Journal of Engineered Fibers and Fabrics* 4, 24-35.

[41] Teli, M.D., and Sheikh, J., (2013). "Modified bamboo rayon-copper nanoparticle composites as antibacterial textiles." *International Journal of Biological Macromolecules.* 61, 302-07.

[42] Bikiaris, Dimitrios N. and Triantafyllidis, K.S. "HDPE/Cu-nanofiber nanocomposites with enhanced antibacterial and oxygen barrier properties appropriate for food packaging applications." *Materials Letters.* 93, 1-4.

[43] Barua, Shaswat, Pronobesh Chattopadhyay, Mayur M. Phukan, Bolin K., Konwar and Niranjan K., (2014). "Hyperbranched epoxy/MWCNT-CuO-nystatin nanocomposite as a high performance, biocompatible, antimicrobial material." *Materials Research Express*. 4, 045402.

[44] Perelshtein, I., Applerot, G., Perkas, N., Wehrschuetz-Sigl, E., Hasmann, A., Guebitz, G. and Gedanken A., (2009). "CuO-cotton nanocomposite: Formation, morphology, and antibacterial activity." *Surface and Coatings Technology.* 204, 54-57.

[45] España-Sánchez, Beatriz, L., Carlos, A., María G., Neira-Velázquez, Pablo González-Morones, José A., Rodríguez-González, Ernesto Hernández-Hernández, Enrique Díaz Barriga-C., Luis Yate, and Ronald F. Ziolo (2014). "Enhanced Antibacterial Activity of Melt Processed Poly[propylene]Ag and Cu Nanocomposites by Argon Plasma Treatment." *Plasma Processes and Polymers* 11, 353-65.

[46] Martins, M. L., Ignazzi R., Eckert J., Watts., Ramon Kaneno,Willian F. Zambuzzi, Daemen L., Saeki M. J, and Bordallo H. N., (2016). "Restricted mobility of specific functional groups reduces anticancer drug activity in healthy cells." *Scientific Reports*. 6, 1-9.

[47] Raman M., Devi V, and Doble M., (2015) "Biocompatible ι-carrageenan-γ-maghemite nanocomposite for biomedical applications-synthesis, characterization and *in vitro* anticancer efficacy." *Journal of Nanobiotechnology* 13, 1-14.

[48] Khan M., Khan M., Marri A., Al-Warthan A., Alkhathlan H. Z., Siddiqui M. Z., Nayak M. L., Kamal A, and Adil S. F., (2016). "Apoptosis inducing ability of silver decorated highly reduced graphene oxide nanocomposites in A549 lung cancer." *International Journal of Nanomedicine* 11,873-883.

[49] Gurunathan S., Han J. W., Park J. H., Kim E., Kwon D. N, and Kim J. H., (2015). "Reduced graphene oxide-silver nanoparticle nanocomposite: a potential anticancer nanotherapy." *International Journal of Nanomedicine* 10, 6257-6276.

[50] Dorniani D., Saifullah B., Barahuie F., Hussein Z. B., Fakurazi S, and Lwyman L. J., (2016). "Graphene Oxide-Gallic Acid Nanodelivery System for Cancer Therapy." *Nanoscale Research Letters* 11, 1-9.

[51] Abdallah N. M., Noaman E., Eltahawy N .A., Badawi A. M., Kandil E., Mansour N. A, and Mohamed H. E., (2016). "Anticancer and Radiosensitization Efficacy of Nanocomposite *Withania somnifera* Extract in Mice Bearing Tumor Cells." *Asian Pac J Cancer Prev.* 17, 4367-4375.

[52] Namvar, F., Azizi, S., Rahman, H. S., Mohamad, R., Rasedee, A., Soltani, M, and Rahim, R. A. (2016). "Green synthesis, characterization, and anticancer activity of hyaluronan/zinc oxide nanocomposite." *Onco Targets and Therapy.* 9, 4549-4559.

[53] Cirillo, G., Vittorio, O., Hampel, S., Iemma, F., Parchi, P., Cecchini, M., Puoci, F, and Picci, N. (2013). "Quercetin nanocomposite as novel anticancer therapeutic: improved efficiency and reduced toxicity." *European Journal of Pharmaceutical Sciences* 49, 359-365.

[54] Chekhun, V. F., Yurchenko, O. V., Naleskina, L. A., Demash, D. V., Lukianova, N. Y, and Lozovska, Y. V., (2013). "*In vitro* modification of cisplatin cytotoxicity with magnetic fluid." *Experimental Oncolology.* 35, 15-19.

[55] Santos, D. P., Ruiz, M. A., Gallardo V., Zanoni M. V, and Arias J. L., (2011). "Multifunctional antitumor magnetite/chitosan-L-glutamic acid [core/shell] nanocomposites." *J. Nanopart Res.* 13, 4311-4323.

[56] Salahuddin, N., Elbarbary, A. A, and Alkabes H A., (2016). "Antibacterial and anticancer activity of loaded quinazolinone Polypyrrole/chitosan silver chloride nanocomposite." *International Journal of Polymeric Materials* 66, 307-316.

[57] Min, J., Moon, H., Yang, H. J., Shin, H., Hong, S. Y, and Kang S., (2014). "Development of P22 Viral Capsid Nanocomposites as Anti-Cancer Drug, Bortezomib [BTZ), Delivery Nanoplatforms." *Macromolecular Bioscience* 14, 557-564.

[58] Wang Peng, (2016). "Nanotechnology: A New Opportunity in Plant Sciences." *Trends in Plant Science.* 21, 699-712.

[59] Ghormade V (2011). "Perspectives for nano-biotechnology enabled protection and nutrition of plants." *Biotechnology Advances.* 29, 792-803.

[60] Cui, H.X., (2010). "Applications of nanotechnology in agrochemical formulation, perspectives, challenges and strategies." Sao Pedro, Brazil, (2010), 28-33.

[61] Yoel, S., (2007). "Nanosuspensions: emerging novel agrochemical formulations." *Springer, Berlin.* 1-39.

[62] Raliya, R., (2016). "Quantitative Understanding of Nanoparticle Uptake in Watermelon Plants." *Frontiers in Plant Science.* 7, 1-10.
[63] Tsuji, K., (2001). "Microencapsulation of pesticides and their improved handling safety." *Journal of microencapsulation.* 18, 137-147.
[64] Khandelwal, N., (2016). "Budding trends in integrated pest management using advanced micro- and nano-materials: Challenges and perspectives." *Journal of Environmental Management.* 184, 157-169.
[65] Lamsal, K., (2011). "Inhibition Effects of Silver Nanoparticles against Powdery Mildews on Cucumber and pumpkin." *Mycobiology.* 39, 26-32.
[66] Chakravarthy, A.K., "Bio efficacy of inorganic nanoparticles CdS, Nano-Ag and Nano-TiO_2 against *Spodoptera litura* [Fabricius] (Lepidoptera: Noctuidae)." *Current Biotica.* 6, 271-281.
[67] Esteban, T. L., "Antibacterial and antifungal activity of a soda-lime glass containing copper nanoparticles." *Nanotechnology.* 20, 1-6.
[68] Jun, H.P., (2006). "A new composition of nanosized silica-silver for control of various plant diseases." *The plant pathology journal.* 22, 295-302.
[69] He, L., (2011). "Antifungal activity of zinc oxide nanoparticles against *Botrytis cinerea* and *Penicillium expansum.*" *Microbiological Research.* 166, 207-215.
[70] Goswami, A., (2010). "Novel applications of solid and liquid formulations of nanoparticles against insect pests and pathogens." *Thin Solid Films.* 519, 1252-1257.
[71] Burman, U., (2013). "Effect of zinc oxide nanoparticles on growth and antioxidant system of chickpea seedlings." *Toxicological and Environmental Chemistry.* 95, 605-612.
[72] Sekhon, S., Bhupinder, (2014). "Nanotechnology in agri-food production: an overview." *Nanotechnology, Science and Applications* 7, 31-53.
[73] Khodakovskaya, Mariya, V., "Carbon Nanotubes Induce Growth Enhancement of Tobacco Cells." *ACS Nano.* 6, 2128-2135.
[74] Mondal, A., (2011). "Beneficial role of carbon nanotubes on mustard plant growth: an agricultural prospect". *Journal of Nanoparticle Research.* 13, 4519-4528.
[75] Shah, V. and Irina, B., (2009). "Influence of Metal Nanoparticles on the Soil Microbial Community and Germination of Lettuce Seeds." *Water Air and Soil Pollution.* 197, 143-148.
[76] Ngo, B., (2014). "Effects of nanocrystalline powders [Fe, Co and Cu] on the germination, growth, crop yield and product quality of soybean (Vietnamese species DT-51)." *Advances in Natural Sciences: Nanoscience and Nanotechnology.* 5, 1-7.

[77] Yang, F., (2007). "The Improvement of Spinach Growth by Nano-anatase TiO_2 Treatment Is Related to Nitrogen Photoreduction." *Biological Trace Element Research.* 119, 77-88.

[78] Ombódi, A. and Masahiko, S., (2000). "Broadcast application versus band application of polyolefin-coated fertilizer on green peppers grown on andisol." *Journal of Plant Nutrition* 23, 1485-1493.

[79] Rameshaiah, G.N., (2015) "Nano fertilizers and nano sensors - An attempt for developing smart agriculture." *International Journal of Engineering Research and General Science.* 3, 314-320.

In: Modified Biopolymers: Challenges and Opportunities ISBN: 978-1-53612-116-2
Editors: D. Pathania, G. Sharma and A. Kumar

Chapter 9

BIOBASED-NANOCOMPOSITES FOR FOOD PACKAGING APPLICATIONS

Shweta Sharma[1], Gaurav Sharma[1,*], Inamuddin[2,3,4]
Abeer Nasser Al-Romaizan[3] and Abdullah M. Asiri[3,4]
[1]School of Chemistry,
Shoolini University of Biotechnology and Management Sciences, Solan, India
[2]Advanced Functional Materials Laboratory, Department of Applied Chemistry,
Faculty of Engineering and Technology, Aligarh Muslim University, Aligarh, India
[3]Chemistry Department, Faculty of Science,
King Abdulaziz University, Jeddah, Saudi Arabia
[4] Centre of Excellence for Advanced Materials Research (CEAMR),
King Abdulaziz University, Jeddah, Saudi Arabia

ABSTRACT

The ever-increasing environmental problems due to plastic wastes have raised the need for the preparation of new packaging materials that are degradable or renewable in nature such that prepared from lipids, proteins, and polysaccharides, etc. But these biopolymers have low mechanical properties and hydrophilic nature so these cannot be used for practical applications. This chapter covers the importance of hybrid biopolymers for food packaging applications. Biobased- nanocomposites have enormous applications in the field of food packaging due to improved gas and water barrier properties, antimicrobial nature, high mechanical strength and thermal stability. Present chapter also focus on the importance of multilayer packaging and impact of biobased- nanocomposites on human health.

[*]Corresponding Author Email: gaurav8777@gmail.com.

Keywords: natural biopolymer, nanocomposite, films, active packaging, multi- layer packaging, anti- microbial packaging, shelf life

1. Introduction

Food packaging provide protection to food from different damages such as physical and chemical caused by air, light, micro-organisms, dirt and pressure, etc. Packaging materials provide physical security to the products that are essential for obtaining a satisfactory shelf life. The packaging system having high gas & water vapour barrier and mechanical properties helps in reducing the product deterioration due to physicochemical or biological factors and maintains the inclusive superiority during storage and handling [1]. After their valuable time period, it is necessary for the packaging materials to biodegrade within particular time without causing environmental problems. Different materials have been used for packaging such as paper, metal, plastic and glass, etc. Maximum use of petroleum based packaging materials was marked in the mid of twentieth century. It was due to their low cost, easy availability, high strength, high durability and excellent physicochemical properties, etc.

Inspite of their high positive aspects, they cause serious environmental deterioration due to their non-degradable nature [2, 3]. According to a survey, about 200 million tonnes of synthetic plastic is produced per year throughout the world and out of this 12 million tonnes are used only by the packing industries annually [4]. The use of plastic packaging materials is increasing the global concern and thus, motivating the researchers to pay more attention towards the preparation of renewable or biodegradable packaging materials. The development of biodegradable packaging materials specially prepared from biopolymers such as lipids, carbohydrates and proteins, etc. are perfect substitute for the conventionally used plastic packaging materials [5-9].

2. Biopolymers

Biopolymers are the naturally obtained polymers that are degradable in nature. Under particular conditions of temperature, moisture and oxygen content, their degradation takes place and the resultant products are non-toxic in nature. They actually have exceptional mechanical properties that make them fit for packaging applications [10-13]. They have the ability of acting as a barrier for the solute or gas particles and also helps in preserving the quality of the packed food. They have not only good barrier properties but also acts as an active carrier system for the controlled release of antimicrobial and antifungal agents integrated into their matrix for food preservation. Due to theirbiodegradable, renewable and abundant nature, they act as an alternative to the petroleum-based plastic materials. Different biopolymer-based films can be used as a base for these packaging materials [14]. These films are made up of different polymers, forming a thin layered structure. Classification of biopolymers is shown in Figure 1.

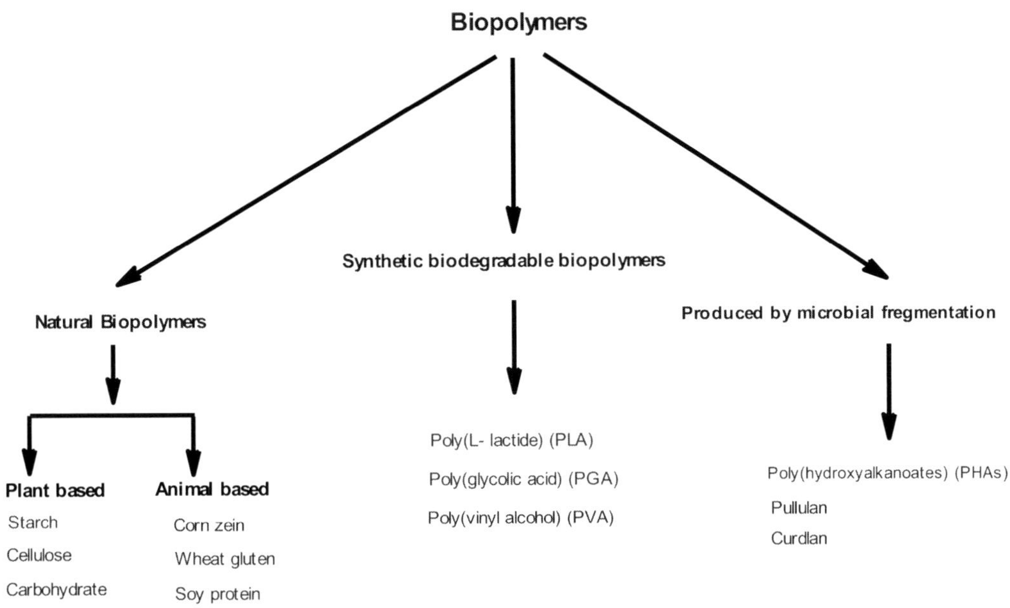

Figure 1. Classification of Biopolymers.

2.1. Polysaccharide Films

Polysaccharide films are also called carbohydrate-based films. Different polysaccharide films are formed by the breakage of polymer segments and remodelling these segments into films that are linked by hydrogen bonding and are hydrophilic in nature. Due to their low cost and abundant availability, these were the primarily used films for packaging applications [15-18]. Retrogradation reactions are the major reasons for the chemical spoilage of polysaccharide based films. Different types of polysaccharide-based films are used such as:

a) **Starch-based films:** Chief sources of starch include wheat, maize, rice and potato. The constituting units of starch are amylose and amylopectin. The film-forming tendency of starch is due tothe amylose segment. The structure of two segments, amylose and amylopectin has been shown in Figure 2. Its major properties include oil resistant ability, hydrophilicity and flexibility [19]. The degradation products of starch arenon- toxic in nature. One major property of starch is gelatinization that encompasses the swelling of starch granules and immediate loss of crystallinity.

The degree of gelatinization depends on numerous factors such as water activity, temperature and amount of shear, etc. [20]. It shows many properties similar to plastic packaging materials in the sense that both are odourless, semi- permeable, tasteless and have resistant ability towards oxygen [21].

b) **Alginate based films:** Alginates are usually obtained from seaweeds. These are used for film formation due to their distinctive property of colloidal formation which helps in thickening, stabilising, gel formation and film formation. Gelling agents such as Mg^{2+}, Ca^{2+}, Fe^{2+}, etc. help in the synthesis of alginate films. These films have high mechanical strength and poor water resistance due to hydrophilic nature [22].

c) **Pectin based films:** Pectin is a polysaccharide found in fruits and vegetables. It is mainly composed of β-1,4-linked D-galacturonic acid residues. It is an anionic electrolyte and its ionic interactions has excessive influence on film forming capability. These films do not work well at high moisture content [23].

2.1.1. Applications of Polysaccharide Films

- These represent a new class of materials that reduce the packaging related problems prevailing in the environment [24].
- They have the ability of reducing the gas transfer rate.
- These help in increasing the durability of the food products.

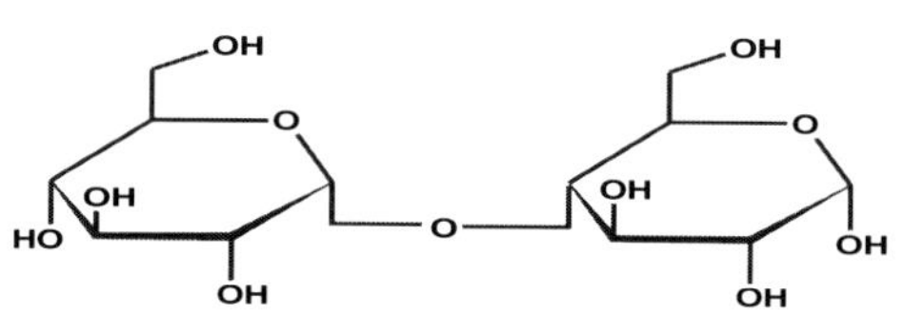

Structure of Amylose

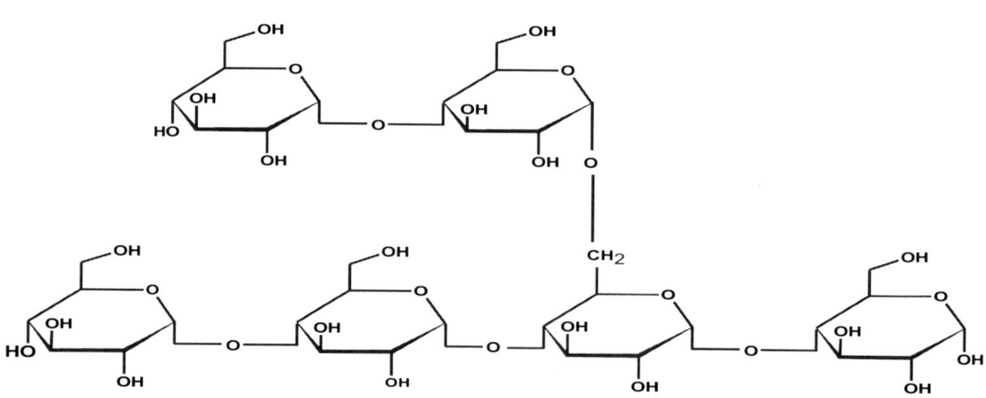

Structure of Amylopectin

Figure 2. Structure of Amylose and Amylopectin.

Low elasticity value and hydrophilic nature are major drawbacks of polysaccharide based films. Their hydrophilic nature does not obstruct the moisture and thus, degrade the packed food products very soon.

2.2. Protein Films

Protein films are composed of repeating units of various amino acids extracted from different animal and plant sources. High elasticity and cross- linking ability make them eligible for packaging applications [25]. Protein films are prepared from the solution of proteins when the solvent evaporates which is generally water or ethanol. Proteins in the solution are denatured by heat treatment or acids so as to form the extended structure prerequisite for film formation [26]. The extended protein structure get linked through covalent, hydrogen or ionic bonding and form the required protein films. The properties of these films are affected by the pH, temperature, relative humidity, additives and raw material, etc. Different types of protein films are used such as:

a) **Collagen films:** Collagen is the structural protein found mainly in bone, cartilage and hide, etc. Biocompatible, improved immunological properties and non- toxic nature makes it appropriate for use as packaging materials.
b) **Gelatin films:** Gelatin is mainly obtained by the controlled hydrolysis of collagen. It is mainly composed of glycine, proline and hydroxyproline. These films provide protection against moisture and light [27].
c) **Wheat gluten films:** Wheat gluten is a globular protein comprised mainly of polypeptide molecules. These films are quite hard and show high barrier ability towards oxygen, light and air contaminants. These are hydrophilic in nature so show poor water resistance properties [28].

2.2.1. Applications of Protein-Based Films

- They act as an alternative to the traditionally used synthetic plastic packaging materials.
- These help in improving the mechanical handling properties of the packaging materials.
- These even restrict the absorption of oxygen and thus reduce the chance of food degradation.
- Helps in reducing the lipid oxidation.

3. Modification of Biopolymer Films towards Better Properties

The required packaging material should possess enough stiffness, durability, gas and moisture barrier abilities, renewable and easily degradable nature. Films prepared from polysaccharides (e.g., starch and pectin) and proteins (e.g., collagen and gelatin) have much of these properties but do not have water resistance properties and absorb more moisture. As a result of this, their mechanical properties decreases which thus, limits their practical use [29, 30].

Researchers are trying to modify these biopolymers so as to reduce their water absorbing tendency and also to improve their mechanical properties. For the modifications of polysaccharide and protein-based films different methods have been employed such as chemical modification, salt addition, heat gelatinization, solvent changes, hydrolysis, addition of other proteins and polysaccharides, enzymatic treatment, coating, composite preparation and grafting, etc. [31-33]. They can be modified using physical methods also such as over- coating, ultrasound treatment, UV-radiation, annealing and addition of surfactants or emulsions, etc. [34, 35]. For example, researchers worked on increasing the mechanical properties and decreasing water absorbing tendency of starch by modifying it by amalgamation with synthetic or natural polymers and by crosslinking.

Although, the above-reported methods have made many significant changes but the results are still not satisfactory. However, in context to nanotechnology, researchers suggested that the use of biopolymer based nanocomposites can be a best substitute for the preparation of packaging materials.

3.1. Biopolymer Based Nanocomposites

It has been recommended that the combination of biopolymers and nanoparticles offers an alternate way to overcome the major problems faced by packaging industries. Nanocomposites actually demonstrate 'averaged' properties of the constituting materials. They show extraordinary properties such as improved barrier properties, high mechanical strength and enhanced chemical &heat resistance, etc. Nanocomposites used in food packaging increases the shelf- life of the product, helps in easy transportation and have ability to endure unfavourable environmental conditions [36].

Researchers have performed a number of experiments to prepare the required packaging materials. In the same context, Vartiainen in 2011, prepared chitosan – nanoclay based nanocomposite films which he cross-linked with glyoxal, glutaraldehyde and genipin. The resultant nanocomposite was quite fruitful because it showed high water barrier properties. The oxygen transmission ability at high humidity value was also quite low [37]. Another was performed by Saarikoski et al., in which they tried to improve the properties and prepare a consistent blend by enzymatically treating cellulose with mixed NaOH/ZnO solution and poly(ethylene-co-acrylic acid) or poly(acrylamide-co-acrylic acid) polymers. Cellulose/poly(ethylene-co-acrylic acid) based films showed high oxygen and water vapour barrier properties [38]. Österberg et al., tried to improve the properties of the CNF films in high humidity conditions by using paraffin wax coating. They dipped the films into melted wax and the excess wax was removed. The treated films showed improved water vapour and oxygen barrier abilities [39]. Wang et al., (2005) prepared a chitosan-based nanocomposite, chitosan/MMT using solvent intercalation method. They found that the nanocomposite was formed at low MMT concentration with high thermal stability and increased hardness.

Chitosan is a biodegradable and renewable polymer having antimicrobial properties. Inspite of great application diversity, its hydrophilic nature decreases its applicability. Researchers tried to combine chitosan with other polymers such as poly(3-hydroxybutyric acid) (PHB) [40], polycaprolactone (PCL)[41, 42] and poly(lactic acid) (PLA) [43] so as to modify its properties according to desired applications.

Table 1. Water Vapor Permeability of different biopolymer films

Different biopolymer films	Water Vapor Permeability (g×μm/m²/day)	Reference
Pectin	113,000	[50]
Pectin+nanoclay	87,050	[50]
Mung bean protein films	-	[51]
Chitosan	21,600	[52]
Chitosan +nanoclay	18,000	[52]
Collagen	-	[51]
Collagen + Carboxamide	-	[51]
Crosslinked Chitosan	11,000	[53]
Crosslinked Chitosan + nanoclay	7200	[53]
Corn zein films + fatty acids	-	[51]
CNF bleached softwood	20,000	[54]
Cellulose triacetate	-	[51]
Methylcellulose	-	[51]
PVA-CNF	31,500	[55]
CNF bleach birch pulp (heat treated)	15,000	[56]
Hydroxyethylcellulose	-	[51]

Table 2. Oxygen penetrability of different biopolymer films

Different biopolymer films	Oxygen penetrability	Reference
Pectin	1,000	[50]
Pectin + nanoclay	600	[50]
PLA	-	[51]
Carboxymethylcellulose	1400	[50]
Hydroxypropyl methylcellulose	-	[51]
Cellulose acetate butyrate	-	[51]
Chitosan	9800	[52]
Crosslinked Chitosan	980	[53]
Hydroxyethyl methylcellulose	-	[51]
PVA-CNF	0.5	[55]
PVA-CNF + 50% nanoclay	0.5	[55]
CNF bleach birch pulp	<1	[56]

It has been noted that mechanical properties of the nanocomposites changes by varying the filler amount. Huang and Yu prepared starch/MMT nanocomposite with varying filler concentrations between 0-11% to the starch. It was observed that tensile strength was increased to marked level with increase in filler concentration upto 8% and then decreased with further increase [44]. Huang et al., measured the change in tensile strength of corn starch/ MMT nanocomposite on addition of 5% clay. They found that the tensile strength increases to about 450% [45].

Rhim et al., conveyed that Cloisite 30B was quite effective in reducing the water vapour permeability (WVP) of chitosan-based nanocomposites. They also reported that WVP of agar/clay nanocomposite films decreases as the clay concentration increases from 0 – 20% [46, 47]. Some other experiments were performed in which the cellulose whiskers were filled in starch. The filling was done so as to decrease the water sensitivity. It was noted that their addition offered a decrease in WVP value from 2.66 to 1.67 g.mm/kPa.h.m^2 [48]. Table 1 summarises the water vapor permeability of different biopolymer films and Table 2 reviews the oxygen permeability of biopolymer films at different reaction conditions [49].

3.1.1. Properties of Bio- Nanocomposite Films

3.1.1.1. Antimicrobial ability

Micro- organisms such as bacteria, viruses and fungi spoils the food very rapidly. Products such as vegetables, meats, milk and bakery products are mainly prone to their attack. Table 3 gives a short survey of critical storage variables of some major food classes [57-59].

Table 3.Critical storage variables of some major food classes [57-59]

Food Classes	Critical storage variables
Milk products	Oxygen, temperature
Ice cream	Temperature (freeze/thaw), oxygen
Fresh poultry	Temperature, oxygen
Seafood	Temperature, oxygen
Fruit	Temperature, light, humidity, oxygen, handling
Vegetables	Temperature, light, oxygen, humidity
Bread	Humidity, temperature, oxygen
Cornflakes	Humidity, temperature, oxidation, handling
Soft bakery products	Humidity, temperature, oxygen
Oily products	Humidity, temperature, oxygen, handling, light
Chocolate	Humidity, temperature, oxidation
Sweets	Temperature, humidity
Caffeine products	Oxygen, light, moisture
Frozen meats	Oxygen, temperature, humidity

The growth of micro- organisms is greatly affected by pH, temperature, nutrient content and light, etc. Temperature is considered as one of the major factor for microbial growth and death. Depending upon the amount of bearable temperature [60-62], micro-organisms are categorised into three types (i) Mesophiles (20°C – 45°C) (ii) Psychrotrophs (20°C – 30°C) (iii) Thermophiles (55°C – 65°C). As the temperature increases above 65°C, micro- organisms start to die. In addition to temperature, pH level also affects the micro-organisms. Microbial growth is best recorded at neutral pH (7.0). At low pH > 3.5, only certain lactic acid bacteria and yeasts can grow [63]. Micro- organisms require carbon, water, nitrogen, certain minerals and vitamins for their growth which they get from the food. For growth, bacteria require maximum nutrients and moulds require minimum. Yeasts have the requirement between bacteria and moulds [64].

In food packaging industries, one of the most important properties is the ability of material to resist the attack of microorganisms so to protect the packed food. New materials have been tested by the researchers that can form a protective layer around the food product and resist the attack of micro- organisms.

Nanocomposite antimicrobial systems are quite effective due to large surface area. The presence of nano- sized antimicrobial agents help to prevent the attack of micro- organisms. Different antimicrobial systems have been tested such as metal ions, metal oxides, natural biopolymers and synthetic antimicrobial agents [65-67]. Metal ion nanoparticles such as silver, copper and zinc, etc. can be used as antimicrobial agents against a number of micro-organisms such as bacteria, yeast and fungi, etc. The tendency of the antimicrobial activity depends upon a number of factors such as particle size, particle size distribution, metal ion concentration and interaction between the metal ion and the surface, etc. So for complete sightsee, metal ions should be completely and properly dispersed on the surface otherwise, their antimicrobial properties gets affected to a larger extent [68].

Silver nanoparticles are one of the most studied nanoparticles due to their non- toxic nature and ability to show wide range of properties such as catalytic, electrical, thermal, optical, and mechanical. Silver nanoparticles can be used for packaging applications due to their antimicrobial nature. Silver ions actually interact with the negatively charged disulphide and sulfhydryl groups present in enzymes and other nucleic acids. This interaction leads to the structural changes and distortion of the micro- organism's cell wall and membrane which ultimately leads to cell death [69, 70]. Increased surface area and potential to release Ag^+ ions increases the antimicrobial activity of the silver based nanocomposites.

Recent inventions showed that the silver ions can be integrated into a number of materials such as zeolites, silicates and nanoclay which increase their antimicrobial activity. Silver ions incorporated with zeolites by replacing the Na^+ ions of zeolites with Ag^+ makes them effective against a number of microorganisms such as bacteria, yeast and mycelium fungi. The only drawback of this composite is that it is effective only against vegetative cells. However, when Ag/zeolite was loaded on chitosan, the resulting composite was effective against gram positive and gram negative bacteria [71]. Intense microbial activity was observed against *Pseudomonas* species when (Ag + MMT) film was prepared by replacing the Na^+ ions with Ag^+ ions.

Researchers have suggested the use of carbohydrates such as sucrose and waxy corn starch to fix the antimicrobial property of silver nanoparticles. Sucrose and other carbohydrates act as a reducing and stabilising agent for the silver nanoparticles and increases their antimicrobial nature. Metal ions such as Ag^+, Zn^{2+} and Cu^{2+} imprinted on chitosan acts as a strong antimicrobial agents especially against *E.coli*, *Salmonella choleraesuis* and *Staphylococcus aureus*[72]. These nanocomposite films have intense potential as antimicrobial agents and therefore, can be used for food packaging applications.

In comparison to silver ions, copper is not that much effective but it can abolish micro-organisms and viruses. Copper ions do not accumulate in the animal cells and are thus, not harmful [73]. Researchers have tried to load copper ions on the polymers and studied their antifungal and antibacterial properties. The results were quite impressive. But due to toxic nature and high catalytic nature of copper ions, these are generally not used for food packaging applications.

TiO_2 is among the most studied metal oxides due to non- toxic, inert, eco- friendly and low-cost nature. TiO_2 based films can be prepared using different techniques such as sol-gel, chemical vapour deposition, laser ablation, chemical spray pyrolysis, evaporation and magnetron sputtering, etc. It is an effective antimicrobial agent against a number of micro-organisms [74].To further increase its application field, researchers load TiO_2 on various materials including lipids, carbohydrates and proteins, etc. To increase its catalytic properties, it is doped with Ag^{+}, Fe^{3+} and Zn^{2+} etc. TiO_2 nanoparticles, loaded on polymer matrix can be used for packaging applications because the resulting nanocomposite provides a shield against micro- organisms, radiation, water vapors and light, etc. [75, 76]. Zhou etal., load TiO_2 nanoparticles on the biopolymer and found that the results were in good agreement with the expectations.

Various experiments proved that the silver nanoparticles and silver loaded on chitosan or starch have excellent properties with high antimicrobial ability. They also protect the food from its spoilage and increase itsshelf life.

3.1.1.2. Oxygen Inhibitors

In addition to antimicrobial, the material should also possess oxygen barrier properties. Addition of oxygen to food results in its browning and sour taste. High oxygen barrier do not allow its passage into the food and maintains low O_2 level in the packing which protect it from deterioration [77].

Xiao-e et al., tried to prepare a nanocomposite packaging material by the combination of TiO_2 and different polymers. The prepared composite could be used for the packaging of food products which are oxygen sensitive. However, TiO_2 nanoparticles work only under UV light which is one of its major drawbacks.

4. Food Packaging Applications

The use of improved packaging materials that can minimise the deterioration of the food has always been the main emphasis of the packaging industries. Increasing population has also increased the demand for the fresh and best quality food products and thus, more efforts have been made by the researchers to meet the ever increasing demands for safer food packaging. Active packaging is a way out solution for the problems faced by the packaging industries. Active packaging offers the ability to increase the shelf life of the products and also to change the sensory properties of the packaging material without harming the packed food. Biopolymer based nanocomposites offer a number improved properties such as:

a) Thermal stability
b) Mechanical strength
c) Light and gas barrier
d) Dimensional stability
e) Heat resistance
f) Antimicrobial and antifungal properties

One of the major problems faced by the packed products is their spoilage due to the attack of micro- organisms. Antimicrobial packaging has offered us a pioneering way to reduce the attack of micro- organisms. They actually forms a protective covering around the food and restrict the contact of micro- organisms with the food. The active constituent of these films can be either organic or inorganic. As already discussed, silver nanoparticles have attractive antimicrobial properties [78]. These materials are attracting huge attention because these offer an extended and quality life to the food packed such as meat, cheese, fruits and vegetables. These packaging films are generally made from biopolymer based nanocomposites which gave us the power to reduce the use of synthetic plastic packaging materials [79]. These films need less material requirement which leads to overall reduced weight, low cost and reduced wastage of the material.

Multilayer food packaging is also used so as to provide better barrier properties. Nanocomposites based on polymer/clay have been substantially explored due to their barrier properties. Multilayer packaging is required for products that are quite sensitive to the external environment such as beer bottles and carbonated beverages.

The presence of oxygen in the packed food leads to many corrosive reactions such as nutrient loss, colour change and flavour change, etc. In order to prevent the interaction of oxygen with the food, many oxygen inhibitors are used. TiO_2 based packaging systems have been used for oxygen inhibition so to protect food. In addition to oxygen inhibitors, multilayer packaging or nanocomposites are also useful in this context.

Researchers are paying more stress on preparing smart packaging. These packaging materials will be able to respond to the adverse external environmental conditions and also provide information about the shelf life. These smart packaging materials are also able to aware the consumer about the nutrition level and food quality. To prepare them,non- toxic nanoparticles of different metals are required [80].

Multi- functional materials have the ability to integrate antioxidants, antimicrobial agents and antifungal agents, etc. into their matrix so as to form new packaging materials with desired properties. These multi- functional materials can be used for a number of applications such as self- cleaning, nanosensors and freshness indicator, etc.

5. Impression on Human Health

Although the beneficial aspects of the nanocomposites are well known but their toxicity is also raising a lot of questions in context to mankind. The extremely small size of the nanoparticles makes them more toxic and reactive. Nanoparticles can enter into the human body through dermal contact and inhalation from the food. It is well-known that bulk ZnO materials are non-toxic in nature but their nanoparticles have been tested to be toxic in nature. These particles can cause pulmonary problems to human body.

Researchers have found that the nanoparticles can easily cross the cellular level and leads to the oxidative damage of the cell [81]. They have noted that TiO_2 and silver nanoparticles enter in the body through gastro- intestinal tract. The impact of these particles depends upon their size, size distribution and on the state in which they enter in the body. These nanoparticles increase the blood circulation drastically when the particles are water loving in nature and their surface is positively charged. Thus, Biobased- nanocomposite packaging materials are better as compared to nanoparticle and biopolymer based packaging materials.

CONCLUSION

In the present chapter, we have tried to compile the developments occurred in the food packaging. Most commonly used biopolymers for the nanocomposite preparation are starch, cellulose, gelatin, etc. Biobased nanocomposite materials offer improved barrier properties, mechanical strength, thermal stability and heat resistance abilities. Biodegradable and renewable nature increases their utility compared to petroleum-based plastic materials. Their antimicrobial properties can be increased by the use of silver nanoparticles, TiO_2 and copper nanoparticles, etc. Utilisation of bio- based nanocomposites offer the less production of waste materials and also environment protection. In addition, they also help in increasing the shelf- life and in easy transportation of food. Now, emphasis is laid on the preparation on smart packaging materials that could not only help in protection of food but can also act as an alarm about the nutritive value of the packed food.

REFERENCES

[1] Echegoyen, Y., (2015).Nano-developments for Food Packaging and Labeling Applications.*Nanotech in Food Agri.* 141-166.

[2] Brody, A., (2003).Nanocomposite technology in food packaging.*Food Technol.* 61, 80-83.

[3] Brody, A. I., Bugusu, B., Han, J. H., Sand, C.K., and Mchugh, T.H., (2008).Innovative Food Packaging Solutions.*J.of Food sci.* 73, R107-R116.

[4] Zhang, L., Ruan, D. and Gao, S., (2002). Dissolution and regeneration of cellulose in NaOH/thiourea aqueous solution.*J. Polym.Sci.: Part B: Polym. Phys.* 40, 1521–1529.

[5] Zhao, R., Torley, P. and Halley, P.J., (2008).Emerging biodegradable materials: starch- and protein based bio-nanocomposites.*J Mater.Sci.* 43,3058-3071.

[6] Marsh, K. and Bugusu, B., (2007).Food packaging – roles, materials, and environmental issues.*J Food Sci* 72, R38-R55.

[7] Cuq, B., Gontard, N. and Guilbert, S., (1998).Proteins as agricultural polymers for packaging production. *Cereal Chem.* 75, 1–9.

[8] Bourtoom, T., (2008).Edible films and coatings: characteristics and properties.*International Food Res. J.* 15, 237-248.

[9] Kester, J.J. and Fennema, O.R., (1986).Edible films and coatings: a review. *Food Technol.* 40, 47–59.

[10] Wong, D.W., Camirand, W.M. and Pavlath, A.E., (1994). Development of edible coatings for minimally processed fruits and vegetables. *Edi.Coat.and films imp. food qua.*65-88.

[11] Modi, J. and Sunny, (2010). Assessing the Feasibility of Poly-(3-hydroxybutyrate-co-3-hydroxyvalerate)(PHBV) and Poly-(lactic acid) for Potential Food Packaging Applications. PhD diss., The Ohio State University.

[12] Imran, H., Revol-Junelles, A.M., Martyn, A., Tehrany, E.A., Jacquot, M., Linder, M. and Desobry, S., (2010).Active food packaging evolution: transformation from Micro- to nanotechnology, *Crit Rev Food SciNutr.* 50, 799-821.

[13] Han, J.H., (2000).Antimicrobial food packaging, *Food Technol.* 54, 56–65.

[14] Baldwin, E.A., (1994).Edible coatings for fresh fruits and vegetables: past, present and future. *Edi.Coat. Films Imp.Food Qua.* 1, 25–64.

[15] Krochta, J.M., and Mulder-Johnston, C.D., (1997).Edible and biodegradable polymer films: Challenges and opportunities. *Food Technol.* 51, 61–74.

[16] Pathania, D., Gupta, D., Kothiyal, N.C., Sharma, G., Eldesoky, G.E. and Naushad, M., (2016). Preparation of a novel chitosan-g-poly(acrylamide)/Zn nanocomposite hydrogel and its applications for controlled drug delivery of ofloxacin. *Inter. J Bio.Macro.* 84, 340-348.

[17] Han, J.H. and Gennadios, A., (2005).Edible films and coatings: a review.*Elsevier Sci. Tech. Books.*

[18] Zobel, H.F. and Stephen, A.M., (1995). Starch: structure, analysis, and application. *Food sci tech* 19-19.

[19] Skurtys, O., Acevedo, C., Pedreschi, F., Enrione, J., Osorio, F. and Aguilera, J.M., (2010).Food hydrocolloid edible films and coatings.*Food Hydrocolloids: Characteristics, Properties, Nova Science Publishers, Inc.*

[20] Kapetanakou, A.E., Manios, S.E. and Skandamis, P.N., (2014).Application of edible films and coatings in foods.*Novel FoodPreservation and Microbial Assessment Techniques* 14, 237-73.

[21] Freire, C., Podczeck, F., VeigaF. andSousa, J., (2009) Starch-based coatings for colon-specific delivery. Part II: Physicochemical properties and in vitro drug release from high amylose maize starch films. *Eur. J Pharm.Biopharm.*72, 587-94.

[22] Borchard, W., Kenning, A., Kapp, A. and Mayer, C., (2005).Phase diagram of the system sodium alginate/water: A model for biofilms.*Inter. J bio.Macro.* 35, 247-56.

[23] Baldwin, E.A.,Nisperos-Carriedo, M.O. and Baker, R.A., (1995).Use of edible coatings to preserve quality of lightly (and slightly) processed products. *Cri.Reviews Food Sci. Nutri.*35, 509-24.

[24] Vargas, M., Pastor, C., Chiralt, A., McClements, D.J. and Gonzalez-Martinez, C., (2008).Recent advances in edible coatings for fresh and minimally processed fruits, *Cri. reviews food sci. nutria.* 48, 496-511.

[25] Cuq, B., Gontard, N. and Guilbert, S., (1998). Proteins as agricultural polymers for packaging production, *Cereal chem.* 75, 1-9.

[26] Hernández-Muñoz, P., Kanavouras, A., Ng, P.K. and Gavara, R., (2003).Development and characterization of biodegradable films made from wheat gluten protein fractions, *J Agri. Food Chem.* 51, 7647-54.

[27] Marquie, C., Aymard, C., Cuq, J.L. and Guilbert, S., (1995). Biodegradable packaging made from cottonseed flour: formation and improvement by chemical treatments with gossypol, formaldehyde, and glutaraldehyde, *J AgriFoodChem.* 43, 2762-2767.

[28] McHugh, T.H. and Krochta, J.M., (1994).Water vapor permeability properties of edible whey protein-lipid emulsion films. *J American Oil Chemists' Society* 71, 307-12.

[29] Debeaufort, F., Quezada-Gallo, J.A. and Voilley, A., (1998) Edible films and coatings: tomorrow's packagings: a review, *Cri.Reviews in Food Sci.* 38, 299-313.

[30] Liu, L., Kerry, J.F. and Kerry, J.P., (2005).Selection of optimum extrusion technology parameters in the manufacture of edible/biodegradable packaging films derived from food-based polymers.*J. Food Agri. Environ.* 3, 51.

[31] Were, L., Hettiarachchy, N.S. and Coleman, M., (1999).Properties of cysteine-added soy protein-wheat gluten films, *J food sci.* 64, 514-8.

[32] Yildirim, M. and Hettiarachchy, N.S., (1997).Biopolymers produced by crosslinking soybean 11S globulin with whey proteins using transglutaminase.*J. Food Sci.* 62, 270–275.

[33] Juvonen, H., Smolander, M., Boer, H., Pere, J., Buchert, J. and Peltonen, J., (2011) Film Formation and Surface Properties of Enzymatically Crosslinked Casein Films.*Journal of Applied Polymer Science* 119, 2205-2213.

[34] Micard, V., Belamri, R., Morel, M.H. and Guilbert, S., (2000).Properties of chemically and physically treated wheat gluten films.*J Agri. Food Chem.* 48, 2948-53.

[35] Pathania, D., Gupta, D., Al-Muhtaseb, A.H., Sharma, G., Kumar, A., Naushad, M., Ahamad, T. and Alshehri, S.M., (2016). Photocatalytic degradation of highly toxic dyes using chitosan-g-poly(acrylamide)/ZnS in presence of solar irradiation. *J. Photochem. Photobio. A: Chem.*329, 61-68.

[36] Giannelis, and Emmanuel P., (1998). “Polymer-layered silicate nanocomposites: Synthesis, properties and applications.”*AppliedOrganometallic Chem.* 12, 675-680.
[37] Vartiainen, J. and Harlin, A., (2011).Crosslinking as an Efficient Tool for Decreasing Moisture Sensitivity of BiobasedNanocomposite Films.*Mat. Sci. Applica.* 2, 346-354.
[38] Saarikoski, E., Rautkoski, H., Rissanen, M., Hartman, J. and Seppälä, J., (2014).Cellulose/Acrylic Acid Copolymer Blends for Films and Coating Applications.*J. Applied Polymer Sci.* 131, 40286-40295.
[39] Österberg, M., Vartiainen, J., Lucenius, J., Hippi, U., Seppälä, J., Serimaa, R. and Laine, J., (2013).A Fast Method to Produce Strong NFC Films as a Platform for Barrier and Functional Materials. *American Chemical Society. ACS Applied Materials andInterfaces* 5, 4640-4647.
[40] Ikejima, T. and Inoue, Y., (2000).Crystallization behavior and environmental biodegradability of the blend films of poly(3-hydroxybutyric acid) with chitin and chitosan, *Carbohyd. Polym.* 41, 351–356.
[41] Olabarrieta, I., Forsström, D., Gedde, U.W. and Hedenqvist, M.S., (2001).Transport properties of chitosan and whey blended with poly (ε-caprolactone) assessed by standard permeability measurements and microcalorimetry.*Polymer* 42, 4401-8.
[42] Senda, T., He, Y. and Inoue, Y., (2002) Biodegradable blends of poly (ε-caprolactone) with α-chitin and chitosan: specific interactions, thermal properties and crystallization behavior. *Polym international* 51, 33-9.
[43] Suyatma, N.E., Copinet, A., Tighzert, L. and Coma, V., (2004).Mechanical and barrier properties of biodegradable films made from chitosan and poly (lactic acid) blends, *J PolymEnv.* 12, 1-6.
[44] Huang, M. and Yu, J., (2006).Structure and properties of thermoplastic corn starch/montmorillonite biodegradable composites. *J ApplPolymSci* 99, 170-176.
[45] Huang, M., Yu, J. and Ma, X., (2005).High mechanical performance MMT-urea and formamide - plasticized thermoplastic cornstarch biodegradable nanocomposites. *CarbohydrPolym*, 62, 1-7.
[46] Rhim, J.W., (2011).Effect of clay contents on mechanical and water vapor barrier properties of agar-based nanocomposite films.*CarbohydrPolym*, 86, 691-699.
[47] Rhim, J.W., Lee, S.B. and Hong, S.I., (2011).Preparation and characterization of agar/clay nanocomposite films: The effect of clay type, *J Food Sci.* 76, N40-N48.
[48] Dufresne, A., Dupeyre, D. and Vignon, M.R., (2000).Cellulose microfibrils from potato tuber cells: Processing and characterization of starch–cellulose microfibrilcomposites.*JApplPolym Sci.* 76, 2080-2092.
[49] Vartiainen, J., Vähä-Nissi, M. and Harlin, A., (2014).Biopolymer Films and Coatings in Packaging Applications—A Review of Recent Developments.*Mat. Sci. Appl.* 5, 708-718.

[50] Vartiainen, J., Tammelin, T., Pere, T.J., Tapper, U. and Harlin, A., (2010).Biohybrid Barrier Films from Fluidized Pectin and Nanoclay.*Carbohy.Polym*, 82, 989-996.

[51] Malhotra,B., Keshwani, A. and Kharkwal, H., (2015). Natural polymer based cling films for food packaging. *Int J Pharm PharmSci*, Vol 7, Issue 4, 10-18.

[52] Vartiainen, J., Tuominen, M. and Nättinen, K., (2010).Bio-Hybrid Nanocomposite Coatings from Sonicated Chitosan and Nanoclay.*J Applied Polym Sci.* 116, 3638-3647.

[53] Vartiainen, J. and Harlin, A., (2011).Crosslinking as an Efficient Tool for Decreasing Moisture Sensitivity of BiobasedNanocomposite Films.*Materials Sci. Appl.* 2, 346-354.

[54] Naushad, M., Ahamad, T., Sharma, G., Al-Muhtaseb, A.H., Albadarin, A.B., Alam, M.M., ALOthman, Z.A., Alshehri, S.M. and Ghfar, A.A., (2016). Synthesis and characterization of a new starch/SnO_2nanocomposite for efficient adsorption of toxic Hg^{2+} metal ion. *Chem. Eng. J.* 300, 306-316.

[55] Spoljaric, S., Salminen, A., Luong, N., Lahtinen, P., Vartiainen, J., Tammelin, T. and Seppälä, J., (2013).Nanofibrillated Cellulose, Poly(Vinyl Alcohol), Montmorillonite Clay Hybrid Nanocomposites with Superior Barrier and ThermomechanicalProperties.*Polym Composites* 35, 1117-1131.

[56] Vartiainen, J., Kaljunen, T., Kunnari, V., Lahtinen, P., Salminen, A., Seppälä, J. and Tammelin, T., (2013).NanocelluloseFilms: Towards Large Scale and Continuous Production.*26th IAPRI Symposium on Packaging* 197-209.

[57] Virtanen, S., Vartiainen, J., Setälä, H., Tammelin, T. and Vuoti, S., (2014).Modified Nanofibrillated Cellulose-Polyvinyl Alcohol Films with Improved Mechanical Performance.*The Royal Society of Chemistry, RSC Advances* 4, 11343- 11350.

[58] Kilcast, D. and Subramaniam, P., (2000).The stability and shelf-life of food.*Cambridge: CRC press.*

[59] Labuza, T.P. and Szybist, L.M., (2008).Open dating of foods, *John Wiley & Sons.*

[60] Steele, R., (2004).Understanding and measuring the shelf-life of food, *Woodhead Publishing.*

[61] Baird-Parker, A.E., Boothroyd, M. and Jones, E., (1970). The effect of water activity on the heat resistance of heat sensitive and heat resistant strains of salmonellae. *J. Appl. Bacteriol.* 33, 515-522.

[62] Beaman, T.E. and Gerhardt, P., (1986).Heat resistance of bacterial spores correlated with protoplast dehydration, mineralization, and thermal adaptation. *Appl. Environ. Microbiol.* 52, 1242-1246.

[63] Brock, T.D., (1967).Life at high temperatures. *Science* 158, 1012-1019.

[64] Jay, J.M., (2008).High-Temperature Food Preservation and Characteristics of Thermophilic Microorganisms. *Modern FoodMicrobio.*341-362.

[65] Rahaie, S., Gharibzahedi, S.M.T., Razavi, S.H. and Jafari, S.M., (2014).Recent developments on new formulations based on nutrient-dense ingredients for the production of healthy-functional bread: a review. *J Food Sci Technol.* 51, 2896–2906.

[66] Damm, C., Münsted, H. and Rösch, A., (2008).The antimicrobial efficacy of polyamide 6/silver nano- and microcomposites. *Mater Chem Phys.*108, 61-66.

[67] Carlson, C., Hussain, S.M., Schrand, M., Braydich-Stolle, L.K., Hess, K.L., Jones, R.L. and Schlager, J.J., (2008).Unique cellular interaction of silver nanoparticles: Size-dependent generation of reactive oxygen species, *J PhysChem*: B 112, 13608-13619.

[68] Dallas, P., Sharma, V.K. and Zboril, R., (2011).Silver polymeric nanocomposites as advanced antimicrobial agents: Classification, synthetic paths, applications, and perspectives. *Adv Colloid Interface Sci.* 166, 119-135.

[69] Russell, A.D. and Hugo, W.B., (1994).Antimicrobial activity and action of silver, *Prog Med Chem.* 31, 351-370.

[70] Feng, Q.L., Wu, J., Chen, G.Q., Cui, F.G., Kim, T.N. and Kim, J.O., (2000).A mechanistic study of the antibacterial effect of silver ions on Escherichia coli and Staaphylococcusaureus. *J Biomed Mater Res.* 52, 662-668.

[71] Sondi, I. and Salopek-Sondi, B., (2004).Silver nanoparticles as antimicrobial agent: a case study of E.coli as a model for Gram-negative bacteria. *J Colloid Inteface Sci.* 275, 177- 182.

[72] Fernández, A., Soriano, E., Hernández-Munoz, P. and Gavara, R., (2010).Migration of antimicrobial silver from composite of polylactide with silver zeolite. *J Food Sci.* 75, E186- E193.

[73] Du, W.L., Niu, S.S., Xu, Y.L., Xu, Z.R. and Fan, C.L., (2009).Antibacterial activity of chitosan tripolyphosphate nanoparticles loaded with various metal ions. *CarbohydrPolym.* 75, 385-389.

[74] Chandra, I. and Rustgi, R., (1998). Biodegradable polymers. *ProgPolym Sci.*23, 1273-1335.

[75] Anpo, M., Kishiguchi, S., Ichihashi, Y., Takeuchi, M., Yamashita, H., Ikeue, K., Morin, B., Davidson, A. and Che, M., (2001).The design and development of second-generation titanium oxide photocatalysts able to operate under visible light irradiation by applying a metal ion-implantation method.*Research on Chemical Intermediates*27, 459-67.

[76] Kumar, T.S., Rahuman, A.A., Jayaseelan, C., Rajakumar, G., Marimuthu, S., Kirthi, A.V., Velayutham, K., Thomas, J., Venkatesan, J. and Kim, S.K., (2014).Green synthesis of titanium dioxide nanoparticles using Psidiumguajava extract and its antibacterial and antioxidant properties. *Asian Pacific J tropical medicine* 7, 968-76.

[77] Choi, W., Termin, A. and Hoffmann, M.R., (1994).The role of metal ion dopants in quantum-sized TiO2: correlation between photoreactivity and charge carrier recombination dynamics.*The J Physical Chem.* 98, 13669-79.

[78] Mills, A., Doyle, G., Peiro, A.M. and Durrant, J., (2006).Demonstration of a novel, flexible, photocatalytic oxygen-scavenging polymer film. *JPhotochemPhotobio. A: Chemistry* 177, 328-31.

[79] Hatzigrigoriou, N.B. and Papaspyrides, C.D., (2011).Nanotechnology in plastic food-contact materials. *J Applied Polym. Sci.* 122, 3719-38.

[80] Kerry, J.P., O'Grady, M.N. and Hogan, S.A., (2006).Past, Current and potential utilization of active and intelligent packaging systems for meat and muscle-based products: A review.*Meat Sci.* 74, 113-130.

[81] Carneiro, J.O., Teixeira, V., Carvalho, P., Azevedo, S. and Manninen, N., (2011).Self-cleaning smart nanocoatings. *Nanocoatings and Ultra-Thin Films: Techn. Appli.* 14, 397-413.

In: Modified Biopolymers: Challenges and Opportunities ISBN: 978-1-53612-116-2
Editors: D. Pathania, G. Sharma and A. Kumar

Chapter 10

NATURAL FIBRE REINFORCED BIODEGRADABLE COMPOSITE MATERIALS

***V. K. Gupta*[1] *and Bhanu Priya*[2,*]**

[1]Distinguished Professor, Sarchi Chair, Applied Chemistry Department
University of Johannesburg, Doornfontein Campus,
Corner Beit and Nind Street John Orr Building, Doornfontein,South Africa
[2]Faculty of Science and Technology, The ICFAI University,
Baddi, Solan, Himachal Pradesh, India

ABSTRACT

Natural biomass such as natural fibres, have been utilized by humans for many centuries for household or other conventional applications. The prime interest in the use of natural fibres results from its biodegradability, low cost, easy availability, low density, nonabrasive nature, recyclability, no health hazards, and high specific properties. Moreover, natural fibres exhibit good thermal and acoustic insulation. Green composites, consisting of an association of a biodegradable polymer matrix and a natural fibre as reinforcement have been of intense interest due to both increasing social and economic pressure to conserve petroleum resource.

Green composites offer environmental benefits such as biodegradability, less greenhouse gas emissions, and renewability of the base material. A variety of natural fibres obtained from the different parts of the plants have been used by material scientists as reinforcement to prepare the green composites. These include bast fibres (jute, flax, hemp, ramie and kenaf), leaf fibres (agave, sisal, henequen etc.) and fruit fibres such as coir etc. Most of the natural fibres and reinforcements used for composites are hydrophilic in nature, whereas synthetic polymers are hydrophobic. Poor adhesion between the natural fibres and polymer matrix often prevents the possibility of natural fibres to act as fillers, resulting in poor dispersion, inadequate reinforcement, and low mechanical properties. Therefore,

[*]Corresponding Author Email: bhanupriya.kanwar@gmail.com.

natural fibres require the addition of coupling agents or the chemical modification for industrial application in composites.

Keywords: natural fibres, chemical modifications, biodegradable polyesters, fibre reinforced biopolymer based composites

1. Introduction

The consumer driven society today demands plastic for the manufacture of millions of products. Packaging materials largely contribute to the high demand for plastics. Our fast paced lifestyle prefers convenience and pre-packaged foods. Plastic is made from crude oil, a non-renewable resource. Although the plastic can be recycled but the amount of solid waste generated by plastic is becoming a problem. In recent years, natural renewable resources have successfully been used to produce plastic that is biodegradable under certain temperature and humidity conditions. Biodegradable plastic made from renewable resources is an important material innovation because it decreases the dependence on petroleum and reduces the amount of waste material, while still generating the product that provides similar benefits of traditional plastics.

Over the last fifty years, the production of plastic has reached enormous levels. Approximately two hundred billion pounds of plastic has been produced worldwide every year. About thirty to forty percent of plastic produced has been used for packaging. It has been recorded that nearly sixty billion tons of plastic are thrown away every year.

Most commercial polymers used today are prepared by using non-biodegradable constituents. Most of these polymers are also derived from petroleum, a non-sustainable resource. These materials cannot be easily degraded and can cause the significant damage to the environment. Approximately 30 to 15 million tons of non-biodegradable polymeric waste have been generated each year. At the end of their life majority of the conventional plastics are traditionally disposed of in landfills where they may persist for several decades without degrading and making the land unusable for any other applications. About 6-7% of the petroleum produced today is used for making polymers, fibres and other chemicals. It has been estimated that the petroleum is being consumed at an 'unsustainable' rate of 100,000 times faster than its production. At the current rate of consumption, it is estimated that the current petroleum reserzves will last only for 50-60 years. These concerns have resulted in the development of environmentally friendly biobased and biodegradable polymers and their composites in recent years.

The development of natural fibre reinforced biodegradable composites has promoted the use of environmentally friendly materials in the diverse applications. The use of green materials provides alternative way to solve the problems associated with waste biomass. They can be obtained in abundance, low cost, and are also renewable sources of biomass.

Among large amount of residues, only a small quantity is applied as household fuel or fertilizer and the rest is burned in the field. As a result, it gives a negative effect on the environment due to the air pollution. The vital alternative to solve this problem is to use the agricultural residues and natural fibre as reinforcement for the development of composites. The natural fibres have been combined with polymer matrix derived from petroleum or renewable resources to produce a useful product for our daily applications as shown in Figure1.1 [1].

The interest in natural fibre reinforced polymer composite materials has been rapidly growing both in terms of their industrial applications and fundamental research. They are renewable, cheap, completely or partially recyclable, and biodegradable. The plants, such as flax, cotton, hemp, jute, sisal, kenaf, pineapple, ramie, bamboo, banana, etc., as well as wood are used as a source of lignocellulosicfibres and often applied as the reinforcement of composites. Their availability, renewability, low density, and price as well as satisfactory mechanical properties make them an attractive ecological alternative to glass, carbon and man-made fibres used for the manufacturing of biocomposites. The natural fibre based composites are more environmentally friendly, and used in transportation (automobiles, railway coaches, aerospace), military applications, building and construction industries (ceiling panelling, partition boards), packaging, consumer products, etc.

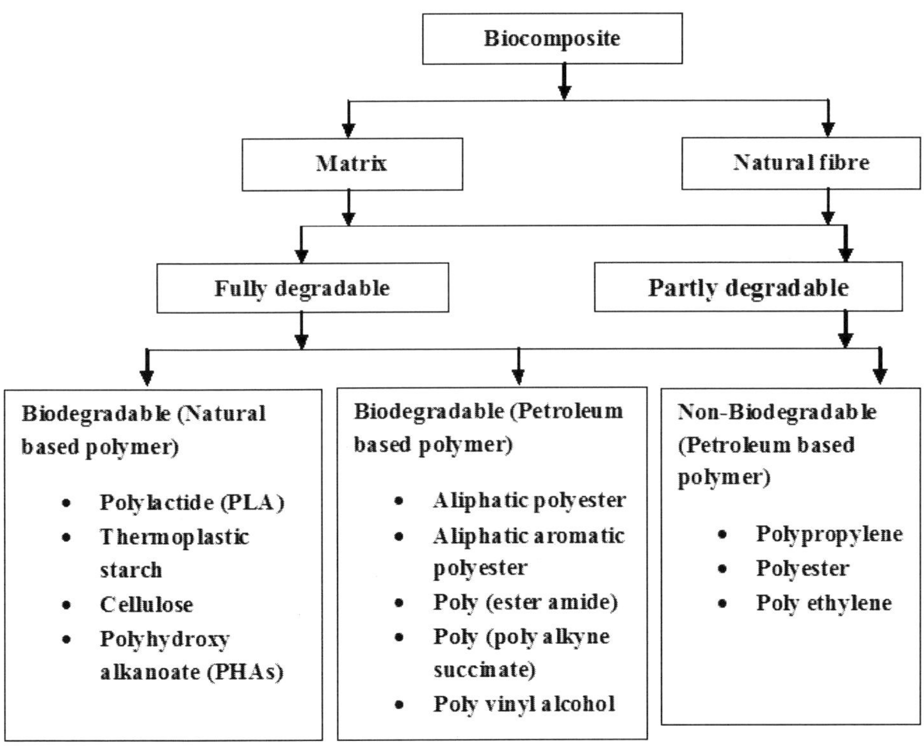

Figure 1.1.Classification of biocomposites.

2. Natural Fibres

During the past decade, increasing environmental awareness, new global agreements and international governmental policies and regulations have been the driving force behind the renewed interest in the natural fibres. Natural fibres are gaining huge attention for their diversified applications in engineering and their uses, such as building materials and structural parts for the automotive application, where light weight is required. Low cost and less tool wear during processing are among the known advantages of the plant-fibres.

Natural fibres are widely available in most developing countries from wood or annual plants. They present several interesting advantages particularly their low density, low cost, renewable, nonabrasive, low energy consumption, high specific properties, fast absorption/desorption of water, no skin irritations, no residues when incinerated, biodegradability and their availability everywhere at modest cost and in a variety of morphologies and aspect ratios.

Fibres are a class of hair like materials that are continuous filaments, similar to pieces of thread. They can be spun into filaments, thread, or rope. They can be used as a component of biocomposites materials. Biocomposites consisting of a polymeric matrix reinforced with natural fibre possess several advantages such as high modulus and strength, relative to traditional materials such as glass fibre reinforced polymer composites. These composites have the advantage of low density, low cost, recyclability, and biodegradability. Their uses are growing in areas where the lightweight and strong structures are required, for example, in the aerospace, automobile, and building construction industries. Fibres are of two types such as natural fibre and manmade or synthetic fibre. Figure 1.2 shows the classification of natural fibres [2].

2.1. Classification of Natural Fibres

Natural fibres include the fibres that are obtained from plant, animals and mineral sources. Natural fibres can be classified into three different categories according to their origin.

2.1.1. Animal Fibres

Animal fibres are comprised of proteins which include animal hair silk fibre and avian fibre. It includes wool, silk, alpaca hair, goat hair, angora, horse hair silk fibre from silk worms and feather fibre.

2.1.2. Mineral Fibres

These are naturally occurring fibres or slightly modified fibres which are obtained from the minerals. These can be categorized into the three categories such as asbestos, ceramic fibres (Glass fibres, aluminium oxide, silicon carbide, and boron carbide) and metal fibres (aluminiumfibres).

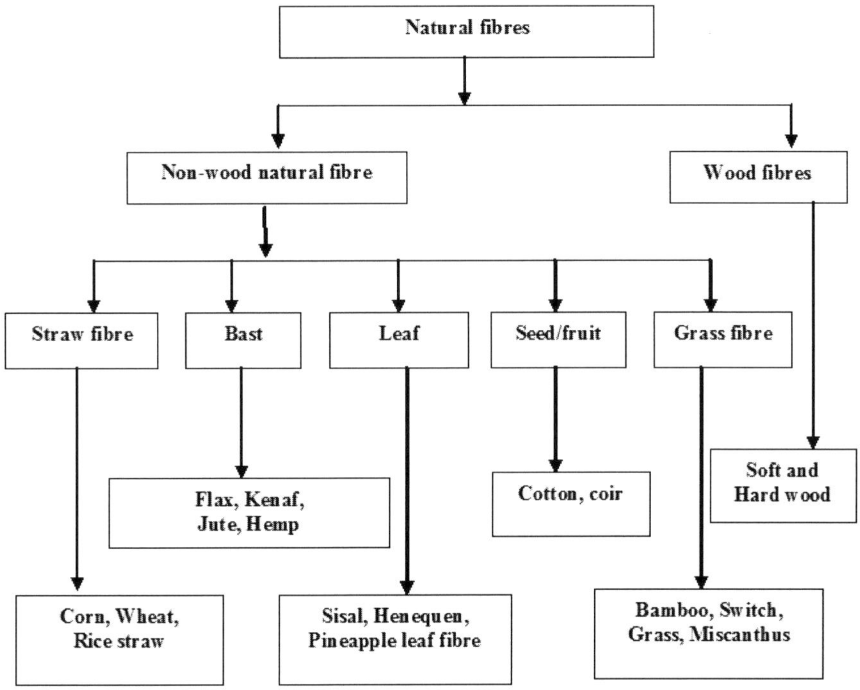

Figure 1.2. Classification of natural fibres.

2.1.3. Plant Fibres

The fibres obtained from plants are called plant fibres. These are obtained from cotton, jute, flax, ramie, sisal and hemp. Cellulose fibres are used in the manufacture of paper, cloth and household things. Plant fibre can be further categorized into seed fibre (cotton and kapok), leaf fibre (sisal and agave), skin fibre (hemp, flax), fruit fibre(coconut, coir) fibre) and stalk fibre (wheat straws, rice, barley, bamboo and grass). These fibres have higher tensile strength therefore used for durable yarn, fabric, packaging, and paper due to high tensile strength.

2.2. Composition of Natural Fibres

Plant fibres are primarily made up of four components such as cellulose, hemicelluloses, lignin and pectin [3]. Cellulose and lignin have received most attention in the recent years due to their world wide applications. Figure 1.3 illustrates the structure of natural fibre material. The different components of plant fibre are cellulose, hemicelluloses, lignin and pectin. Cellulose, the major constituent of all plant materials, forms about half to one third of plant tissues and is constantly replenished by photosynthesis. In particular, cellulose is the main constituent of higher plants, including sugarcane bagasse, rice husk and rice straw. Chemically, cellulose is a linear polymer D-anhydroglucose unit [4]. Cellulose is an organic compound with the formula $(C_6H_{10}O_5)_n$, a polysaccharide consisting of a linear chain of $\beta(1\rightarrow4)$ linked D-glucose units. The degree of polymerization (DP) in cellulose is around 10,000. Cellulose is the structural component of the primary cell wall of green plants, many forms of algae and oomycetes. Cellulose is the most common organic compound on earth. All plants are composed of about 33% cellulose [5]. For industrial use, cellulose is mainly obtained from wood pulp and cotton. It has been used to produce paperboard, paper and a wide variety of derivative products such as cellophane and rayon [6]. Hemicellulose is a complex carbohydrate polymer. It is a polysaccharide with a molecular weight lower than cellulose. It consists of mixture of polysaccharides such as D-xylose, D-mannose, D-galactose, D-glucose, L-arabinose, 4-o-methylglucuronic, D-galacturonic and D-glucuronic acids. It is partially soluble in water and hygroscopic in nature. Sugars are linked together by β-1,4- and occasionally β-1,3-glycosidic bonds. Lignin is a complex hydrocarbon polymer with both aliphatic and aromatic constituents. It is formed by removal of water from sugars to create aromatic structures [7]. They are insoluble in most of the solvents and cannot be broken down to monomeric units. Lignin is totally amorphous and hydrophobic in nature. It gives rigidity to the plants. Lignin is considered to be a thermoplastic polymer exhibiting a glass transition temperature of around 90°C and melting temperature of around 170°C. It is soluble in hot alkali, readily oxidized, and easily condensable with phenol [8-9].

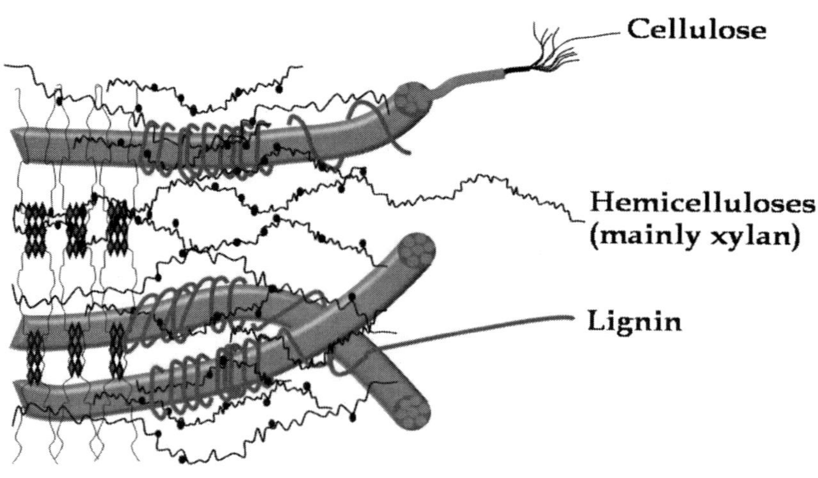

Figure 1.3.Structure of lignocellulosic material.

Pectins are a collective name for heteropolysaccarides. It gives flexibility to plants. It is a natural, non-toxic, and amorphous carbohydrate present in cell walls of all plant tissues, which functions as an intercellular and intracellular cementing material.

2.3. Advantages of Natural Fibre

Natural fibres have attracted the attention due to their advantages over other established materials. They are environmentally friendly, biodegradable, abundantly available, renewable, cheap, low density and possess good mechanical properties as compared to synthetic fibre. Plant fibres are light compared to glass, carbon and aramid fibres. The biodegradability of plant fibre can contribute to a healthy ecosystem while their low cost and high performance fulfils the economic interest of industry [10]. The lower abrasive nature of fibre is an advantage for recycling process of the composite materials [11]. The use of natural fibres derived from different resources in polymeric materials provides positive environmental benefits with respect to ultimate disposability and best utilization. Natural fibres, such as pine needles, hibiscus sabdariffa, flax and jute etc. hold good reinforcing capability when mixed with polymer matrices. Various workers have explored the effective utilization of natural fibres in place of toxic synthetic fibres for various purposes. Synthetic fibres cause not only health problems but are also difficult to recycle. These polymer composite materials find extensive applications in various fields viz. in aerospace, automotive, marine, infrastructure, military, etc. It has been observed that reinforcing power of natural fibres was almost similar to that of synthetic fibres. The major disadvantage in using synthetic fibres is that these fibres are toxic in nature unlike the natural fibres. Therefore, more and more emphasis has been given on natural fibres reinforced polymer composites because of non-toxic and biodegradable nature.

2.4. Limitations of Natural Fibres

Natural fibres have several advantages and also possess good mechanical properties as compared to synthetic fibres. One of the major drawbacks of natural fibres is their hydrophobic character which causes the fibre to swell and ultimately rotting takes place through attack by fungi. These fibres are inherently incompatible with hydrophobic thermoplastics, such as polyolefins [12]. The major limitations of using these fibres as reinforcements in such matrices include poor interfacial adhesion between polar hydrophilic fibre and nonpolar-hydrophobic matrix which leads to poor mechanical properties. Moreover, there is difficulty in mixing because of poor wetting of the fibre with the matrix that leads to composites with weak interface. Degradation at higher temperature, aspect ratio and fibre dispersion of natural fibres also affect the composite properties.

2.5. Surface Modification of Natural Fibres

Natural fibres are used to a limited extant in industries, due to certain limitation associated with surface interaction. Interfacial adhesion between fibre and matrix plays a predominant part in characterizing the mechanical properties of the composites. Poor adhesion of natural fibres across the phase boundary leads to weak dispersion of force which ultimately results poor mechanical properties of the composites [13]. For reinforcement of natural fibre in composites, several problems occur along the interface due to the presence of hydrophilic hydroxyl groups on the fibre surface. Hydrophilic nature of natural fibres hinders effective reaction with the matrix. In addition to this, pectin and waxy substance cover the reactive functional groups of the fibre and act as a barrier to interlocking with the matrix. To optimize effective interfacial bonding between fibre and matrix, the fibre surface needs to be modified with different chemical treatments which enhance the compatibility of natural fibres [14]. Various chemical modifications and treatments such as graft copolymerization, mercerization, silane treatment, peroxide treatment, acetylation, benzoylation, permanganate treatment etc. have been used to overcome the disadvantages associated with natural fibres [15].

2.5.1. Graft Copolymerization

Graft copolymerization is a well established technique used to modify the properties of natural polymers. Modification of fibres by graft copolymerization allows incorporating new and desired properties without drastically affecting the basic properties of the fibre. Depending on the polymer that is grafted onto the cellulose, it is possible to attain the properties such as dimensional stability, resistance to abrasion and ware, wrinkle recovery, oil and water repellence, elasticity, ion exchange capacities, temperature responsiveness, thermal resistance and resistance to microbiological attack. A graft copolymer generally forms a sequence of one monomer which is referred as the backbone polymer with one or more branches (grafts) of long sequences of different monomers. The graft copolymerization is the one of the key way to modify the natural lignocellulosic materials without affecting its basic properties. Figure 1.4 represents a schematic structure of a cellulose graft copolymer.

The history of polymer modification by copolymerization began around 1910 when it was discovered that copolymers of olefins and dienes resulted in better elastomers than polyolefins or polydiens alone. In 1943, first graft copolymer of vinyl and allyl ethers of cellulose copolymerized with maleic acid ester was reported [16]. The concept of graft copolymer was actively promoted by Mark in 1953. During last decades several methods have been suggested for the preparation of graft copolymers by conventional chemical techniques.

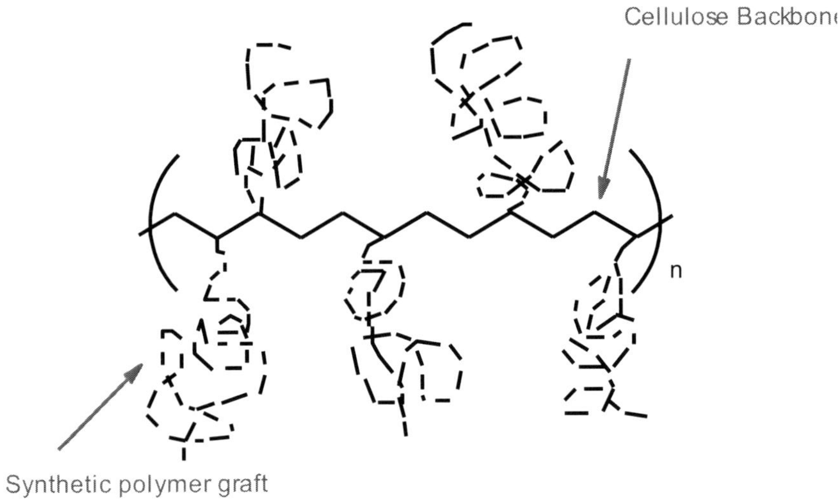

Figure 1.4. Schematic representation of cellulose graft copolymer.

Graft copolymerization of different vinyl monomers has been successfully attempted onto both natural and synthetic polymers. Graft copolymerization of vinyl monomer onto polymeric materials has been the subject of extensive studies for about four decades due to its abundance in nature. The physicochemical properties such as chemical resistance, moisture absorption, water uptake behaviour etc., of natural fibres (jute, hemp, flax, ramie, sisal, pinus, oil palm, coir, bamboo, okra, agave, *Grewiaoptiva*, *Cannabis indica*, *Hibiscus sabdriffa* etc.) have been improved by graft copolymerization [17-18].

Initiator plays an important role on the grafting, and it determines the grafting percentage. The grafting of vinyl monomers onto cellulose or cellulose derivatives is performed by chemical initiators. Redox initiator systems can be used at low temperatures, and they react only with the amorphous region of cellulose because the latter are more reactive than the crystalline phase. The redox initiators such as ceric (IV) ion (ceric ammonium nitrate (CAN) or cerium (IV) sulfate [19-21], ceric ammonium sulfate (CAS) [22], iron(II)-hydrogen peroxide (Fe^{2+}-H_2O_2:Fenton reagent) [23], cobalt (III) acetylacetonate complex salts [24], Co(II) potassium monopersulfate [25], and sodium sulphite-ammonium persulfate [26] have been attempted for the grafting of different vinyl monomers onto the cellulosic fibres. The free radical initiators such as azobisisobutyronitrile ($C_8H_{12}N_4$: AIBN) [27], potassium persulfate ($K_2S_2O_8$: KPS) [28-33], ammonium persulfate ($(NH_4)_2S_2O_8$: APS), and benzoyl peroxide ($C_{14}H_{10}O_4$: BPO) have also been used for the grafting of different monomer onto natural fibres.

Potassium bromated thiourea redox system was used for initiation of graft copolymerization of methyl methacrylate (MMA) onto nylon-66 [34]. The grafting of acrylamide (AAm) or acrylonitrile (AN) onto nylon 6 or nylon-66 to produce N-grafted polyamides *via* redox reaction was investigated [35]. Potassium diperiodatonickelate [Ni(IV)]-casein, novel redox system has been used to initiate the graft copolymerization of butyl acrylate (BA) onto casein under heterogeneous conditions in aqueous medium [36].

The graft copolymerization of MMA onto nylon-6 by using acetyl acetonate complex of Mn(III), Co(III) and Fe(III) as initiators has been observed [37]. It was observed that the rate of polymerization, increased with increase in concentration of initiators in the initial stages and decreased with further increase of complex concentrations. Graft copolymerization of acrylonitrile (AN) onto pineapple leaf fibre (PALF) was studied using $CuSO_4$ and KIO_4 combination as an initiator in an aqueous medium and the grafted samples was recorded with higher thermal stability [38].

The grafting of vinyl monomers onto natural fibres was also performed by radiation or microwave irradiation. The use of microwave irradiation to induce multi component polymerization reactions is a new area of increasing interest. The use of microwave irradiation was used to reduce reaction times as compare to traditional thermal initiated reactions. The improvement in chemical resistant behaviour of methylmethacrylate (MMA) grafted flax fibre under the influence of microwave radiation (MWR) and the improved swelling behaviour of natural *Cannabis indica*fibre from graft copolymerization of acrylic acid onto by using CAN/nitric acid redox initiator system under the influence of microwave radiation (MWR) also have been observed [39-41]. The corn starch was grafted with acrylic acid using potassium persulphate (KPS) as redox initiator by microwave irradiation. It has been observed that the molecular chains of corn starch were introduced to hydrophilic and hydrophobic groups [42].

Gamma irradiation induced graft copolymerisation of acrylonitrile onto cellulosic filter paper has been reported. It has been observed that the amount of homopolymer was reduced by the introduction of a small amount of styrene monomer [43]. The graft polymerization of Glycidyl methacrylate (GMA) onto Kapok fibre under irradiation with ^{60}Co gamma rays was used as ion exchangers and adsorption of toxic gas [44].

Physico-chemical properties such as decrease in humidity, chemical resistance, improved hydrophobicity and mechanical strengthhave also observes by graft copolymerization.The compatibility and adhesion of cellulosic fibres with hydrophobic composites have been enhanced by grafting of hydrophobic monomers such as methyl methacrylate, styrene, acrylonitrile, butadiene, isobutyl vinyl ether, and vinyl acetate [45-47].

The exhibited excellent protein adsorption properties have been observed by the preparation of an anion exchange matrix, poly(vinyl benzyltrimethyl ammonium chloride)-g-cotton cellulose, using ^{60}Co gamma-radiation source. The grafted samples have also used for the removal of heavy metal ions from waste water [48-50].

Several workers have studied the graft copolymerization of binary vinyl monomer mixtures onto cellulosic fibres. Grafting of flax fibre with binary vinyl monomer mixtures such as methyl methacrylate (MMA)-ethyl acrylate (EA), MMA-acrylonitrile (AN), EA-MMA, EA- acrylic acid (AA) and MMA- AA has been studied to improve the thermal stability of the fibre [50-54].The thermal stability of methyl methacrylate (MMA) grafted *Bombyxmori*fibre (natural silk) [55].

The grafted fibre depicted greater tensile strength as compared of untreated fibre. Textile strength of okra bleached fibres was improved by grafting of acrylonitrile (AN) using potassium persulphate as redox initiator [56-58].

Surface modification of natural fibres through graft copolymerization improved the swelling behaviour, water uptake, moisture absorbance, resistance to chemicals, thermal behaviour, tensile strength and percent crystallinity. These grafted fibres have been potentially used as reinforcement for thermosetting, thermoplastic and biodegradable polymer composites [59-72].

2.5.2. Chemical Methods

The interest in using natural fibres for the preparation of composites has increased in recent years due to their lightweight, nonabrasive, combustible, nontoxic, low cost and biodegradable properties. Hydrophilic nature of natural fibres reduces their potential as reinforcing agents due to the low interfacial properties between fibre and polymer matrix. However, lack of good interfacial adhesion, low melting point and poor resistance to moisture absorption, make the use of natural fibre reinforced composites less attractive. The pre treatments of the fibre clean the fibre surface, chemically modify the surface, stop the moisture absorption process, increase the surface roughness and improve the fibre matrix interaction. The chemicals may activate hydroxyl groups or introduce new moieties that can effectively interlock with the matrix. Many chemical treatments such as mercerization, isocyanate treatment, permanganate treatment, acetylation, silane treatment and peroxide treatment with various coupling agents etc., have been attempted for the modification of fibre surface and enhance its properties [73].

2.5.2.1. Alkaline Treatment

Alkali treatment leads to the increase in the amount of amorphous cellulose at the expense of crystalline cellulose. Various alkalies such as KOH, LiOH and NaOH can be used for the treatment of fibre. The important modification occurs due to the removal of hydrogen bonding in the network structure. The following reaction takes place during the alkali treatment [74]:

$$\text{Cellulose—OH} + \text{NaOH} \longrightarrow \text{Cellulose—ONa} + H_2O$$

This treatment changes the orientation of highly packed crystalline cellulose order and forms amorphous region by swelling the fibre cell wall. It provides more access for the penetration of chemicals. Moreover, mercerization increases the number of possible reactive sites and allows better fibre wetting. It has been observed that 2% alkali solution at 200°C and 1.5 MPa pressure for 90 seconds was suitable for degumming and defibrillation to individual fibres [75]. The 10-30% sodium hydroxide (NaOH) solution produced the best effects on flax fibre properties. Flax fibres soaked in different solution of NaOH (2.5, 5, 10, 13, 15, 18, 20, 25, or 30%) were studied for the surface modification. It has been found that 5 and 18% of sodium hydroxide solution were the appropriate concentrations for mercerization of the flax fibres [76].

It has been reported that 5% NaOH treated sisal fibre-reinforced polyester composite had better tensile strength than 10% NaOH treated composites [77]. Alkali treated flax fibre was used as reinforcement in epoxy resins. It was observed that alkaline treatment increased about 30% tensile properties (both strength and modulus) of flax fibre-epoxy composites [78-79].

2.5.2.2. BenzoylationTreatment

In benzoylation treatment, benzoyl chloride is used to decrease the hydrophilicity of the fibre which improves the fibre matrix adhesion. Benzoyl chloride includes benzoyl ($C_6H_5C{=}O$) which is attributed to the decreased hydrophilic nature of the treated fibre. It enhanced the thermal stability of the fibre [80-81]. During benzoylation treatment alkali pretreatment is used to activate the hydroxyl groups of the fibre as shown in Figure 1.4. It also improves the tensile strength and moisture resistance properties of fibres [82].

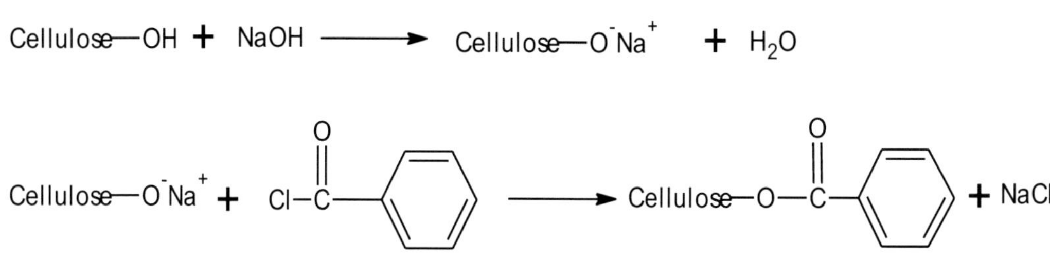

Figure 1.4. Benzoyl chloride treatment of cellulosic fibres.

Benzoylation treatment of the fibre improves the fibre matrix adhesion which considerably increases the strength of composite.Benzoylation decreases the water absorption.It improves the thermal stability of natural fibre. NaOH and benzoyl chloride (C_6H_5COCl) solution have been used for surface treatment of sisal fibres [83-84]. It was observed that the thermal stability of treated composites were higher than the untreated fibre composites.

Benzoylation of *Cannabis indica*fibre was carried out by treating raw fibre with varying concentrations (1, 2 3, 5, 7 and 9%) of benzoyl chloride solution. The chemical resistance and hydrophobic behaviour of surface modified *Cannabis indica* fibres was increased [85]. Flax fibres have been treated with NaOH to activate the hydroxyl groups. It has been observed that the interfacial adhesion of benzoylated flax fibre and polyethylene (PE) matrix was improved after treatment.

2.5.2.3. SilaneTreatment

Silane is used as coupling agents to modify the fibre surface as shown in Figure 1.5. It resulted in several stages of hydrolysis, condensation and bond formation during the treatment process with the fibre. Silanol molecules form in the presence of moisture and hydrolysable alkoxy groups. It reacts with cellulose hydroxyl group of the fibre and improves fibre matrix adhesion to stabilize composite properties [86]. The chemical composition of silane coupling agents (bifunctionalsiloxane molecules) allows the formation a chemical link between the surface of the cellulose fibre and the resin through a siloxane bridge. It provides molecular continuity across the interface region of the composite. It also results in the hydrocarbon chains that restrain fibre swelling into the matrix [87-88]. Natural fibres exhibit micropores on their surfaces and silane coupling agent act as a surface coating which penetrates into the pores and develop the coating on the fibre surface [89].

Treatment of natural fibres with silane solutions in a water and ethanol mixture with concentration between 0.033% and 1% was investigated [90]. Aminopropyltrimeth-oxysilane have been used to modify the surface of henequen fibres and oil palm fibres [91]. Flax fibre surface was modified by using aminopropyltrimethoxysilane with concentration of 1% in a solution of acetone and water (50/50 by volume) for 2 hours [92].

The silane treated fibres were used as reinforcement for the preparation of thermosetting composites. The improvement of the physico-chemical properties of raw fibre after silane treatment has been observed. Silane treated *Grewiaoptiva*fibres were reinforced in urea formaldehyde resin to improve the mechanical and thermal properties [93].

2.5.2.4. Acetylation Treatment

Acetylation treatment is generally known as esterification method for plasticizing of cellulosic fibres. Acetylation improves the hydrophobic nature of the fibre and also improves the dimensional stability and environmental degradation [94]. This treatment provides rough surface topography with less number of void contents that give better mechanical interlocking with the matrix [95-96]. The treatment improves the moisture resistance a property due to the removal of hemicellulose and lignin constituents from the fibre [97].The reaction of acetic anhydride with fibre is shown in Figure 1.6.

Acetylation is beneficial for reducing the moisture absorption of natural fibres. Acetylated fibres have use as reinforcement material for the preparation of biocomposites [98-100]. The acetylation of sisal fibre in 18% NaOH solution, acetic acid, acetic anhydride and sulphuric acid for 1 hour has also been reported.

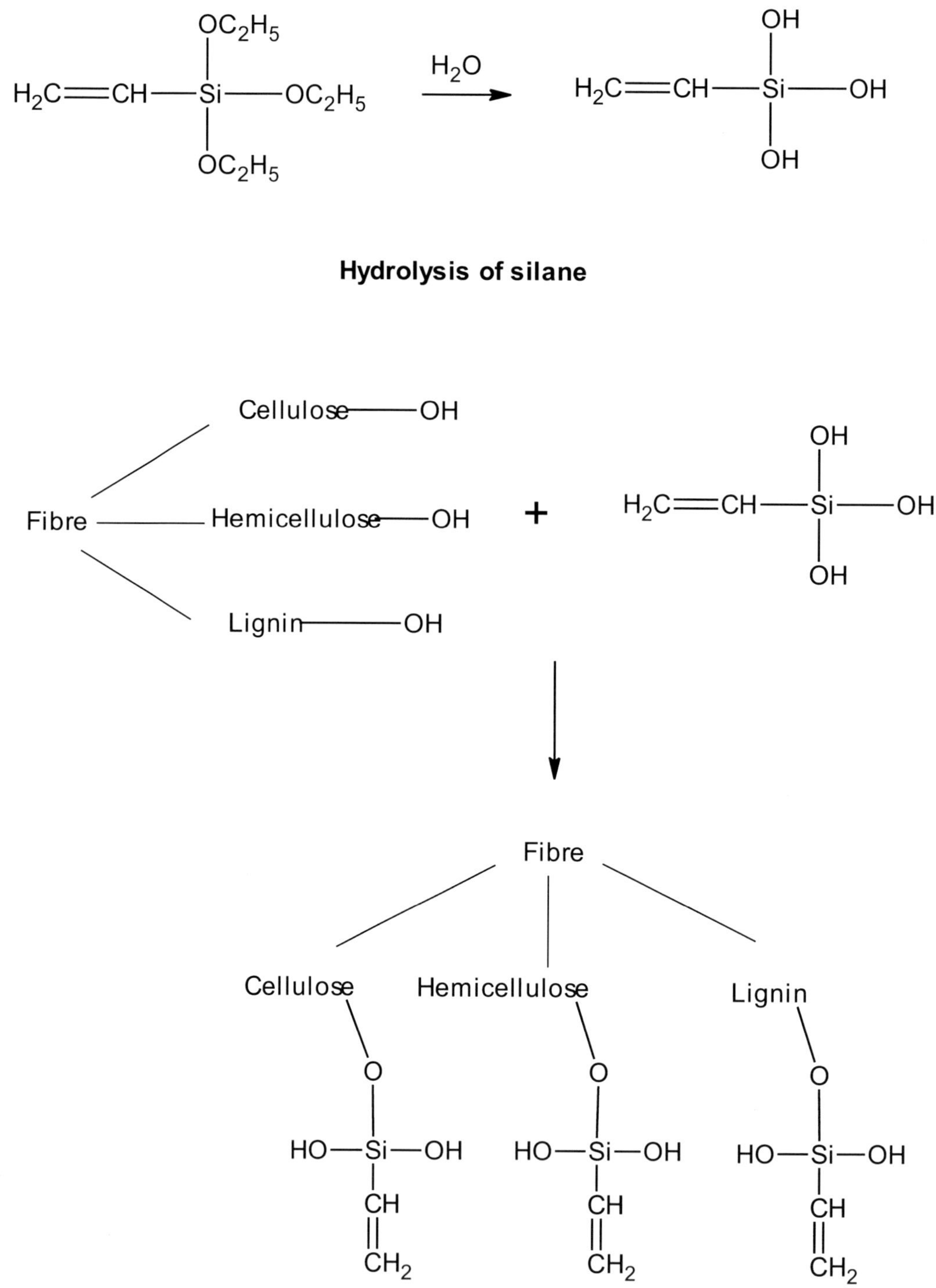

Figure 1.5. Interaction of silane with cellulosic fibres.

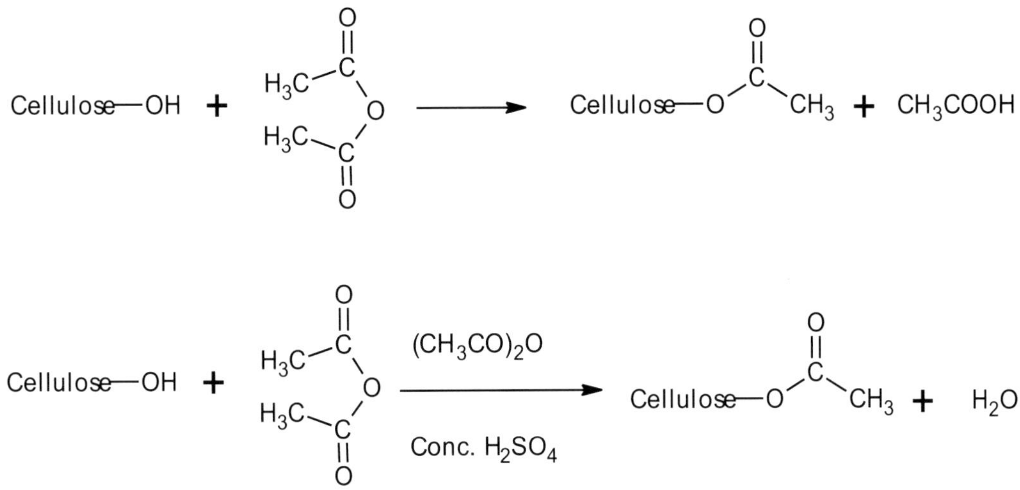

Figure 1.6. Acetylation of cellulosic fibre.

The decrease in the moisture absorption was observed for the treated fibres [101]. Acetylated flax fibre reinforced polypropylene composite [102]. It was observed that the flax fibre surface morphology and moisture resistance properties improved remarkably after acetylation treatment. Tensile and flexural strengths of acetylated flax fibre reinforced polypropylene composite were found to increase as compared to untreated flax fibre reinforced composite.

2.5.2.5. Isocyanate Treatment

An isocyanate is a compound containing the isocyanate functional group –N=C=O, which is highly susceptible to reaction with the hydroxyl groups of cellulose and lignin of the fibres. The reaction between fibre and isocyanate coupling agent is shown below [103] as:

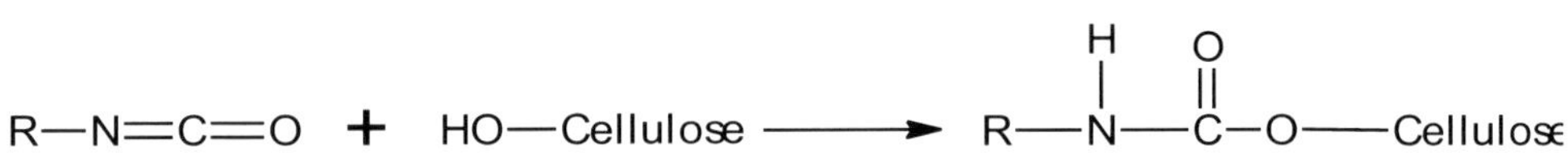

Where, R is the different chemical groups (such as alkyl). Toluene di-isocyanatediisocyanate (CTDIC), phenyl isocyanate (C_6H_5NCO) and polymethylene-polyphenyl-isocyanate ($C_{15}H_{10}N_2O_2$) solutions are used to improve the fibre-matrix interfacial adhesion [104].

The sisal fibres were treated with cardanol derivative of toluene di-isocyanate (CTDIC) as coupling agent [105]. The treated sisal fibres were used as reinforcement materials in LDPE matrix based composites. Pineapple leaf fibres were treated with polymethylene-polyphenyl-isocyanate ($C_{15}H_{10}N_2O_2$) solution at 50°C for 30 minutes to improve the fibre-matrix interfacial adhesion [106]. TDIC treated oil palm fibre were reinforced in phenol formaldehyde matrix. It was observed that the tensile strength of treated composites was higher than untreated fibre reinforced composites [107].

2.5.2.6. Sodium Chlorite Treatment

Sodium chlorite ($NaClO_2$) is usually used for the bleaching of natural fibres. This treatment is also known as delignification treatment. Lignin is removed by this process which makes fibre hydrophobic. Delignification lowered the tensile strength of fibre and also increases the flexural strength because of lower stiffness and more flexible character of fibres after delignification. Flax fibres were treated with sodium chlorite for the surface modification [108]. Tensile strength of bleached sisal fibre-phenol formaldehyde composite was found less than other chemical treated fibre composites. It may be due to the fact that the delignification of the fibre lowered the tensile strength of the fibres. Textile strength of okra bleached fibres has improved by grafting acrylonitrile (AN) using potassium persulphate as redox initiator [109]. This treatment also improves the thermal resistance and crystalline structure of Kenaffibre (*Hibiscus cannabinus*) fibre [110]. Green coconut fruit fibres were treated with alkaline solution and bleached with sodium chlorite.Bleached fibres were used as reinforcement in high impact polystyrene. It was observed that the tensile strength of treated composites was higher than that of untreated fibre reinforced composites [111].

2.5.2.7. Maleated Coupling Agents

Maleated coupling agents have been widely used to strengthen the composites containing fillers and fibre reinforcements [112-113]. Maleic anhydride is not only used to modify the fibre surface but it also modifies the polypropylene matrix to achieve better interfacial bonding and mechanical properties of fibre and matrix as shown in Figure 1.7.

This treatment increases the surface energy of cellulose fibres to a level much closer to the surface energy of the matrix. This results in better wettability and higher interfacial adhesion of the fibre. Maleic anhydride treatment reduces the water absorption to a great extent in banana, hemp and sisal fibre reinforced composites. Mechanical properties like Young's modulus, flexural modulus, hardness and impact strength of plant fibre reinforced composites were increased after maleic anhydride treatment [114-116].

2.5.2.8. Permanganate Treatment

Permanganate contains permanganate group MnO_4^-. Permanganate treatment leads to the formation of cellulose radical through MnO_3^- ion formation. The highly reactive Mn^{3+} ions are responsible for initiating graft copolymerization as shown in Figure 1.8. The most permanganate treatments have been performed by using potassium permanganate ($KMnO_4$) in different concentrations in acetone.

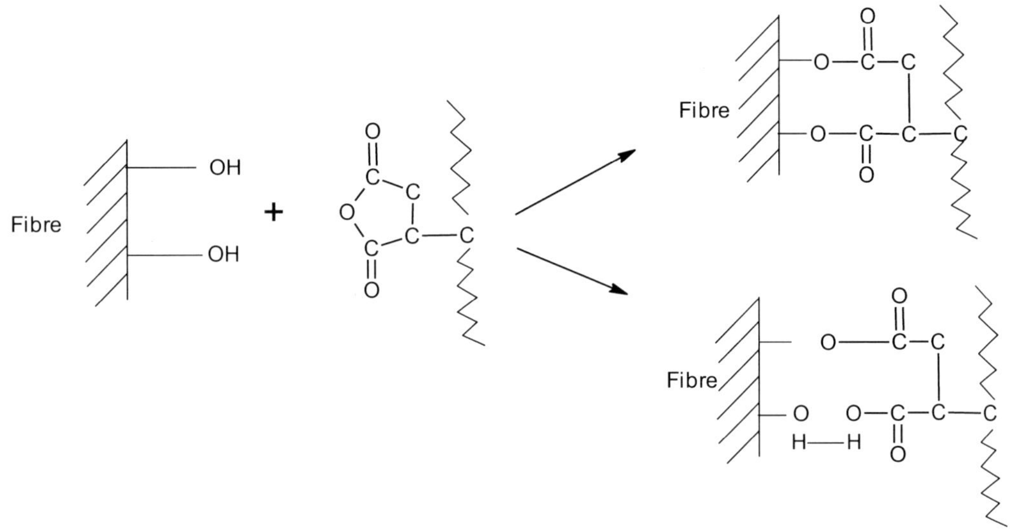

Figure 1.7. The reaction of cellulose fibres with maleic anhydride grafted polypropylene copolymers.

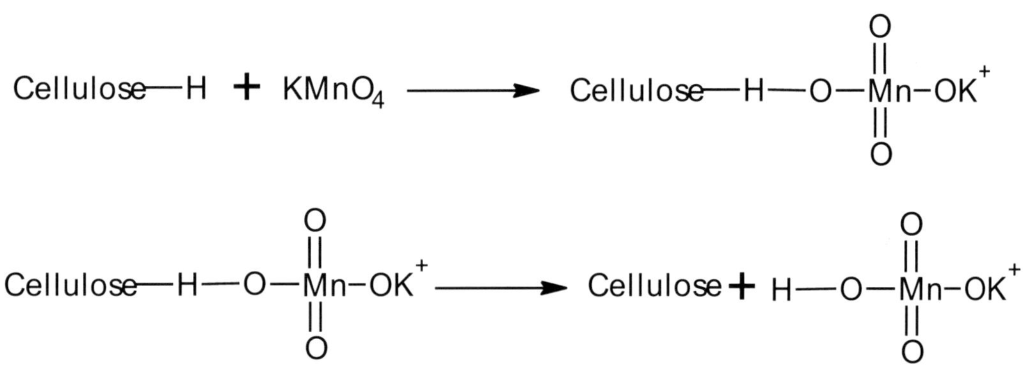

Figure 1.8. Treatment of cellulosic fibre using potassium permanganate ($KMnO_4$).

Permanganate treatment reduces the hydrophilic nature of the fibres and decreases the water absorption of fibre reinforced composite. The hydrophilic tendency of fibre decreased as the $KMnO_4$ concentrations increased. Higher $KMnO_4$ concentrations resulted in the degradation of cellulosic fibre and the formation of polar groups between fibre and matrix.

Permanganate treatment was attempted by using different concentration of potassium permanganate ($KMnO_4$) solution in acetone after alkaline pretreatment [117-119]. Sisal fibres were modified by dipping in permanganate solution of concentrations (0.033, 0.0625 and 0.125%) [120].

2.5.2.9. Peroxide Treatment

Benzoyl peroxide [$(C_6H_5CO)_2O_2$] and dicumyl peroxide [$C_6H_5C(CH_3)_2O)_2$] are the organic peroxide used for natural fibre surface modifications. Interface properties of fibre and matrix can be improved by peroxide treatment. Peroxide initiates free radicals that react with the hydroxyl group of the fibre and matrix. As a result, good fibre matrix adhesion occurs and it shows the higher mechanical properties of composites [121].This treatment also reduces the moisture absorption tendency of the fibre and improves the thermal stability [122-123]. Peroxide is a molecule with the functional group ROOR containing the divalent ion O-O. Organic peroxides decompose easily to free radicals such as $RO^{\cdot}$ and $RO^{\cdot}$ which reacts with the hydrogen group of the cellulose fibres.

$$\text{RO—OR} \longrightarrow 2\text{RO}^*$$

$$\text{RO}^* + \text{Cellulose—H} \longrightarrow \text{R—OH} + \text{Cellulose}'$$

In peroxide treatment fibres are coated with benzoyl peroxide or dicumyl peroxide in the acetone solution. The complete decomposition of peroxide was achieved by heating the solution at higher temperature [124-125]. Peroxide treatment of fibre was performed by using 6% benzoyl peroxide or dicumyl peroxide in acetone solution for about 30 minutes after alkali pre-treatment. Sisal fibres treated with peroxide were used as reinforcement material in low-density polyethylene. Tensile and electrical properties of composites were found higher [126]. Peroxide treated oil palm fibres were reinforced in phenol formaldehyde resin. It has been observed that treated fibre reinforced composites were hydrophobic in nature [127-128]. Flax fibres were treated with dicumyl peroxide after alkali pre treatments [129].

3. BIODEGRADABLE POLYMERIC MATERIALS

Synthetic polymers are considered to be an invaluable gift of modern sciences and technology to mankind. They have become indispensable to our life with range of applications in diverse fields such as packaging, agriculture, food, consumer products, medical appliances, building materials, industry, aerospace materials etc. However, the resistance of synthetic polymers to chemical, physical and biological degradation has become a serious concern when used in areas such as surgery, pharmacology, agriculture etc. [130-131]. The severe environmental problems, including the increasing difficulties of waste disposal and the deepening threat of global warming (due to carbon dioxide released during incineration) caused by the non-biodegradability of a number of polymers such as polyethylene (used in packaging and agriculture field) have raised concerns all over the world. There is the need of development of new biodegradable polymeric compound in order to build a new society and economy free of plastic pollution in the twenty first century. The use of biodegradable counter parts as an alternative to non-biodegradable polymers is evoking considerable interest. Unfortunately, the degradable polymers available until recently have inferior physical properties in terms of strength and dimensional stability and most of them are very expensive and are technically difficult to process. As a result, attempts have been made to solve these problems by including biodegradability into polymers through appropriate modifications of their structure. Biodegradable polymers from renewable resources have attracted much interest in recent years because of its desirable properties of biocompatibility, biodegradability and natural abundance.

The biodegradable polymers can be classified according to their chemical composition, origin and synthesis method, processing method, economic importance, application, etc. Biodegradable polymers are classified according to their origin into two groups as natural and synthetic polymers. The natural fibres are obtained from natural resources and synthetic polymers are obtained from oil. Further these are classified into four families. Figure 1.9 shows a classification of biodegradable polymers [132].

Biodegradable polymers have been stimulated and developed to overcome problems regarding the persistence of plastic wastes in the environment. Biodegradable materials and composites have been used in different fields such as packaging, paper coating and other disposable articles, as well as in biomedical applications such as restorable surgical sutures, implants and controlled drug delivery devices.

Soyparotein resins were blended with ramie fibres to obtain composites. The soyaprotein was mixed with starch and glycerol in the presence of sodium sulphite and extruded to form pellets. The composites showed good mechanical properties [133]. poly(vinyl alcohol)-chitosan blend films have been synthesized by using glutaraldehyde as a crosslinking agent [134]. It has been observed that crosslinker improved the tensile strength and decreased the elongation of the films. PVA content in the blends increased the water uptake while cross-linking with glutaraldehyde decrease the hydrophilicity.

Silk/poly(butylenes succinate) (PBS) biocomposites have been synthasized [135-136]. It was found that the manufacturing processes effects the mechanical properties including tensile strength, fracture toughness and impact resistance, and thermal stability of biocomposites. Polyols (glycerol, sorbitol, and manitol) were used as a plasticizer to improve the rheological and thermal properties of zein resin films [137].

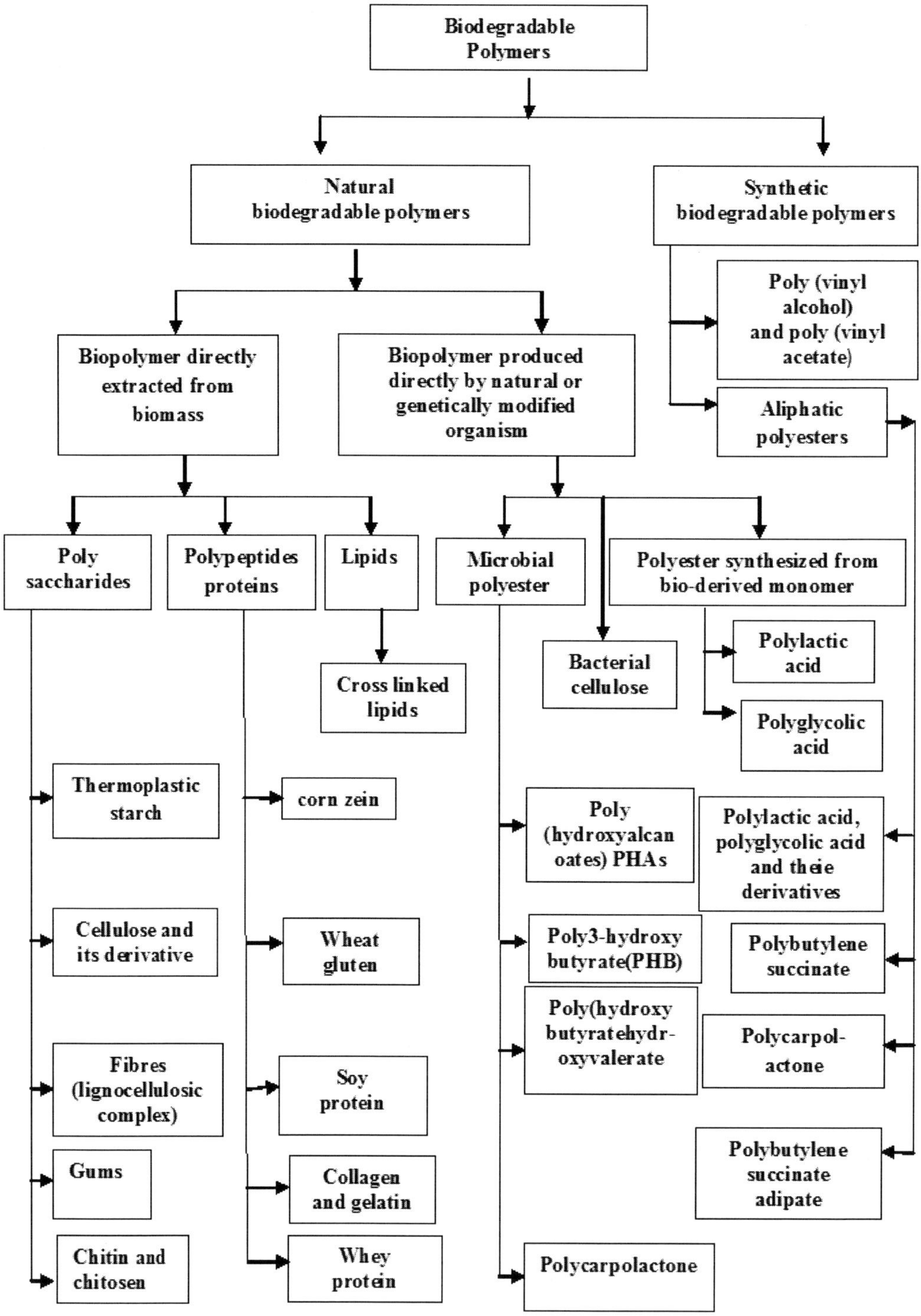

Figure 1.9.Classification of biodegradable polymers.

The effects of lysine based diisocyanate (LDI) coupling agent on the properties of biocomposite of bamboo fibre (BF) reinforced poly(lactic acid) (PLA) and poly(butylene succinate) (PBS) have been investigated [138]. It was observed that tensile properties, water resistance, and interfacial adhesion of both PLA/BF and PBS/BF composites were improved. Crystallization temperature and enthalpy of the composites were increased with increase in LDI content. The decrease in the heat of fusion and thermal degradation temperature in both composites was observed. The kenaffibre reinforced PLA laminated composites by compression molding have been sunthesized [139].

The starch/carboxy methyl cellulose (CMC) composite films have been formed by casting method [140].The effect of CMC addition on the physical properties of blend films was investigated. The blend film composed of 15% w/w CMC/starch showed the lowest water vapour permeability. The moisture absorption and solubility of the blend films were recorded less. The addition of 20% w/w starch to CMC caused the increase in tensile strength (TS) by 59% in comparison to pure starch film.

4. Natural Fibre Reinforced Biopolymer Based Composites

The materials formed by combining natural fibres with biodegradable resins result in a new class of fully biodegradable 'green composites'. The green composites are environmentally friendly, fully degradable and sustainable. At the end of their life they can be easily disposed of or composted without harming the environment. Fibre reinforced biodegradable polymer matrix based composites have successfully proven their value in various applications because of their excellent specific properties, e.g., high strength, flexibility and low weight. They have not only the functional capability to substitute the widely used glass fibres, but they have advantages from the point of view of weight and fibre-matrix adhesion, specifically with polar matrix materials.

These "green" composites exhibit good tensile and flexural properties and may be used for packaging, medicine, furniture, automotive components and indoor panelling. They have good potential for use in waste management due to their biodegradability and much lower production of ash during incineration. Fibre reinforcement to the biodegradable polymer matrix increases the mechanical properties such as tensile strength of composites.The use of biodegradable materials in composites have many advantages such asincreases in awareness of environmentally-friendly manufacturing, reduces the dependency on petroleum products, increases flexibility, invites the possibility of thermal recycling, and reduces health concerns in the manufacturing process. Several studies have been reported on the development of such composites, including blends of starch with polyethylene of low and high density (LDPE, HDPE), starch and poly(vinyl alcohol) (PVA), and poly(3-hydroxybutyrate) with natural fibres such as pineapple leaf, sisal, jute and ramie fibres.

The woven and nonwoven flax fibre reinforced poly(lactic acid) (PLA) biocomposites have been synthasized [141]. Ecofrindlybiocomposites prepared from poly(vinyl alcohol) (PVA) and cellulose microfibres (CMF). It was observed that interactions of -OH groups of PVA with CMF molecular chains lead to increase in the melting point and crystallinity of biocomposites. It was observed that the addition of CMF filler increased the shear strength of samples from 1.55 to 2.41 MPa [142]. The increase in the mechanical and thermal properties of sago pith as a bio-filler (SPB) on poly(vinyl alcohol) (PVA) was studied [143].

The flexural strength, tensile strength and tensile modulus of the soyprotein-jute fibre composites using water as plasticize were observed higher as compare to polypropylene (PP)-jute fibre composites [144]. Starch/polyvinyl alcohol (PVA)/titanium oxide (TiO_2) nanocomposite films have been synthesized by casting method [145]. Sisal fibre (SF) reinforced poly(lactic acid) (PLA) composites prepared by a compression molding technique [146]. Two samples of PLA/SF composites were developed with a fibre content of 10 and 20% by weight.

The biocomposite films have shown transparency, flexibility and batter mechanical properties. It also showed the good biodegradability and better barrier properties [147-148].

The dielectric properties of PVA/ZnOnanocomposites films have been studied. It has been observed that PVA/ZnOnanocomposites behave as lossless materials above 10^6 kHz. The electrical characterization of sisal powder filled poly(vinyl alcohol) (PVA) composites has been investigated. Dielectric measurements of the composite were determined in the temperature ranged from 30 to 150°C [149-150].

REFERENCES

[1] Mohanty, S., Nayak, S.K., Verma, S.K. and Tripathy, S. S. 2004. "Effect of MAPP as a coupling agent on the performance of jute-pp composites." *Journal of Reinforced Plastics and Composites* 23: 625-637.

[2] Mohanty, A.K., Misra, M. and Drzal, L.T. 2001. "Surface modifications of natural fibers and performance of the resulting biocomposites: an overview." *Comp Interfaces* 85: 313-43.

[3] Kayseerilioglu, B. S, Bakir, U., Yilmaz, L., and Akkas, N. 2003. "Use of xylan, an agriculture by product, in wheat gluten based biodegradable films: mechanical, solubility and water vapour transfer rate properties." *Bioresource Technology* 87: 239-246.

[4] Bledzki, A.K. and Gassan, 1999. "Composites reinforced with cellulose based fibres." *Journal of Progress in Polymer Science* 24: 221-274.

[5] Kaith, B.S. and Kalia, S.2008. "Graft copolymerization of MMA onto flax under different reaction conditions: a comparative study." *Express Polymer Letters* 2: 93-100.

[6] Dipa, R., Sarkar, B.K., Rana, A.K. and Bose, N.R. 2001. "Effect of alkali treated jute fibres on composite properties." *Bulletin of Materials Science* 24: 129-135.

[7] Ouajai, S. and Shanks, R.A. (2005) Composition, structure and thermal degradation of hemp cellulose after chemical treatment. *Polymer degradation and stability* 89(2): 327-335.

[8] Bismarck, A., Mishra, S., and Lampke, T. 2005. "Plant fibers as reinforcement for green composites." *Natural fibers, biopolymers and biocomposites*CRC Press: Boca Raton 37-108.

[9] Cantero, G., Arbelaiz, A., Llano-Ponte, R. and Mondragon, I. 2003. "Effects of fibre treatment on wettability and mechanical behaviour of flax/polypropylene composites."*Composite Science and Technology* 63: 1247-1254.

[10] Wambua, P., Ivens, U. and Verpoest, I. (2003) Natural fibers: can they replace glass in fiber-reinforced plastics? *Composite Science and Technology*63:1259-1264.

[11] Larbig, H., Scherzer, H., Dahlke, B, and Poltrock, R. 1998. "Natural Fiber reinforced foams based on renewable resources for automotive interior applications." *Journal of Cellular Plastics*34: 361-379.

[12] Facca, A.G., Kortschot, M.T. and Yan, N. 2007. "Predicting the elastic modulus of natural fiber reinforced thermoplastics." *Composite A: Applied Science and Manufacturing* 37:1660-1671.

[13] Wang, B., Panigrahi, S., Tabil, L. and Crerar, W. 2007. "Pre-treatment of flax fibers for use in rotationally molded biocomposites." *Journal of Reinforced Plastic Composite* 26: 447-463.

[14] Dash, B.N., Rana, A.K., Mishra, S.C., Mishra, H.K., Nayak, S.K. and Tripathy, S.S. 2000. "Novel low-cost jute-polyester composite. II. SEM observation of the fracture surfaces." *Polymer Plastic Technology and Engineering* 39: 333-350.

[15] Paul, A., Joseph, K. and Thomas, S. 1997. "Effect of surface treatments on the electrical properties of low-density polyethylene composites reinforced with short sisal fibers"*Composite Science and Technology* 57: 67-79.

[16] Bardhan, K., Mukhopadhyay, S. and Chatterjee, S.R. 1977. "Grafting of acrylamide onto methyl cellulose by persulfate." *Journal of Polymer Science: Polymer Chemistry* 15:141-148.

[17] Singha, A.S. and Rana, A. K. 2012. "A study on benzoylation and graft copolymerization of lignocellulosic cannabis indica fiber." *Journal of Polymer and Environment* 20: 361-371.

[18] Singha, A.S. and Thakur, V. K. 2008. "Saccaharumcilliarefiber reinforced polymer composites. *E-Journal of Chemistry* 5: 782-791.

[19] Okieimen, E.F. and Ebhoaye, J.E. 1986. "Grafting acrylic acid monomer on cellulosic materials." *Journal of Macromolecular Science: Part A-Chemistry* 23: 349-353.

[20] Bicak, N., Sherrington, D.C. and Senkal, B.F. 1999. "Graft copolymer of acrylamide onto cellulose as mercury selective sorbent." *Reactive and Functional Polymers* 41: 69-76.

[21] Gupta, K.C. and Khandekar, K. 2002. "Graft copolymerization of acrylamide-methylacrylatecomonomers onto cellulose using ceric ammonium nitrate." *Journal of Applied Polymer Science* 86: 2631-2642.

[22] Ibrahim, M.M., Flefel, E.M. and El-Zawawy, W.K. 2002. "Cellulose membranes grafted with vinyl monomers in a homogeneous system." *Polymer Advanced Technology* 13: 548-557.

[23] George, F.F., Robert, C.B. and William, M. 1987. "Graft polymerization of acrylonitrile onto wheat straw." *Journal of Applied Polymer Science* 33: 899-906.

[24] Gupta, K.C. and Sahoo, S. 2000. "Ceric ion initiated grafting on cellulose from binary mixture of acrylonitrile and methylacrylate." *Journal of Macromolecular Science: Part A* 37: 447-468.

[25] Sahoo, P.K., Samantaray, H.S. and Samal, R.K. 1986. "Graft copolymerization with new class of acidic peroxo salts as initiators. I. Grafting of acrylamide onto cotton-cellulose using potassium monopersulfate catalyzed by Co(II)." *Journal of Applied Polymer Science* 32: 5693-5703.

[26] Yang, F., Li, G., He, Y.G., Ren, F.X. and Wang, J.X. 2009. "Synthesis, characterization, and applied properties of carboxymethyl cellulose and polyacrylamide graft copolymer." *Carbohydrate Polymer* 78: 95-99.

[27] Ouajai, S., Hodzic, A. and Shanks, R.A. 2004. "Morphological and grafting modification of natural cellulose fibers." *Journal of Applied Polymer Science* 94: 2456-2465.

[28] Abdel-Razik, E.A. 1997. "Aspects of thermal graft copolymerization of methyl methacrylate onto ethyl cellulose in homogeneous media." *Polymer Plastic Technology and Engineering* 36: 891-903.

[29] Abdel-Razik, E.A. 1990. "Homogeneous graft copolymerization of acrylamide onto Ethylcellulose." *Polymer* 31:1739-1744.

[30] Bardhan, K., Mukhopadhyay, S. and Chatterjee, S.R. 1977. "Grafting of acrylamide onto methyl cellulose by persulfate." *Journal of Polymer Science: Polymer Chemistry 15*:141-148.

[31] Ibrahim, M.D., Mondal, H., Uraki, Y., Ubukata, M., Itoyama, K. 2008. "Graft polymerization of vinyl monomers onto cotton fibres pretreated with amines." *Cellulose* 15: 581-592.

[32] Liu, S. and Sun, G. 2008. "Radical graft functional modification of cellulose with allyl monomers: Chemistry and structure characterization." *Carbohydrte Polymer* 71: 614-625.

[33] Mukhopadhyay, S., Prasad, J. and Chatterjee, S.R. 1975. "Grafting of acrylic acid onto methylcellulose." *Macromolecular Chemistry*, 176: 1-7.

[34] Hebeish, A., Refie, E. I. and Waly, A. I. 1976. "Thiourea-potassium bromate redox system-initiated graft polymerization of methyl methacrylate and methacrylic acid on nylon 6."*Journal of Polymer Science: Part A-Polymer Chemistry*14: 2895-2902.

[35] Lenka, S. and Nayak, P.L. 1982. "Grafting vinyl monomers onto nylon 6. VI. Graft copolymerization of methyl methacrylate onto nylon 6 using acetylacetonate complex of Mn(III), Co(III), and Fe(III)." *Journal of Applied Polymer Science*27:1959-65.

[36] Liu, Y., Zhang,Y., Liu, Z. and Deng, K. 2002. "Graft copolymerization of butyl acrylate onto casein initiated by potassium diperiodatonickelate(IV) in alkaline medium." *European Polymer Journal* 38: 1619-1625.

[37] Lenka, S. and Nayak, P.L. 1982. "Grafting vinyl monomers onto nylon 6. VI. Graft copolymerization of methyl methacrylate onto nylon 6 using acetylacetonate complex of Mn(III), Co(III), and Fe(III). *Journal of Applied Polymer Science*27(6):1959-65.

[38] Mohanty, A.K., Tripathy, P.C., Misra, M., Parija, S. and Sahoo, S. 2000. "Chemical modification of pineapple leaf fiber: Graft copolymerization of acrylonitrile onto defatted pineapple leaf fibers." *Journal of Applied Polymer Science* 7: 3035-3043.

[39] Singh, I., Bajpaia, P.K., Malika, D., Sharma, A.K. and Kumar, P.2011. "Feasibility study on microwave joining of 'green' composites." *Akademeia* 1: 1923-1504.

[40] Singha, A.S. and Rana, A.K. 2012. "Ce(IV) ion-initiated and microwave radiation-induced graft copolymerization of acrylic acid onto lignocellulosic fibers." *International Journal of Polymer Analysis and Characterization*17: 72-84.

[41] Kaith, B.S., Singha, A.S. and Kalia, S. 2007. "Grafting MMA onto flax under the influence of microwave radiation and the use of flax-g-poly(MMA) in preparing PF composites." *Autex Research Journal* 7: 119-129.

[42] Da-Wei, H., Jian-Kun, W., Hai-Rong, L. and Ming-Ming, X. 2011. "The graft copolymerization of corn starch by microwave irradiation with rheological phase reaction." *Journal of Applied Polymer Science* 121: 1481-1487.

[43] Badawy, S.M., Dessouki, A.M. and El-Din, H.M.N. 2001. "Direct pyrolysis mass spectrometry of acrylonitrile-cellulose graft copolymer prepared by radiation-induced graft polymerization in presence of styrene as homopolymer suppressor."*Radiation Physics and Chemistry* 61: 143-148.

[44] Kang,P.H., Jeun, J.P., Chung, B.Y., Kim, J.S. and Nho, Y.C. 2007. "Preparation and characterization of glycidyl methacrylate (GMA) grafted kapok fiber by using radiation induced-grafting technique." *Journal of Industrial and Engineering Chemistry*13: 956-958.

[45] Rout, J., Misra, M., Tripathy, S.S., Nayak, S.K. and Mohanty, A.K. 2002. "Surface modification of coir fibers. II. Cu(II) initiated graft copolymerization of acrylonitrile onto chemically modified coir fibers." *Journal of Applied Polymer Science* 84: 75-82.

[46] Ibrahim, N.A., Wan-Yunus, W.M.Z., Abu-Ilaiwi, F.A.F., Rahman, M.Z.A., Ahmad, M.B. and Dahlan, K.Z.M. 2003. "Optimized condition for grafting reaction of poly(butyl acrylate) onto oil palm empty fruit bunch fibre." *Polymer International*52: 1119-1124.

[47] Wang, L. and Xu, Y. 2006. "Preparation and characterization of graft copolymerization of ethyl acrylate onto hydroxypropyl methylcellulose in aqueous medium." *Cellulose* 13: 191-200.

[48] Kumar, V., Bhardwaj, Y.K., Jamdar, S.N., Goel, N.K. and Sabharwal, S.J. 2006. "Preparation of an anion - exchange adsorbent by the radiation - induced grafting of vinylbenzyltrimethylammonium chloride onto cotton cellulose and its application for protein adsorption." *Applied Polymer Science* 102: 5512-5521.

[49] Liu, Y., Zhang,Y., Liu, Z. and Deng, K. 200. "Graft copolymerization of butyl acrylate onto casein initiated by potassium diperiodatonickelate(IV) in alkaline medium." *European Polymer Journal* 38: 1619-1625.

[50] Yea, J.H., Donga, J.J., Lub, J.L., Zhenga, X.Q., Jina, J., Chena, H. and Lianga,Y.R. 2010."Effect of graft copolymerization of fir sawdust lignocellulose with N-vinylpyrrolidone on adsorption capacity to tea catechins." *Carbohydrate Polymers* 81: 441-447.

[51] Kaith, B.S. and Kalia, S. 2007. "Synthesis and characterization of graft co-polymers of flax fiber with binary vinyl monomers."*International Journal of Polymer Analysis and* Characterization 12: 401-412.

[52] Mittal, H., Kaith, B.S. and Jindal, R. 2010. "Synthesis, characterization and swelling behaviour of poly(acrylamide-comethacrylic acid) grafted Gum ghattibased superabsorbent hydrogels." *Advances in Applied Science Research* 1: 56-66.

[53] Kalia, S., Kumar, A, Kaith, B.S. 2011. "Sunn hemp cellulose graft copolymers polyhydroxybutyrate composites: morphological and mechanical studies." *Advance Material Letter*2: 17-25.

[54] Gupta, K.C. and Sahoo, S. 2000. "Ceric ion initiated grafting on cellulose from binary mixture of acrylonitrile and methylacrylate." *Journal of Macromolecular Science: Part A* 37: 447-468.

[55] Ojah, R. and Dolui, S.K. 2006. "Graft copolymerization of methyl methacrylate onto Bombyxmori initiated by semiconductor-based photocatalyst." *Bioresource Technology* 97: 1529-1535.

[56] Khan, G.M.A.T.2007."Dyeing of grafted jute fibre with reactive dyes and its improved properties."*International Journal of Science and Technology* 12: 8-13.

[57] Khan, G.M.A., Saheruzzanam, M., Abdur, R.S.M., Islam, M.S., Alam, M.S. and Islam, M.M. 2009. "Grafting of acrylonitrile monomer onto bleached okra fiber and its textile properties." *Indian journal of fibre and textile research* 34: 321-327.

[58] Singha, A.S. and Rana, A. K. 2012. "A study on benzoylation and graft copolymerization of lignocellulosic cannabis indica fiber." *Journal of Polymer and Environment* 20: 361-371.

[59] Kaith, B.S. and Kalia, S.2008. "Graft copolymerization of MMA onto flax under different reaction conditions: a comparative study." *Express Polymer Letters* 2(: 93-100.

[60] Kaith, B.S., Singha, A.S., Kumar, S. and Kalia, S. 2008. "Mercerization of flax fiber improves the mechanical properties of fiber-reinforced composites." *International Journal of Polymeric Materials* 57: 54-72.

[61] Kalia, S and Kaith, B.S. 2009. "Synthesis of flax-g-copolymers under pressure for use in phenolic composites as reinforcement." *Journal of Chilean Chemical Society* 54: 108-112.

[62] Kalia, S, Kaith, B.S. 2008. "Mechanical properties of phenolic composites reinforced with flax-g-copolymers prepared under different reaction conditions - a comparative study." *E-Journal of Chemistry*5: 177-184.

[63] Singha, A.S. and Thakur, V. K. 2008. "Saccaharumcilliarefiber reinforced polymer composites." *E-Journal of Chemistry* 5: 782-791.

[64] Singha, A.S. and Thakur, V.K 2008. "Effect of fibre loading on urea-formaldehyde matrix based green composites."*Iranian Polymer Journal*17: 861-873.

[65] Thakur, V.K. Singha, A.S. and Thakur, M.K. 2012. "Modification of natural biomass by graft copolymerization." *International Journal of Polymer Analysis and Characterization*17: 547-555.

[66] Thakur, V.K., Singha, A.S. and Nath, M.B. 2011. "Graft copolymerization of methyl methacrylate onto cellulosic biofibers."*Journal of Applied Polymer Science* 122: 532-544.

[67] Ekebafe, L.O., Imanah, J.E., Ekebafe, M.O. and Ugbesia, S.O. 2010. "Graft polymerization of polyacrylonitrile onto rubber (hevea brasiliensis) seed shell-cellulosic and its utilization potential for heavy metal uptake from aqueous medium." *The Pacific Journal of Science and Technology*11:488-498.

[68] Singha, A.S. and Thakur, V. K. 2008. "Saccaharumcilliarefiber reinforced polymer composites." *E-Journal of Chemistry* 5: 782-791.

[69] Singha, A.S. and Thakur, V.K 2008. "Effect of fibre loading on urea-formaldehyde matrix based green composites." *Iranian Polymer Journal*17: 861-873.

[70] Analysis and characterization of microwave irradiation induced graft copolymerization of methyl methacrylate onto deilignifiedgrewiaoptivafibre. Amar Singh Singha, BhanuPriya Thakur, Deepak Pathania, *International Journal of Polymer Analysis and Characterization*, 19(2): 115-123, 2014.

[71] Priya, B., Singha, A.S., Pathania, D. 2014"Synthesis and Kinetics of Ascorbic Acid Initiated Graft Copolymerized Delignified Cellulosic Fibre." *Polymer Engineering and Science,* 2: 211-214.

[72] Priya. B., Gupta, V. K., Pathania, D., Singha, A. S. 2014. "Synthesis, characterization and antibacterial activity of biodegradable starch/PVA composite films reinforced with cellulosic fibre." *Carbohydrate Polymers,* 109: 171-179.

[73] Thakur, V.K., Singha, A.S. and Thakur, M.K. 2012. "Pressure induced synthesis of EA grafted saccaharumcilliare fibers." *International Journal of Polymeric Materials and Polymeric Biomaterials* 63: 17-22.

[74] Wu, Z.H., Pittman, C.U. and Gardner, S.D. 1996. "Chemical treatments of natural fiber for use in natural fiber-reinforced composites: A review." *Carbon* 34:59-62.

[75] Garcia-Jaldon, C., Dupeyre, D. and Vignon, M.R. 1998. "Fibres from semi-retted hemp bundles by steam explosion treatment." *Biomass and Bioenergy* 14: 251-260.

[76] Sreekala, M.S., Kumaran, M.G., Joseph, Jacob, S.M. and Thomas, S. 2000. "Oil palm fibre reinforced phenol formaldehyde composites: influence of fibre surface modifications on the mechanical performance." *Applied Composite Materials* 7: 295-329.

[77] Mishra, S., Mohanty, A.K., Drzal, L.T., Mishra, M., Parija, S., Nayak, S.K. and Tripathy, S.S. 2003. "Studies on mechanical performance of biofibre/glass reinforced polyester hybrid composites." *Composite Science and Technology* 63:1377-1385.

[78] Van de Weyenberg, I., Ivens, J., De Coster, A., Kino, B., Baetens, E. and Vepoes, I. 2003. "Influence of processing and chemical treatment of flax fibres on their composites." *Composite Science and Technology* 63:1241-1246.

[79] Jacob, M., Thomas, S. and Varughese, K.T. 2004. "Mechanical properties of sisal/oil palm hybrid fiber reinforced natural rubber composites." *Composite Science and Technology* 64: 955-965.

[80] Manikandan Nair, K.C., Thomas, S and Groeninckx, G. 2001. "Thermal and dynamic mechanical analysis of polystyrene composites reinforced with short sisal fibres." *Composite Science and Technology* 61: 2519-2529.

[81] Xue, L., Lope, G.T. and Satyanarayan, P. 2007. "Chemical treatment of natural fibre for use in natural fibre-reinforced composites: A review." *Polymer Environment* 15: 25-33.

[82] Joseph, K., Thomas, S. and Pavithran, C. 1996. "Effect of chemical treatment on the tensile properties of short sisal fibre reinforced polyethylene composites." *Polymer* 37: 5139-5149.

[83] Joseph, K., Filho, R.D.T., James, B., Thomas, S. and Carvalho, L.H. 1999. "A review on sisal fiber reinforced polymer composites."*RevistaBrasileira de Engenharia Agricola e Ambiental* 3:367-376.

[84] Nair, K.C.M., Thomas, S. and Groeninckx, G. 2001. "Thermal and dynamic mechanical analysis of polystyrene composite reinforced with short sisal fibers." *Composite Science and Technology* 61: 2519-2529.

[85] Singha, A.S. and Thakur, V.K. 2009. "Grewiaoptivafiber reinforced novel, low cost polymer composites." *E-Journal of Chemistry*6:71-76.

[86] Abdelmouleh, M., Boufis, S., Belgacem, M.N. and Dufresne, A. 2007. "Short natural-fibre reinforced polyethylene and natural rubber composites: Effect of silane coupling agents and fibre loading." *Composite Science and Technology* 67: 1627-1639.

[87] Wang, B., Panigrahi, S., Tabil, L. and Crerar, W. 2007."Pre-treatment of flax fibers for use in rotationally molded biocomposites." *Journal of Reinforced Plastic Composite* 26: 447-463.

[88] George, J., Sreekala, M.S. and Thomas, S. 2001. "A review of interface modification and characterisation of natural fibre reinforced plastic composites." *Polymer Engineering and Science* 41: 1471-1485.

[89] Valadez-Gonzalez, A., Cervantes-Uc, J.M., Olayo, R. and Herrera-Franco, P.J. 1999. "Chemical modification of henequen fibres with an organosilane coupling agent." *Composite Part B: Engineering* 30: 321-331.

[90] Valadez-Gonzalez, A., Cervantes-Uc, J.M., Olayo, R. and Herrera-Franco, P.J. 1999. "Chemical modification of henequen fibres with an organosilane coupling agent." *Composite Part B: Engineering* 30: 321-331.

[91] Agrawal, R., Saxena, N.S., Sharma, K.B., Thomas, S. and Sreekala, M.S. 2000."Activation energy and crystallization kinetics of untreated and treated oil palm fibre reinforced phenol formaldehyde composites." *Material Science and Engineering: A* 277:77-82.

[92] Sever, K., Sarikanat, M., Seki, Y., Erkan, G.K. and Erdog, U.M. 2010. "Mechanical properties of γ methacryl-oxypropyltrimethoxysilane-treated jute/ polyester composites." *Composite Materials* 44: 1913-1924.

[93] Rong, M.Z., Zhang, M.Q., Liu, Y., Yang, G.C. and Zeng, H.M. 2001. "Effect of fiber treatment on the mechanical properties of unidirectional sisal-reinforced epoxy composites." *Composite Science and Technology* 61:1437-1447.

[94] Sreekala, M.S., Kumaran, M.G., Joseph, S., Jacob, M. and Thomas, S. 2000. "Oil palm fibre reinforced phenol formaldehyde composites: influence of fibre surface modifications on the mechanical performance." *Applied Composite Materials* 7: 295-305.

[95] Tserki, V., Zafeiropoulos, N.E., Simon, F. and Panayiotou, C. 2005. "A study of the effect of acetylation and propionylation surface treatments on natural fibres." *Composites: Part A* 36: 1110-1118.

[96] Bledzki, A.K., Mamun, A.A., Lucka-Gabor, M. and Gutowski, V.S.2008. "The effects of acetylation on properties of flax fibre and its polypropylene composites." *Express Polymer Letters* 2: 413-422.

[97] Rowell, R.M., Young, R.A. and Rowell, J.K. 2000. "Paper and composites from agro-based resources." *Carbohydrate Polymers* 41: 69-78.

[98] Paul, A., Joseph, K. and Thomas, S. 1997. "Effect of surface treatments on the electrical properties of low-density polyethylene composites reinforced with short sisal fibers." *Composite Science and Technology* 57: 67-79.

[99] Sreekala, M.S. and Thomas, S. 2003. "Effect of fibre surface modification on water-sorption characteristics of oil palm fibres." *Composite Science and Technology* 63: 861-869.

[100] Ouajai, S. and Shanks, R.A. 2005. "Composition, structure and thermal degradation of hemp cellulose after chemical treatment." *Polymer degradation and stability* 89: 327-335.

[101] Mishra, S., Misra, M., Tripathy, S.S., Nayak, S.K. and Mohanty, A.K. 2002. "The influence of chemical surface modification on the performance of sisal polyester biocomposites." *Polymer Composites* 23:164-170.

[102] Bledzki, K., Reihmane, S. and Gassan, J. 1996. "Properties and modification methods for vegetable fibers for natural fiber composites."*Journal of Applied Polymer Science* 59: 1329-1336.

[103] Frederick, T.W. and Norman, W. 2004. "Natural fibers plastics and composites." *Kluwer Academic Publishers,* New York.

[104] George, J., Janardhan, R., Anand, J.S., Bhagawan, S.S. and Thomas, S. 1996. "Melt rheological behaviour of short pineapple fibre reinforced low density polyethylene composites."*Polymer* 37: 5421-5431.

[105] Joseph, P.V., Joseph. K., Thomas, S., Pillai, C.K.S., Prasad, V.S., Groeninckx, G. and Sarkissova, M. 2003. "The thermal and crystallization studies of short sisal fibre reinforced polypropylene composites." *Composite A: Applied Science and Manufacturing* 34: 253-266.

[106] Gassan, J. and Bledzki, A.K. 1997."The influence of fiber-surface treatment on the mechanical properties of jute-polypropylene composites." *Composite A: Applied Science and Manufacturing* 28:1001-1005.

[107] Viviana, P.C., Claudia, V., Jose, M.K. and Analia, V. 2004. "Effect of chemical treatment on the mechanical properties of starch based blends reinforced with sisal fibre." *Composite Material* 38: 1387-1399.

[108] Mustata, A. 1997. "Factors influencing fiber-fiber friction in the case of bleached flax." *Celluose Chemistry and Technology* 31:405-413.

[109] Taj, S., Munawar, M.A. and Khan, S. 2007. "Review: Natural fibre reinforced polymer composites." *The Proceedings of the Pakistan Academy of Sciences* 44: 129-144.

[110] Jonoobi, M., Harun, J., Shakeri, A., Misra, M., &Oksman K. 2009. "Chemical composition crystanility and thermal degradation of bleached and unbleached kanefbast (Hibiscus Cannabinus)." *Bioresource*4: 626-639.

[111] Bledzki, A.K. and Gassan, 1999. "Composites reinforced with cellulose based fibres." *Journal of Progress in Polymer Science* 24: 221-274.

[112] Joseph, S., and Thomas, S., (2008) Electrical properties of banana fiber-reinforced phenol formaldehyde composites.*Journal of Applied Polymer Science*109: 256-263.

[113] Mishra, S., Misra, M., Tripathy, S.S., Nayak, S.K. and Mohanty, A.K. 2002. "The influence of chemical surface modification on the performance of sisal polyester biocomposites." *Polymer Composites* 23:164-170.

[114] Mishra, S., Naik, J.B. and Patil, Y.P. 2000. "The compatibilising effect of maleic anhydride on swelling and mechanical properties of plant-fiber-reinforced novolac composites." *Composite Science and Technology* 60:1729-1735.

[115] Mohanty, A.K., Misra, M. and Drzal, L.T. 2001. "Surface modifications of natural fibers and performance of the resulting biocomposites: an overview. *Comp Interfaces* 8: 313-43.

[116] Chand, N., Nigrawal,A. and Jain, D. 2008. "Dielectric behavior of maleic anhydride grafted polypropylene (MAgPP) modified sisal fiber reinforced pp composites. *Journal of Natural Fibers* 5: 270-288.

[117] Sreekala, M.S., Kumaran, M.G. and Thomas, S. 2002. "Water sorption in oil palm fiber reinforced phenol formaldehyde composites." *Composite A: Applied Science and Manufacturing* 33: 763-777.

[118] Leonard, Y.M. and Ansell, M.P. 1999. "The effect of chemical treatment on the properties of hemp, sisal, jute and kapok fibres for composite reinforcement." *Macromolecular Chemistry* 272: 109-110.

[119] Leonard, Y.M. and Martin, P.A. 2002."Chemical modification of hemp, sisal, jute and kapok fibres by alkalisation." *Journal of Applied Polymer Science* 84: 2222-2234.

[120] Miah, M.J., Khan, M.A. and Khan, R.A. 2011. "Fabrication and characterization of jute fiber reinforced low density polyethylene based composites: effects of chemical treatment." *Journal of Scientific Research*3: 249-259.

[121] Sapieha, S., Allard, P. and Zang, Y.H. 1990. "Dicumylperoxidemodified cellulose/LLDPE composites." *Applied Polymer Science* 41: 2039-2048.

[122] Wambua, P., Ivens, U. and Verpoest, I. 2003. "Natural fibers: can they replace glass in fiber-reinforced plastics?" *Composite Science and Technology*63:1259-1264.

[123] Kalaprasad, G., Francis, B., Radhesh, C., Pavithran, C., Groeninckx, G. and Thomas, S. 2004. "Effect of fibre length and chemical modifications on the tensile properties of intimately mixed short sisal/glass hybrid fibre reinforced low density polyethylene composites." *Polymer International* 53: 1624-1638.

[124] Kaith, B.S., Singha, A.S., Kumar, S. and Kalia, S. 2008. "Mercerization of flax fiber improves the mechanical properties of fiber-reinforced composites." *International Journal of Polymeric Materials* 57: 54-72.

[125] Perpeir, P.H.F., Voorward, H.C.J., Cioffi, M.O.H., Mulinati, D.R., Luz, S.M.D. and Silva, M.L.C.P.D. 2011. "Sugarcane bagasses pulping and bleaching: thermal and chemical characterization." *Bioresource* 6: 2471-2482.

[126] Pickering, K.L., Li, Y., Farrell, R.L. and Lay, M. 2007. "Interfacial modification of hemp fibre reinforced composites using fungal and alkali treatment." *Biobased Material and Bioenergy* 1: 109-117.

[127] Prasad, S.V., Pavithran, C. and Rohatgi, P.K. 1983. "Alkali treatment of coir fibres for coir-polyester composites." *Material Science* 18: 1443-1454.

[128] Ray, D., Sarkar, B.K., Rana, A.K. and Bose, N.R. 2001. "Effect of alkali treated jute fibres on composite properties." *Bulletin Material Science* 24: 129-135.

[129] Dipa, R., Sarkar, B.K., Rana, A.K. and Bose, N.R. 2001. "Effect of alkali treated jute fibres on composite properties." *Bulletin of Materials Science* 24: 129-135.

[130] Vert, M, Santos, I.D., Ponsart, S., Alauzet, N., Morgat, J.L., Coudane, J. and Garreau, H. 2002."Degradable polymers in a living environment: where do you end up?" *Polymer International* 51: 840-844.

[131] Wollerdorfer, M. and Bader, H. 1998. "Influence on natural fibers on the mechanical properties of biodegradable polymers." *Industrial Crops and Products* 8: 105-112.

[132] Averous, L., Fringant, C. and Moro, L. 2001. "Plasticized starch-cellulose interactions in polysaccharide composites." *Polymer* 42: 6571-6578.

[133] Reddy, N. and Yang, Y. 2011. "Completely biodegradable soyprotein-jute biocomposites developed using water without any chemicals as plasticizer." *Industrial Crops and Products*33: 35-41.

[134] Bahrami, S. B., Kordestani S.S., Mirzadeh, H and Mansoori, P. 2003. "Poly (vinyl alcohol)-chitosan blends: preparation, mechanical and physical properties." *Iranian Polymer Journal* 12: 139-146.

[135] Katori, S. and Kimura, T. 2002. "Injection moulding of silk fiber reinforced biodegradable composites." In Brebbia C.A. and de Wilde W.P. (ed), High Performance Structures and Composites. WIT Press, Boston, 2: 97-105

[136] Lee S.M., Cho D., Park W.H., Lee S.G., Han S.O. and Drzal L.T. 2005. "Novel silk/poly (butylenes succinate) biocomposites: the effect of short fiber content on their mechanical and thermal properties." *Composites Science and Technology* 65: 647-657.

[137] Ghanbarzadeh, B., Oromiehie A., Musavi, M., Razmi, E. and Milani, J. 2006. " Effect of polyolic plasticizers on rheological and thermal properties of zein resins. *Iranian Polymer Journal* 15: 779-787.

[138] Lee, S.H. and Wang, S. 2006. "Biodegradable polymers/bamboo fiber biocomposite with bio-based coupling agent." *Composites: Part A* 37: 80-91.

[139] Lee, B.H., Kim, H.J. and Yu, W.R. 2009. "Fabrication of long and discontinuous natural fibre reinforced polypropylene biocomposites and their mechanical properties." *Fiber and Polymers* 10: 83-90.

[140] Cheung, H.Y., Lau, K.T. and Tao, X.M. 2008. "A potential material for tissue engineering: silkworm silk/PLA biocomposite. *Composite Part. B: Engineering*36: 1026-1033.

[141] Lipsa, R., Tudorachi, N. and Vasile, C. 2008. "Poly(vinyl alcohol)/poly(lactic acid) blends biodegradable films doped with colloidal silver." *Revue Roumaine de Chimie*53(5): 405-413

[142] Majid, R.A., Ismail, H. and Taib, R.M. 2010. "Effects of polyethylene-g-maleic anhydride on properties of low density polyethylene/thermoplastic sago starch reinforced kenaffibre composites." *Iranian Polymer Journal*19: 501-510.

[143] Yee, T.W., Rahman, W.A.W.A. and Sin, L.T. 2011. "Properties and morphology of poly(vinyl alcohol) blends with sago pith bio-filler as biodegradable composites." *Journal of Vinyl andAdditive Technology* 17: 184-189.

[144] Tripathi, S., Mehrotra, G.K. and Dutta, P.K. (2009) Physicochemical and bioactivity of cross-linked chitosan-PVA film for food packaging applications. *International Journal of Biological Macromolecules* 45: 372-376.

[145] Park, H.R., Chough, S.H., Yun, Y.H. and Yoon, S.D. (2005) Properties of starch/PVA blend films containing citric acid as additive. *Journal of Polymers and the Environment* 13(4): 375-382.

[146] Yun, Y.H., Youn, Y.N., Yoon, S.D. and Lee, J.U. 2012. "Preparation and physical properties of starch-based nanocomposite films with the addition of titanium oxide nanoparticles." *Journal of Ceramic Processing Research* 13: 59-64

[147] Ramirez, M.G.L., Muniz, G.I.B., Satyanarayana K.G. and Tanobe, V. 2010. "Preparation and characterization of biodegradable composites based on brazilian cassava starch, corn starch and green coconut fibers." *Matéria (Rio de Janeiro) 15*: 330-337.

[148] Sariffuddin, N., Ismail, H. and Ahmad, Z. 2012. "Effect of fibre loading on properties of Thermoplastic sago starch (TPSS)/kenaf core fibre (KF)."*Bioresources* 7: 4294-4306.

[149] Dhal, J.P and Mishra S.C. (2013) Processing and dielectric properties of bio-fiber reinforced polymer composite using brown grass flower broom. *Journal of Polymer Composites* 1(1): 28-32.

[150] Nigrawal, A. and Chand, N. 2012. "Preparation, dielectric and A.C. conductivity studies on chemically treated sisal powder filled polyvinyl alcohol biocomposites."*International Journal of Engineering. Science and Technology* 4: 191-209.

In: Modified Biopolymers: Challenges and Opportunities ISBN: 978-1-53612-116-2
Editors: D. Pathania, G. Sharma and A. Kumar

Chapter 11

BIO-INSPIRED POLYMER COMPOSITES: ROBUST BIOMEDICAL APPLICATION PODIUM

Enza Torino[2], Deepika Jamwal[1], Kritika Sood[1], Virender Pratap Singh[1], Pardeep Singh[1] and Pankaj Thakur[1,2]
[1]School of Chemistry, Faculty of Basic Sciences, Shoolini University, Solan, India
[2]Center for Advanced Biomaterials for Healthcare (CABHC)
Istituto Italiano Di Tecnologia, Naples, Italy

ABSTRACT

It is an exciting time for chemists to be involved in all aspects of biomedical research. The word "molecular" has always been an integral part of chemistry, but it has gained a new appreciation from physicians and biologists, who have named emerging fields "molecular medicine," "molecular genetics," "molecular biology," "molecular bioengineering," "molecular cardiology," etc. Clearly, "chemistry" or "molecular science" has become the centerpiece of the modern biomedical research. "Drug discovery" has historically been a stronghold of chemistry, especially natural products chemistry and synthetic organic chemistry. Chemistry is thus inherently well-positioned to have a major impact on drug discovery, since other disciplines cannot create novel small molecules. In other words, "drug discovery and development" cannot happen without chemists. Classical natural products-based drug discovery, involving extraction, assay-based functional fractionation, and isolation, characterization, and target validation has been gradually replaced by molecular target-based drug discovery. High-throughput screening of large libraries of compounds (including computer-based in silico screening), lead identification from hits (often using the in silico docking analysis of the protein X-ray crystal structures), and lead optimization have been in the mainstream for the last two decades. However, simple random high-throughput screening of a huge number of small molecule libraries generated by combinatorial chemistry has not produced expected results: there has been only one FDA-approved drug (sunitinib for renal carcinoma) from the high-throughput screening of combinatorial chemistry libraries followed by the optimization of hits. Thus, natural products- based drugs (parent compounds, derivatives, analogs, and mimics) are still major entities among the FDA-approved drugs (57.7% of all drugs). Nevertheless, combinatorial chemistry in the form of parallel synthesis or diversity-oriented synthesis (DOS) for the optimization of highly promising lead compounds has been successful in many drug discovery and development cases. The focused library approach is especially beneficial for clearing ADME/Tox (absorption, distribution, metabolism, excretion, and toxicology) requirements.

Keywords: molecular science, natural products, drug discovery, molecular bioengineering

1. Introduction

The use of polymers and polymeric structures in the treatment, diagnosis, and monitoring of biological systems is continuing to increase in sophistication. Research exploring the use of nanoparticles for drug delivery applications has included the exploration and development of nanocolloids, liposomes, micelles, nanogels, polymer-drug conjugates, and a variety of other polymeric nanoparticles that are composed of organic polymers with specific physical or chemical properties that make them relevant delivery vehicles. One of the current trends is the fabrication of tunable, mono-dispersed, polymeric nanostructures to be used as nanoparticle drug delivery platforms. The Polymer printer technology used to fabricate monodisperse particles from a wide range of particle matrix materials including biocompatible poly(ethylene glycol) and bioabsorbable poly(D-lactic acid) is described along with an abbreviated overview of other polymeric drug delivery platforms. Nanoscience has been variously defined at different fora, books, journals and the web, yetone thing is common; it involves the study of the control of matter on an atomic and molecular scale. This molecular level investigation is at a range usually below 100 nm. In simple terms, a nanometer is one billionth of a meter and the properties of materials at this atomic or subatomic level differ significantly from properties of the same materials at larger sizes. Although, the initial properties of nanomaterials studied were for its physical, mechanical, electrical, magnetic, chemical and biological applications, recently, attention has been geared towards its pharmaceutical application, especially in the area of drug delivery. This is because of the challenges with use of large size materials in drug delivery, some of which include poor bioavailability, *in vivo* stability, solubility, intestinal absorption, sustained and targeted delivery to site of action, therapeutic effectiveness, generalized side effects, and plasma fluctuations of drugs. Of recent, several research in nanodrug delivery have been designed to overcome these challenges through the development and fabrication of nanostructures. It has been reported that, nanostructures have the ability to protect drugs from the degradation in the gastrointestinal tract, the technology can allow target delivery of drugs to various areas of the body. The technology enables the delivery of drugs that are poorly water soluble and can provide means of bypassing the liver, thereby preventing the first pass metabolism Nanotechnology increases oral bioavailability of drugs due to their specialized uptake mechanisms such as absorptive endocytosis and are able to remain in the blood circulation for a long time, releasing the incorporated drug in a controlled fashion, leading to less plasma fluctuations and minimized side-effects. Nanoscale size nanostructures are able to penetrate tissues and are easily taken up by cells, allowing for efficient delivery of drugs to target sites of action. Uptake of nanostructures has been reported to be 15-250 times greater than that of microparticles in the 1-10 um range. Nanotechnology improves performance and acceptability of dosage forms by increasing 70 recent advances in novel drug carrier systems their effectiveness, safety, patient adherence, as well as ultimately reducing health care costs. It may also enhance the performance of drugs that are unable to pass clinical trial phases. Nanotechnology definitely promises to serve as drug delivery carrier of choice for the more challenging conventional drugs used for the treatment and management of chronic diseases such as cancer, asthma, hypertension, HIV and diabetes. It has been reported that 1 in 10 marketed drugs have solubility problems, over a third of pipeline drugs are poorly soluble, and almost 2/3 of drugs coming from early pre-clinical development have low solubility. As such, nearly 40% of all possible drug targets fail early due to poor solubility characteristics and insufficient pharmacokinetics [1]. Biological agents, such as siRNA have a short half-life, are quickly degraded and due to these factors need to be protected during circulation, delivered to the desired tissue, and then released intracellularly into the cytosol to be used effectively as a therapeutic. Proteins and peptides face similar hurdles and in addition, they can also trigger an immunological response. The delivery of these therapeutic agents, as well as the delivery of detection and imaging agents for the diagnosis and treatment of disease, has improved somewhat over the years with the development of nano-carriers such as liposomes, micelles, dendrimers, polymer particles, polymer conjugates, and colloidal precipitates. However, only a handful of drugs and imaging agents delivered using these classical approaches have made it into the clinic to treat patients [1-2]. A myriad of opportunities exist for new methodologies for targeted drug delivery and controlled release applications. Polymer chemistry in its many forms has the chance as well as the ability to improve the effectiveness of drug delivery by targeting pharmaceutically relevant cargo to specific sites, managing drug pharmacokinetics and pharmacodynamics, and improving on nonspecific toxicity and immunogenicity [3]. Nanoparticles, nanoscopic "vessels" and polymer-drug conjugates that are the most effective drug delivery vehicles are those that are engineered to be biocompatible, site-specific, have optimal capability to carry relevant cargo, and can demonstrate controlled release of that cargo. Research exploring the use of nanoparticles for drug delivery applications has included the exploration and development of nanocolloids, liposomes, micelles, nanogels, polymer-drug conjugates, and a variety of other polymeric nanoparticles that are composed of organic polymers that have specific physical or chemical properties that make them relevant delivery vehicles. Micelles and liposomes, in particular, are the subject of great interest for drug delivery applications [4-7]. They possess distinctive advantages over usual "free" drug administration in that they have the ability to protect the cargo from premature chemical or physical breakdown, and the surface can be modified such that the liposome is targeted to a specific receptor which can also lead to a decrease in toxic side effects of the drug. In fact, conventional liposomes have been used to carry and transport antimicrobials [8], chemotherapeutics [9], as well as DNA and proteins [10-11]. There are disadvantages to conventional liposomal carriers, however, the most prominent being the limited range of types of cargo that can be loaded due to the solubility and partitioning issues as well as chemical and physical instability [12]. Apart from conventional liposomes, there have been recent advancements in the development of stabilized liposomes using a polyethylene glycol (PEG) coating to create long-circulating liposomal carriers [13-15]. Placing targeting ligands on the surface of the PEG-coated liposome to effect site-specific delivery and release of cargo is also under investigation [16-19]. The functionalized liposomes show some improvement over conventional liposomes, although disadvantages still exist because liposomes are not shape or size specific leading to difficulty in precise load ability [20]. Additional directions in the search for the ideal organic nanoparticle are the design and synthesis of polymer conjugates. Polymer–drug (or protein) conjugates are hybrid structures that tend to be water soluble (due to the control of the chemical composition of the polymer), can be tumor specific via the enhanced permeability and retention (EPR) effect and targeting ligands decorating the polymer portion of the conjugate, and can be captured via the endosome [21-22]. These conjugates offer additional alternatives in the field of nanomedicine but they are plagued with some challenges. The conjugates need to be comprised of high molecular mass, biodegradable polymeric carriers so they can better exploit the EPR effect. Additionally, the conjugates need to move away from the heterogeneity of the carrier and to a more uniform structure. Dendrimers can supply a valuable architecture to these conjugates but lack control over shape and the ability to load large amounts of therapeutics, targeting moieties, and imaging modalities [23]. Using emerging materials science technologies, synthetic polymer nanoparticles have been developed as a more effective drug delivery method. Considerable research has gone into the advancement of this architecture for drug transport. For instance, recent research has yielded nanoparticles composed of biodegradable polymer matrices [24-25]. These show great potential for use as delivery agents for drugs, oligonucleotides for antisense therapy, and DNA in gene therapy through increased control over release. These degradable polymers can be signaled to degrade by changes in pH or temperature [26-28]. Furthermore, using charge or surface modification with targeting ligands, researchers have yielded ways to target cells for uptake [29-31]. An ideal therapeutic carrier would be biocompatible, shape and size specific, monodisperse, composed of virtually any material, amenable to functionalization, and gentle enough for fragile biological cargo. In addition, the fabrication of the carrier must be amenable to industrial scale development. Current research in this important area is driven by the desire to create a methodology to effectively deliver biologically relevant cargo, reduce side effects, and more successfully diagnose and treat disease. By developing lithographic technologies that are able to fabricate transistors that are smaller and smallerto under 100 nm (approximately the size of an individual virus particle) using lithographic techniques. As such, it is now timely to envision and propose using lithographic techniques found in the microelectronics industry to fabricate carriers for use in nano-medicine, because the state-of-the-art today in the microelectronics industry has finally reached the size scale relevant for the size of an ideal drug delivery carrier. Indeed transistors in today's microelectronic devices are even smaller than the traditional drug delivery carriers that have been studied over the last 25 years: virus particles, micelles, liposomes, and polymer particles [32-34].

2. Biopolymer Oriented Smart Drug Delivery Systems

Ideally, nanoparticulate drug delivery system should selectively accumulate in the required organ or tissue and at the same time, penetrate target cells to deliver the bioactive agent. It has been suggested [35-36] that, organ or tissue accumulation could be achieved by the passive or antibody-mediated active targeting [37-38] while the intracellular delivery could be mediated by certain ligands [39-40] or by cell-penetrating peptides [41-42]. Thus, a drug delivery system (DDS) should be multifunctional and possess the ability to switch on and switch off certain functions when necessary. Another important requirement is that different properties of the multifunctional DDS are coordinated in an optimal fashion. Thus, for example, if the system is to be constructed that can provide the combination of the longevity allowing for the target accumulation and specific cell surface binding allowing, two requirements must be met; the half-life of the carrier in the circulation should be long enough and second, the internalization of the DDS by the target cells should proceed fast enough not to allow for the carrier degradation and drug loss in the interstitial space. Intracellular transport of bioactive molecules is one of the key problems in drug delivery. Nanoparticulate DDS, such as liposomes and micelles, are frequently used to increase the efficacy of drug and DNA delivery and targeting [43-44]. So far, very few successful attempts have been made to deliver various drug carriers directly into the cell cytoplasm, bypassing the endocytic pathway, to protect drugs and DNA from the lysosomal degradation, thus enhancing drug efficiency and DNA incorporation into the cell genome. Within the multifunctional DDS, it has been postulated that the development of a DDS built in such a way that during the first phase of delivery, a nonspecific cell-penetrating function is shielded by the organ/tissue-specific delivery will be possible. Upon accumulating in the target, protecting polymer or antibody attached to the surface of the DDS via the stimuli-sensitive bond should detach under the action of local pathological conditions such as abnormal pH or temperature and expose the previously hidden second function allowing for the subsequent delivery of the carrier and its cargo inside cells. While such DDS should be stable in the blood for a long time to allow for an efficient target accumulation, it has to lose the protective coat inside the target almost instantly to allow for fast internalization thereby minimizing the washing away of the released drug or DNA. Intracellular trafficking, distribution, and fate of the carrier and its cargo can be additionally controlled by its charge and composition, which can drive it to the nuclear compartment or toward other cell organelles. It has been reported within the past few years, that specific proteins and peptides (such as TAT peptide) can enter cell cytoplasm directly and even target cell nuclei [45-46].

Selected proteins and peptides have also been used for the intracellular delivery of small drug molecules, large molecules (enzymes, DNA), and nanoparticulates (quantum dots, iron oxide nanoparticles, liposomes) [47-52]. The mechanism of this phenomenon is currently a subject of investigation, although important progress has been made. Indeed, some reports show that electrostatic interactions and hydrogen bonding lay behind certain proteins and peptides-mediated direct transduction of small molecules [53-54], while the energy dependent macropinocytosis is responsible for certain proteins and peptides-mediated Intracellular delivery of large molecules and nanoparticulates with their subsequent enhanced release from endosomes into the cell cytoplasm [55-58]. One of the most outstanding achievements in the drug delivery field was the development of smart drug delivery systems (SDDSs), also called stimuli-sensitive delivery systems. The concept is based on rapid transitions of a physicochemical property of polymer systems upon a stimulus. This stimulus includes physical (temperature, mechanical stress, ultrasound, electricity, light), chemical (pH, ionic strength), or biological (enzymes, biomolecules) signals and such stimuli can either be internal, resulting from changes in the physiological condition of a living subject, or "external" signals, artificially induced to provoke desired events. SDDS provides a programmable and predictable drug release profile in response to various stimulation sources. Figure 1 below shows a typical smart drug delivery system; Depending on the desired applications, one may design different drug delivery systems for enhanced therapeutic efficiency with low systemic toxicity and side effects. SDDS has several advantages compared to conventional drug delivery systems. The conventional controlled release systems are based on the predetermined drug release rate irrespective of the environmental condition at the time of application. On the other hand, SDDS is based on the release-on-demand strategy, allowing a drug carrier to liberate a therapeutic drug only when it is required in response to a specific stimulation. The best example of SDDS has been self-regulated insulin delivery systems that can respond to changes in the environmental glucose level [59-61]. One of the most widely used SDDSs has been polymeric micelles. Many polymeric micelles consisting of hydrophobic and hydrophilic polymer blocks have been developed. They have been found to dissolve water-insoluble drugs, such as doxorubicin or paclitaxel, at high concentrations. When administered to the body, drug release from polymeric micelles usually depends on simple diffusion, degradation of the micelle blocks, or disruption of the micelles by body components.

Drug delivery is the administration of pharmaceutical agent(s) to achieve a therapeutic effect in patients. Conventional drug-delivery strategies often require repetitive drug administration, and this protocol can lead to strongly fluctuating drug levels that may pose undue physiological discomfort to the recipient [62-63]. Furthermore, certain drugs at high concentrations can sometimes trigger undesirable toxicity or immunological responses in the patient, which can be detrimental [64-65]. Proper drug administration requires a fine balance between time, dosage, and mode of delivery, and, more importantly, should aid in improving patient compliance and clinical outcomes. These factors have consequently led to the emergence of various drug delivery systems, such as liposomes, micelles, emulsions and micro/nanoparticles [66-67]. However, monolithic particulate systems may suffer from burst release due to the presence of undissolved drug trapped on the particle surface. In addition, these monolithic particulate systems do not allow for the loading of multiple drugs from a single particulate carrier. Drug release profiles are usually altered through the direct tailoring of the chemistry or structural composition (e.g., multi-layered structure, porosity, composite particles) of the polymeric matrix [68-69]. Subsequently, the release of the drugs is achieved either by diffusion through the polymer matrix, by gradual dissolution of the polymer, by catalytic action from enzymes in-vivo or a combination of the above [70-74]. A great deal of work had been done in engineering a variety of different particle types, encompassing various shapes, sizes, morphologies, and in evaluating how these can affect drug release kinetics [75-76].

3. Biopolymer-Nanocomposites for Drug Delivery

Nanoparticles made from organic or inorganic materials have attracted increasing attention during recent years for the delivery of drugs, peptides/proteins, genes, and vaccines [77-82]. In general, submicron particles are used for drug delivery, and the active agents can be dissolved, entrapped, or encapsulated within the particles, or adsorbed/conjugated onto the particles' surfaces. Based on the inner structure, nanoparticles may be classified as having a matrix or vesicular system (Figure 2). Matrix - type nanoparticles are homogeneous systems in which the active agents are dissolved/dispersed throughout the whole particles. The vesicular - type nanoparticles (which are also termed nanocapsules) are heterogeneous systems, where the active agents are confined within an aqueous or oily cavity enclosed by the organic or inorganic membrane [77-78]. Depending on the fabrication process and the physico - chemical properties of the components which make up the nanoparticles, the active agents may exist as a solution, a solid solution, or as a crystalline or amorphous solid in the nanoparticles. Organic nanoparticles have been studied extensively due to their good biocompatibility, good biodegradability, and high loading efficiency. These include polymeric nanoparticles, submicron liposomes, solid lipid nanoparticles (SLN), micelles, and dendrimers.

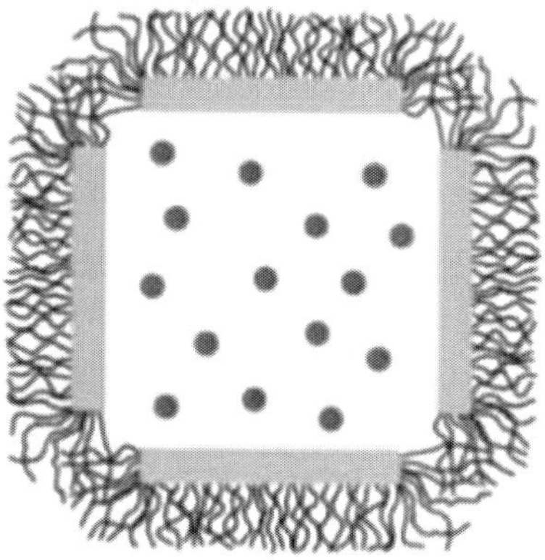

Figure 1. Smart drug delivery system-Gold nanocage covered with polymer.

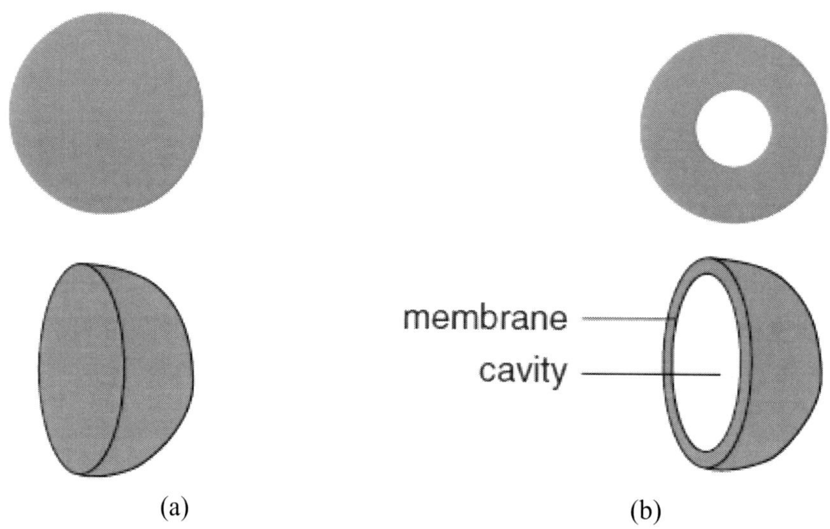

Figure 2. (a) Matrix - type and (b) vesicular - type nanoparticles.

By comparison, inorganic nanoparticles have been investigated to a lesser degree, with the majority of studies being performed with silica nanoparticles and iron oxide nanoparticles. Both size and surface chemistry play important roles in determining the performance of the administered drug - loaded nanoparticles and the eventual therapeutic efficiency. Due to their submicron size, intravenously administered nanoparticles possess an Enhanced Permeation and Retention effect (EPR), and tend to accumulate within the sites of the pathological changes (this is known as " passive targeting") [83]. The extremely small nanoparticles (i.e., those <100 nm in size) are capable of avoiding uptake by the reticuloendothelial system (RES), and consequently exhibit an extended blood circulation time [84]. When orally and/or locally administered (e.g., topical, nasal, corneal), the nanoparticles - due to their large surface area - usually demonstrate a prolonged residence time and provide high local drug concentrations, leading to an increased bioavailability [85]. The importance of size is also reflected at the cellular level. It has been reported that the internalization of nanoparticles by cells (tumor cells, intestinal cells, etc.) is size - dependent the smaller the particles, the greater the extent of internalization [86-87]. Consequently, those nanoparticles loaded with the active agents may show a higher level of cytotoxicity against tumor cells due to their greater cellular uptake [88]. On the other hand, nanoparticles - on the basis of their small size - are able to cross physiological barriers (e.g., gastrointestinal, blood - brain), and this results in an improved bioavailability [89]. The surface chemistry of nanoparticles is equally important. For example, intravenously administered "naked" hydrophobic nanoparticles tend to be captured by the macrophages and to accumulate in the RES (liver, spleen, and lung). This phenomenon called opsonization results in a short circulation time that is induced by the adsorption of blood proteins (opsonins) onto the particles' surface (the process of opsonization). Nanoparticles with a hydrophilic coating, such as poly(ethylene glycol) (PEG; often referred to as "stealth nanoparticles") have the ability to effectively suppress the opsonization process, which leads to an extended blood residence time [90]. Nanoparticle drug delivery systems possess many unique advantages that may result in a reduced toxicity and increased therapeutic efficacy. The term "nanoparticles" when used in relation to drug delivery, should, in fact, be replaced by "composited nanoparticle" although this is not usually made clear in the relevant literature. This substitution should be made because:

- Nanoparticles, due to their high surface energy, have a natural tendency to aggregate. Stabilizers are generally required from the initial synthesis to the final *in vivo* applications.
- The surfaces of the drug - loaded nanoparticles are often decorated with specific molecules so as to impart novel functions, such as a long - circulation time, precise targeting, and stimulus - sensitivity. A surface layer is sometimes also required to reduce the toxicity of the nanoparticles and increase their biocompatibility. Iron oxide nanoparticles are an example of this.

Nanotechnology has been exploited extensively to enhance the pharmacokinetic properties and therapeutic index of a myriad of drugs [91-94]. In the past two decades, a large number of nanoparticle-based therapeutic products have entered clinical development or been approved for clinical use [95]. Polymeric nanoparticles (NPs) and liposomes represent two dominant classes of nanocarriers capable of efficiently encapsulating and delivering a variety of drug classes [95-96]. Polymeric NPs which are suitable for systemic administration may be formed by self-assembly of biodegradable copolymers consisting of two or more polymer blocks with different hydrophobicities. The use of amphiphilic polymers results in the formation of NPs with a hydrophobic core and a hydrophilic shell.

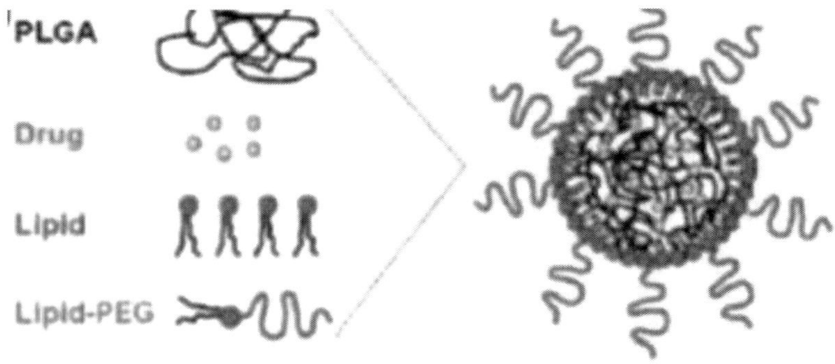

Figure 3. Schematic illustration showing the formulation of lipid polymer hybrid NPs. The NPs comprise a hydrophobic core, a hydrophilic shell, and a lipid monolayer at the interface of the hydrophobic core and the hydrophilic shell.

The core-shell structure of polymeric NPs allows them to encapsulate and carry poorly water-soluble drugs, decrease the biofouling of the NPs resulting in long circulation half-life, release drugs at a sustained rate in the optimal range of drug concentration, and be further functionalized with targeting ligands for differential delivery [97-99]. In addition, the preparation of polymeric NPs based on the nanoprecipitation or double emulsion methods is simple and scalable [97]. For example, Genexol-PM is a NP formulation of paclitaxel and poly(D, L-lactide)-bpolyethylene glycol-methoxy, which has been approved for metastatic breast cancer therapy in Korea [100]. Liposomes are spherical lipid vesicles with a bilayer structure of natural or synthetic amphiphilic lipid molecules. Liposomes have been widely used as drug delivery vehicles because of the favorable safety profile, ease of surface modification, and long systemic circulation half-life that can reach days after being surface modified with hydrophilic polymers such as polyethylene glycol (PEG) [101-102]. Several liposomal drug formulations have been approved for clinical use, including Doxil (doxorubicin liposomes) [103] which was first to be approved in 1995, AmBisome (amphotericin B liposomes), DaunoXome (daunorubicin liposomes), DepoCyt (cytarabine liposomes), DepoDur (morphine liposomes), and Visudyne (verteporfin liposomes) [95, 102]. Given the clinical success of polymeric NPs and liposomes, we hypothesized that a lipid-polymer hybrid NP might be developed that takes advantage of the unique strengths of polymeric NPs and liposomes. The hybrid NPs can be a robust drug delivery platform with high drug encapsulation yield, tunable and sustained drug release profile, excellent serum stability, and potential for differential targeting of cells or tissues.

4. Situate Explicit or Selective Targeting

Site-directed targeting to cells or organs is desired to reduce the concomitant negative effects due to the action of the drug in sites other than necessary in the organism. As already mentioned, by encapsulating the drug in a carrier, the distribution process depends on the carriers' characteristics. Moreover, the carrier can be modulated to provide selective targeting to the cell or organ of interest, thus minimizing unwanted systemic side effects. For this purpose, two mechanisms may be used; passive targeting and active targeting. Passive targeting takes place due to the action of the reticuloendothelial system (RES) in the common response of the organism to destroy foreign materials. Passive targeting is mainly dictated by the physical properties of the carrier and its interaction with plasma proteins. This form of targeting can be used to target diseases that affect the RES, for instance this is used to activate the immune system to destroy tumor cells. Passive targeting usually occurs by phagocytosis of the particle or carrier by the mononuclear phagocytic system, belonging to the RES. In particular when the disease occurs in cells other than the RES, this kind of uptake must be avoided. In such cases, active targeting needs to be used. Active targeted drug delivery occurs when the drug or carrier are directed to specific sites, in particular receptors located on the cell's membrane or tissue of interest, with the aid of a homing device (antibody, ligand, epitope) [104]. For instance, to provide recognition to specific target cells, antibodies were covalently attached to the surface of liposomes [105-111]. Active targeting usually involves the attachment of a ligand to the surface of the carrier to achieve specific ligand-receptor interaction. Once the targeted interaction takes place, the cell's mechanisms, mainly receptor-mediated endocytosis, provides the proper conditions for internalization of the carrier.

5. Biopolymer Functionalized Magnetic Nanoparticles

Magnetic nanoparticles (MNPs) represent a class of non-invasive imaging agents that have been developed for magnetic resonance (MR) imaging. These MNPs have traditionally been used for disease imaging via passive targeting, but recent advances have opened the door to cellular-specific targeting, drug delivery, and multi-modal imaging by these nanoparticles. As more elaborate MNPs are envisioned, adherence to proper design criteria (e.g., size, coating, molecular functionalization) becomes even more essential. This review summarizes the design parameters that affect MNP performance *in vivo*, including the physicochemical properties and nanoparticle surface modifications, such as MNP coating and targeting ligand functionalization that can enhance MNP management of biological barriers. A careful review of the chemistries used to modify the surfaces of MNPs is also given, with attention paid to optimizing the activity of bound ligands while maintaining favorable physicochemical properties.

With recent advances in nanotechnology and molecular biology, development of magnetic nanoparticles (NPs) with specific functional properties has gained much attention. Magnetic nanoferrites have super fluidity of application in myriad fields of technology. Superparamagnetic nanoparticles have been studied extensively from past few decades, due to their unique magnetic properties and ability to function at the cellular and molecular levels makes them a frontier in the field of biotechnology and clinical science [112-116]. The most prominent applications are the treatment of cancer [112-113], drug delivery [114], magnetic resonance imaging (MRI) [115] and magnetic separation [116].

6. Magnetic Nanoferrites Based Hyperthermia

Hyperthermia means generation of 'heat' and in context to cancer it means treatment based on heat generation on the required tumor site [117]. The basic idea hyperthermia is the heating of a magnetic nanoparticle in the presence of an applied a. c. field and this way to convert dissipated magnetic energy (hysteresis losses) into heat energy increases the possibilities for the use of magnetic nano ferrites in the treatment of cancer. Use of super paramagnetic Nano ferrites in the hyperthermia treatment has a number of advantages in comparison to conventional way of giving hyperthermia treatment. Some of the challenges in traditional hyperthermia treatment are (a) restrained penetration of heat into body tissues by laser, ultrasound and microwave radiations [118], (b) Uncontrolled heating of healthy tissue resulting in burns, blisters and unusual pain [119] and (c) Often issue of recurrent tumor growth due to thermal under-dosage in the target region in the case of presence of bone of pelvis or scull above targeted region which shield deep tissues from being destroyed. In clinical aspect of hyperthermia, always efforts are made to enhance the thermal homogeneity at 42-60°C inside the tumor tissue by latest thermometry systems. Magnetic materials with their curie temperature (T_c) between 42 and 60°C are the best candidates for effective hyperthermia treatment, since they act as *in vivo* temperature control switches and prevent over-heating. Use of Super Paramagnetic Nano ferrites in the hyperthermia treatment has a number of advantages in comparison to conventional way of giving hyperthermia treatment [120-121]. In the case of super paramagnetic particles, a heating mechanism that governs heating capacity is the induction of magnetic moments and cause of their rapid variation. Magnetic Nano ferrites such as nickel ferrite ($NiFe_2O_4$), manganese ferrite ($MnFe_2O_4$) and cobalt-nickel ferrite ($Co_xNi_{(1-x)}Fe_2O_4$) [122-123] response to applied a. c. frequency due to two main mechanisms of Neel and Brownian relaxation and in addition to this MNP can capture free floating cancer cells, which can then be carried out of the body and deliver therapeutic heat directly to cancer cells.

6.1. Nanoferrites as Fascinating Carrier for Targeted Drug Delivery

A major disadvantage of cancer treatment is an inevitable condition to limit cancer drug delivery to healthy cells. To limit cytotoxicity of anticancer drug, a suitable way for antitumor drug vehiculation has to be developed. A good solution to this problem is to bind magnetic nanoparticles with the drug molecule. This assembly can perform well under the guided magnetic field for the efficient delivery of the drug. Magnetic Nano ferrites can give remarkable results in these required conditions after making their superparamagnetic nature more effective with required type of surface functionalization. From last few decades a lot of research has been done in this field and magnetic Nano ferrites have been evolved as an efficient carrier because of their large surface to volume ratio and their tendency to agglomerate and adsorb plasma proteins [124].

6.2. Magnetic Resonance Imaging

MRI is a protruding imaging technology that produces three dimensional detailed phenomenal images without the use of harmful radiation. MRI uses powerful magnets which produce a strong magnetic field that forces protons in the body to align with that field. When a radiofrequency current is then pulsed through the patient, the protons are stimulated, and move out of equilibrium, straining against the pull of the magnetic field. When the radiofrequency field is turned off, the MRI sensors are able to detect the energy released as the protons realign with the magnetic field. The time it takes for the protons to realign with the magnetic field, as well as the amount of energy released changes depend on the environment and the chemical nature of the molecules. In general, MR imaging contrast agents must have a strong effect to accelerate longitudinal relaxation (T1) of water and exhibit bright or positive contrast where they are localized, or to accelerate the transverse relaxation (T2) and produce dark or negative-contrast images. Compared with conventional paramagnetic gadolinium complex-based contrast agents such as Magnevist, Prohance, and Dotarem, super paramagnetic nanoparticles (NPs) are emerging as the next generation magnetic probes for MR imaging because of their excellent magnetism and long circulating times [125-127].

6.3. Functionalized Magnetic Nano-Ferrites in Bio-Sensing

Currently we use bulk and thin film forms of magnetic and superconducting materials in the fabrication of biosensors and associated signal processing elements like transducers, etc. Magnetic Nano ferrites based biosensors provide a new horizon for novel functions with a variety of applications in clinical and non-clinical diagnostics and have already made a major impact in various applications, ranging from enzyme electrodes to genoelectronics. The useful properties of Magnetic Nano ferrites suggest that future interdisciplinary research is likely to lead to a new generation of electrochemical biosensors. Researchers are now focusing on understanding the various biomolecule-transducer interactions using these interesting nanomaterials. Highly sensitive bio detection protocols based on

Magnetic Nano ferrites open up the possibility of creating biosensors for disease markers, biological and infectious agents in early-stage detection of disease and threats. The ability to functionalize Magnetic Nano ferrites with the desired groups to bind target biomolecules, and doping these with electronically active materials to obtain enhanced charge transfer, may pave the way for new methods for biosensor development [128-129].

REFERENCES

[1] Duncan, R., (2003). The dawning era of polymer therapeutics. *Nature reviews Drug discovery*, 2(5), 347-360.

[2] Pison, U., Welte, T., Giersig, M. and Groneberg, D.A., (2006). Nanomedicine for respiratory diseases. *European Journal of Pharmacology*, 533(1), 341-350.

[3] Barenholz, Y., (2001). Liposome application: problems and prospects. *Current opinion in colloid and interface science*, 6(1), 66-77.

[4] Lavasanifar, A., Samuel, J. and Kwon, G.S., (2002). Poly (ethylene oxide)-block-poly (L-amino acid) micelles for drug delivery. *Advanced drug delivery reviews*, 54(2), 169-190.

[5] Torchilin, V.P., (2004). Targeted polymeric micelles for delivery of poorly soluble drugs. *Cellular and molecular life sciences*, 61(19), 2549-2559.

[6] Kwon, G.S. and Kataoka, K., (1995). Block copolymer micelles as long-circulating drug vehicles. *Advanced drug delivery reviews*, 16(2), 295-309.

[7] Pinto-Alphandary, H., Andremont, A. and Couvreur, P., (2000). Targeted delivery of antibiotics using liposomes and nanoparticles: research and applications. *International journal of antimicrobial agents*, 13(3), 155-168.

[8] Drummond, D.C., Meyer, O., Hong, K., Kirpotin, D.B. and Papahadjopoulos, D. (1999). Optimizing liposomes for delivery of chemotherapeutic agents to solid tumors. *Pharmacological reviews*, 51(4), 691-744.

[9] Nishikawa, M. and Hashida, M., (2002). Nonviral approaches satisfying various requirements for effective *in vivo* gene therapy. *Biological and Pharmaceutical Bulletin*, 25(3), 275-283.

[10] Rao, M. and Alving, C.R., (2000). Delivery of lipids and liposomal proteins to the cytoplasm and Golgi of antigen-presenting cells. *Advanced drug delivery reviews*, 41(2), 171-188.

[11] Soltero, R. (2005). Oral Protein and Peptide Drug Delivery in Drug Delivery: Principles and Applications.

[12] Allen, T.M., Hansen, C.B. and de Menezes, D.E.L., (1995). Pharmacokinetics of long-circulating liposomes. *Advanced Drug Delivery Reviews*, 16(2-3), 267-284.

[13] Gregoriadis, G., (1995). Engineering liposomes for drug delivery: progress and problems. *Trends in biotechnology*, 13(12), 527-537.

[14] Torchilin, V.P., (1998). Polymer-coated long-circulating microparticulate pharmaceuticals. *Journal of microencapsulation*, 15(1), 1-19.

[15] Park, Y.S., (2002). Tumor-directed targeting of liposomes. *Bioscience reports*, 22(2), 267-281.

[16] Maruyama, K., (2002). PEG-immunoliposome. *Bioscience reports*, 22(2), 251-266.

[17] Maruyama, K., Ishida, O., Takizawa, T. and Moribe, K., (1999). Possibility of active targeting to tumor tissues with liposomes. *Advanced drug delivery reviews*, 40(1), 89-102.

[18] Papahadjopoulos, D., Kirpotin, D.B., Park, J.W., Hong, K., Shao, Y., Shalaby, R., Colbern, G. and Benz, C.C., (1998). Targeting of drugs to solid tumors using anti-HER2 immunoliposomes. *Journal of Liposome Research*, 8(4), 425-442.

[19] Torchilin, V.P., (2005). Recent advances with liposomes as pharmaceutical carriers. *Nature reviews Drug discovery*, 4(2), 145-160.

[20] Vicent, M.J. and Duncan, R., (2006). Polymer conjugates: nanosized medicines for treating cancer. *Trends in biotechnology*, 24(1), 39-47.

[21] Malik, N., Evagorou, E.G. and Duncan, R., (1999). Dendrimer-platinate: a novel approach to cancer chemotherapy. *Anti-cancer drugs*, 10(8), 767-776.

[22] Lee, C.C., MacKay, J.A., Fréchet, J.M. and Szoka, F.C., (2005). Designing dendrimers for biological applications. *Nature biotechnology*, 23(12), 1517.

[23] Diwan, M., Elamanchili, P., Lane, H., Gainer, A. and Samuel, J., (2003). Biodegradable nanoparticle mediated antigen delivery to human cord blood derived dendritic cells for induction of primary T cell responses. *Journal of drug targeting*, 11(8-10), 495-507.

[24] Shen, H., Ackerman, A.L., Cody, V., Giodini, A., Hinson, E.R., Cresswell, P., Edelson, R.L., Saltzman, W.M. and Hanlon, D.J., (2006). Enhanced and prolonged cross-presentation following endosomal escape of exogenous antigens encapsulated in biodegradable nanoparticles. *Immunology*, 117(1), 78-88.

[25] Langer, R., (2000). Biomaterials in drug delivery and tissue engineering: one laboratory's experience. *Accounts of Chemical Research*, 33(2), 94-101.

[26] Bogunia-Kubik, K. and Sugisaka, M., (2002). From molecular biology to nanotechnology and nanomedicine. *Biosystems*, 65(2), 123-138.

[27] Galindo-Rodriguez, S.A., Allemann, E., Fessi, H. and Doelker, E., (2005). Polymeric nanoparticles for oral delivery of drugs and vaccines: a critical evaluation of *in vivo* studies. *Critical Reviews™ in Therapeutic Drug Carrier Systems*, 22(5).

[28] Shenoy, D., Little, S., Langer, R. and Amiji, M., (2005). Poly (ethylene oxide)-modified poly (β-amino ester) nanoparticles as a pH-sensitive system for tumor-targeted delivery of hydrophobic drugs: part 2. *In vivo* distribution and tumor localization studies. *Pharmaceutical research*, 22(12), 2107-2114.

[29] Ramanan, R.M.K., Chellamuthu, P., Tang, L. and Nguyen, K.T., (2006). Development of a temperature-sensitive composite hydrogel for drug delivery applications. *Biotechnology progress*, 22(1), 118-12.

[30] Fernandez, V.V.A., Tepale, N., Sanchez-Diaz, J.C., Mendizabal, E., Puig, J.E. and Soltero, J.F.A., (2006). Thermoresponsive nanostructured poly (N-isopropylacrylamide) hydrogels made via inverse microemulsion polymerization. *Colloid and Polymer Science*, 284(4), 387-395.

[31] Park, E.K., Kim, S.Y., Lee, S.B. and Lee, Y.M., (2005). Folate-conjugated methoxy poly (ethylene glycol)/poly (ε-caprolactone) amphiphilic block copolymeric micelles for tumor-targeted drug delivery. *Journal of controlled release*, 109(1), 158-168.

[32] Fukumori, Y. and Ichikawa, H., (2006). Nanoparticles for cancer therapy and diagnosis. *Advanced powder technology*, 17(1), 1-28.

[33] McAllister, K., Sazani, P., Adam, M., Cho, M.J., Rubinstein, M., Samulski, R.J. and DeSimone, J.M., (2002). Polymeric nanogels produced via inverse microemulsion polymerization as potential gene and antisense delivery agents. *Journal of the American Chemical Society*, 124(51), 15198-15207.

[34] Maeda, H., Wu, J., Sawa, T., Matsumura, Y. and Hori, K., (2000). Tumor vascular permeability and the EPR effect in macromolecular therapeutics: a review. *Journal of controlled release*, 65(1), 271-284.

[35] Palmer, T.N., Caride, V.J., Caldecourt, M.A., Twickler, J. and Abdullah, V., (1984). The mechanism of liposome accumulation in infarction. *Biochimica et Biophysica Acta (BBA)-General Subjects*, 797(3), 363-368.

[36] Jaracz, S., Chen, J., Kuznetsova, L.V. and Ojima, I., (2005). Recent advances in tumor-targeting anticancer drug conjugates. *Bioorganic and medicinal chemistry*, 13(17), 5043-5054.

[37] Torchilin, V.P., (2004). Targeted polymeric micelles for delivery of poorly soluble drugs. *Cellular and molecular life sciences*, 61(19), 2549-2559.

[38] Widera, A., Norouziyan, F. and Shen, W.C., (2003). Mechanisms of TfR-mediated transcytosis and sorting in epithelial cells and applications toward drug delivery. *Advanced drug delivery reviews*, 55(11), 1439-1466.

[39] Gupta, B., Levchenko, T.S. and Torchilin, V.P., (2005). Intracellular delivery of large molecules and small particles by cell-penetrating proteins and peptides. *Advanced drug delivery reviews*, 57(4), 637-651.

[40] Lochmann, D., Jauk, E. and Zimmer, A., (2004). Drug delivery of oligonucleotides by peptides. *European journal of pharmaceutics and biopharmaceutics*, 58(2), 237-251.

[41] Torchilin, V.P., (2005). Lipid-core micelles for targeted drug delivery. *Current drug delivery*, 2(4), 319-327.

[42] Torchilin, V.P., (2005). Recent advances with liposomes as pharmaceutical carriers. *Nature reviews Drug discovery*, 4(2), 145-160.

[43] Maheshwari, A., Mahato, R.I., McGregor, J., Han, S.O., Samlowski, W.E., Park, J.S. and Kim, S.W., (2000). Soluble biodegradable polymer-based cytokine gene delivery for cancer treatment. *Molecular Therapy*, 2(2), 121-130.

[44] Tachibana, R., Harashima, H., Shono, M., Azumano, M., Niwa, M., Futaki, S. and Kiwada, H., (1998). Intracellular regulation of macromolecules using pH-sensitive liposomes and nuclear localization signal: qualitative and quantitative evaluation of intracellular trafficking. *Biochemical and biophysical research communications*, 251(2), 538-544.

[45] Torchilin, V.P., Rammohan, R., Weissig, V. and Levchenko, T.S., (2001). TAT peptide on the surface of liposomes affords their efficient intracellular delivery even at low temperature and in the presence of metabolic inhibitors. *Proceedings of the National Academy of Sciences,* 98(15), 8786-8791.

[46] Wattiaux, R., Laurent, N., Wattiaux-De Coninck, S. and Jadot, M., (2000). Endosomes, lysosomes: their implication in gene transfer. *Advanced drug delivery reviews*, 41(2), 201-208.

[47] Caron, N.J., Torrente, Y., Camirand, G., Bujold, M., Chapdelaine, P., Leriche, K., Bresolin, N. and Tremblay, J.P., (2001). Intracellular delivery of a Tat-eGFP fusion protein into muscle cells. *Molecular Therapy*, 3(3), 310-318.

[48] Vives, E., Brodin, P. and Lebleu, B., (1997). A truncated HIV-1 Tat protein basic domain rapidly translocates through the plasma membrane and accumulates in the cell nucleus. *Journal of Biological Chemistry*, 272(25), 16010-16017.

[49] Fawell, S., Seery, J., Daikh, Y., Moore, C., Chen, L.L., Pepinsky, B. and Barsoum, J.A.M.E.S., (1994). Tat-mediated delivery of heterologous proteins into cells. *Proceedings of the National Academy of Sciences*, 91(2), 664-668.

[50] Rudolph, C., Plank, C., Lausier, J., Schillinger, U., Müller, R.H. and Rosenecker, J., (2003). Oligomers of the arginine-rich motif of the HIV-1 TAT protein are capable of transferring plasmid DNA into cells. *Journal of Biological Chemistry*, 278(13), 11411-11418.

[51] Santra, S., Yang, H., Holloway, P.H., Stanley, J.T. and Mericle, R.A., (2005). Synthesis of water-dispersible fluorescent, radio-opaque, and paramagnetic CdS: Mn/ZnS quantum dots: a multifunctional probe for bio imaging. *Journal of the American Chemical Society*, 127(6), 1656-1657.

[52] Schwarze, S.R., Ho, A., Vocero-Akbani, A. and Dowdy, S.F., (1999). *In vivo* protein transduction: delivery of a biologically active protein into the mouse. *Science*, 285(5433), 1569-1572.

[53] Torchilin, V.P., Levchenko, T.S., Rammohan, R., Volodina, N., Papahadjopoulos-Sternberg, B. and D'Souza, G.G. (2003). Cell transfection *in vitro* and *in vivo* with nontoxic TAT peptide-liposome-DNA complexes. *Proceedings of the National Academy of Sciences*, 100(4), 1972-1977.

[54] Zhao, M., Kircher, M.F., Josephson, L. and Weissleder, R., (2002). Differential conjugation of tat peptide to superparamagnetic nanoparticles and its effect on cellular uptake. *Bioconjugate chemistry*, 13(4), 840-844.

[55] Mai, J. C., Shen, H., Watkins, S. C., Cheng, T., and Robbins, P. D. (2002). Efficiency of protein transduction is cell type-dependent and is enhanced by dextran sulfate. *Journal of Biological Chemistry*, 277(33), 30208-30218.

[56] Vives, E., Richard, J., Rispal, C., and Lebleu, B. (2003). TAT peptide internalization: seeking the mechanism of entry. *Current Protein and Peptide Science*, 4(2), 125-132.

[57] Snyder, E. L., and Dowdy, S. F. (2004). Cell penetrating peptides in drug delivery. *Pharmaceutical research*, 21(3), 389-393.

[58] Wadia, J. S., Stan, R. V., and Dowdy, S. F. (2004). Transducible TAT-HA fusogenic peptide enhances escape of TAT-fusion proteins after lipid raft macropinocytosis. *Nature medicine*, 10(3), 310-315.

[59] Sawant, R. M., Hurley, J. P., Salmaso, S., Kale, A., Tolcheva, E., Levchenko, T. S., and Torchilin, V. P. (2006). "SMART" drug delivery systems: double-targeted pH-responsive pharmaceutical nanocarriers. *Bioconjugate chemistry*, 17(4), 943-949.

[60] Chu, L. Y., Liang, Y. J., Chen, W. M., Ju, X. J., and Wang, H. D. (2004). Preparation of glucose-sensitive microcapsules with a porous membrane and functional gates. *Colloids and Surfaces B: Biointerfaces*, 37(1), 9-14.

[61] Kim, J. J., and Park, K. (2001). Modulated insulin delivery from glucose-sensitive hydrogel dosage forms. *Journal of Controlled Release*, 77(1), 39-47.

[62] Yoshimura, R., Ueda, N., Ikenouch-Sugita, A., Umene-Nakano, W., Hori, H., Kakihara, S., and Nakamura, J. (2009). Fluctuating plasma levels of the active moiety of risperidone is related to occurrence of extrapyramidal symptoms. *International journal of psychiatry in clinical practice*, 13(1), 21-24.

[63] Khung, Y. L., Li Lee, W., Chui, K. L., Liu, Y., Lim, M. P., Huang, C. L., and Loo, S. C. J. (2012). Microencapsulation of Dye-and Drug-Loaded Particles for Imaging and Controlled Release of Multiple Drugs. *Advanced healthcare materials*, 1(2), 159-163.

[64] Pannu, N., and Nadim, M. K. (2008). An overview of drug-induced acute kidney injury. *Critical care medicine*, 36(4), S216-S223.

[65] Aster, R. H., Curtis, B. R., McFarland, J. G., and Bougie, D. W. (2009). Drug-induced immune thrombocytopenia: pathogenesis, diagnosis, and management. *Journal of Thrombosis and Haemostasis*, 7(6), 911-918.

[66] Kenney, B., and Stack, G. (2009). Drug-induced thrombocytopenia. *Archives of pathology and laboratory medicine*, 133(2), 309-314.

[67] Beck-Broichsitter, M., Thieme, M., Nguyen, J., Schmehl, T., Gessler, T., Seeger, W., and Kissel, T. (2010). Novel 'Nano in Nano' Composites for Sustained Drug Delivery: Biodegradable Nanoparticles Encapsulated into Nanofiber Non-Wovens. *Macromolecular bioscience*, 10(12), 1527-1535.

[68] Rothstein, S. N., and Little, S. R. (2011). A "tool box" for rational design of degradable controlled release formulations. *Journal of Materials Chemistry*, 21(1), 29-39.
[69] Lee, W. L., Hong, M., Widjaja, E., and Loo, S. C. J. (2010). Formation and Degradation of Biodegradable Triple-Layered Microparticles. *Macromolecular rapid communications*, 31(13), 1193-1200.
[70] Yuan, Q., Hein, S., and Misra, R. D. K. (2010). New generation of chitosan-encapsulated ZnO quantum dots loaded with drug: synthesis, characterization and *in vitro* drug delivery response. *Acta Biomaterialia*, 6(7), 2732-2739.
[71] Lee, W. L., Widjaja, E., and Loo, S. C. J. (2010). One-Step Fabrication of Triple-Layered Polymeric Microparticles with Layer Localization of Drugs as a Novel Drug-Delivery System. *Small*, 6(9), 1003-1011.
[72] Luo, L., Zhang, Y., Li, F., Si, X., Ding, Y., Deng, D., and Wang, T. (2013). Enzyme mimics of spinel-type $Co_xNi_{1-x}Fe_2O_4$ magnetic nanomaterial for eletroctrocatalytic oxidation of hydrogen peroxide. *Analytica chimica acta*, 788, 46-51.
[73] Bertram, J. P., Rauch, M. F., Chang, K., and Lavik, E. B. (2010). Using polymer chemistry to modulate the delivery of neurotrophic factors from degradable microspheres: delivery of BDNF. *Pharmaceutical research*, 27(1), 82-91.
[74] Klose, D., Siepmann, F., Elkharraz, K., Krenzlin, S., and Siepmann, J. (2006). How porosity and size affect the drug release mechanisms from PLGA-based microparticles. *International journal of pharmaceutics*, 314(2), 198-206.
[75] Lee, W. L., Loei, C., Widjaja, E., and Loo, S. C. J. (2011). Altering the drug release profiles of double-layered ternary-phase microparticles. *Journal of controlled release*, 151(3), 229-238.
[76] Douglas, P., Andrews, G., Jones, D., and Walker, G. (2010). Analysis of *in vitro* drug dissolution from PCL melt extrusion. *Chemical Engineering Journal*, 164(2), 359-370.
[77] Soppimath, K. S., Aminabhavi, T. M., Kulkarni, A. R., and Rudzinski, W. E. (2001). Biodegradable polymeric nanoparticles as drug delivery devices. *Journal of controlled release*, 70(1), 1-20.
[78] Brigger, I., Dubernet, C., and Couvreur, P. (2002). Nanoparticles in cancer therapy and diagnosis. *Advanced drug delivery reviews*, 54(5), 631-651.
[79] Bannon-Peppas, L., and Blanchette, J. O. (2004). Nanoparticle and targeted system for cancer therapy. *Adv. Drug Deliv. Rev.*, 56, 1649-1659.
[80] Hamidi, M., Azadi, A., and Rafiei, P. (2008). Hydrogel nanoparticles in drug delivery. *Advanced drug delivery reviews*, 60(15), 1638-1649.
[81] Liu, Z., Jiao, Y., Wang, Y., Zhou, C., and Zhang, Z. (2008). Polysaccharides-based nanoparticles as drug delivery systems. *Advanced drug delivery reviews*, 60(15), 1650-1662.

[82] Husseini, G. A., and Pitt, W. G. (2008). Micelles and nanoparticles for ultrasonic drug and gene delivery. *Advanced drug delivery reviews*, 60(10), 1137-1152.

[83] Yuan, F., Leunig, M., Huang, S. K., Berk, D. A., Papahadjopoulos, D., and Jain, R. K. (1994). Mirovascular permeability and interstitial penetration of sterically stabilized (stealth) liposomes in a human tumor xenograft. *Cancer research*, 54(13), 3352-3356.

[84] Storm, G., Belliot, S. O., Daemen, T., and Lasic, D. D. (1995). Surface modification of nanoparticles to oppose uptake by the mononuclear phagocyte system. *Advanced drug delivery reviews*, 17(1), 31-48.

[85] Chen, H., and Langer, R. (1998). Oral particulate delivery: status and future trends. *Advanced drug delivery reviews*, 34(2), 339-350.

[86] Desai, M. P., Labhasetwar, V., Walter, E., Levy, R. J., and Amidon, G. L. (1997). The mechanism of uptake of biodegradable microparticles in Caco-2 cells is size dependent. *Pharmaceutical research*, 14(11), 1568-1573.

[87] Win, K. Y., and Feng, S. S. (2005). Effects of particle size and surface coating on cellular uptake of polymeric nanoparticles for oral delivery of anticancer drugs. *Biomaterials*, 26(15), 2713-2722.

[88] Jin, C., Bai, L., Wu, H., Song, W., Guo, G., and Dou, K. (2009). Cytotoxicity of paclitaxel incorporated in PLGA nanoparticles on hypoxic human tumor cells. *Pharmaceutical research*, 26(7), 1776-1784.

[89] Bhardwaj, V., Hariharan, S., Bala, I., Lamprecht, A., Kumar, N., Panchagnula, R., and Ravi Kumar, M. N. V. (2005). Pharmaceutical aspects of polymeric nanoparticles for oral drug delivery. *Journal of Biomedical Nanotechnology*, 1(3), 235-258.

[90] Gref, R., Domb, A., Quellec, P., Blunk, T., Müller, R. H., Verbavatz, J. M., and Langer, R. (1995). The controlled intravenous delivery of drugs using PEG-coated sterically stabilized nanospheres. *Advanced drug delivery reviews*, 16(2-3), 215-233.

[91] Ferrari, M. (2005). Cancer nanotechnology: opportunities and challenges. *Nature Reviews Cancer*, 5(3), 161-171.

[92] Langer, R. (1998). Drug delivery abd targeting. *Nature*, 392(6679), 5-10.

[93] Jiang, W., Kim, B. Y., Rutka, J. T., and Chan, W. C. (2008). Nanoparticle-mediated cellular response is size-dependent. *Nature nanotechnology*, 3(3), 145-150.

[94] Farokhzad, O. C., and Langer, R. (2006). Nanomedicine: developing smarter therapeutic and diagnostic modalities. *Advanced drug delivery reviews*, 58(14), 1456-1459.

[95] Zhang, L., Gu, F. X., Chan, J. M., Wang, A. Z., Langer, R. S., and Farokhzad, O. C. (2008). Nanoparticles in medicine: therapeutic applications and developments. *Clinical pharmacology and therapeutics*, 83(5), 761-769.

[96] Wagner, V., Dullaart, A., Bock, A. K., and Zweck, A. (2006). The emerging nanomedicine landscape. *Nature biotechnology*, 24(10), 1211.

[97] Farokhzad, O. C., and Langer, R. (2006). Nanomedicine: developing smarter therapeutic and diagnostic modalities. *Advanced drug delivery reviews*, 58(14), 1456-1459.
[98] Torchilin, V. P. (2007). Micellar nanocarriers: pharmaceutical perspectives. *Pharmaceutical research*, 24(1), 1.
[99] Tong, R., and Cheng, J. (2007). Anticancer polymeric nanomedicines. *Journal of Macromolecular Science, Part C: Polymer Reviews*, 47(3), 345-381.
[100] Kim, T. Y., Kim, D. W., Chung, J. Y., Shin, S. G., Kim, S. C., Heo, D. S., and Bang, Y. J. (2004). Phase I and pharmacokinetic study of Genexol-PM, a cremophor-free, polymeric micelle-formulated paclitaxel, in patients with advanced malignancies. *Clinical cancer research*, 10(11), 3708-3716.
[101] Moghimi, S. M., and Szebeni, J. (2003). Stealth liposomes and long circulating nanoparticles: critical issues in pharmacokinetics, opsonization and protein-binding properties. *Progress in lipid research*, 42(6), 463-4.
[102] Torchilin, V. P. (2005). Recent advances with liposomes as pharmaceutical carriers. *Nature reviews Drug discovery*, 4(2), 145-160.
[103] Skubitz, K. M. (2003). Phase II Trial of Pegylated-liposomal Doxorubicin (Doxil™) In Sarcoma original article. *Cancer investigation*, 21(2), 167-176.
[104] Barenholz, Y. (2001). Liposome application: problems and prospects. *Current opinion in colloid and interface science*, 6(1), 66-77.
[105] Wassef, N. M., and Alving, C. R. (1987). Complement-dependent phagocytosis of liposomes by macrophages. *Methods in enzymology*, 149, 124-134.
[106] Sato, T., and Sunamoto, J. (1992). Recent aspects in the use of liposomes in biotechnology and medicine. *Progress in lipid research*, 31(4), 345-372.
[107] Leserman, L. D., Barbet, J., Kourilsky, F., and Weinstein, J. N. (1980). Targeting to cells of fluorescent liposomes covalently coupled with monoclonal antibody or protein A. *Nature*, 288(5791), 602-604.
[108] Heath, T. D., Fraley, R. T., and Papahdjopoulos, D. (1980). Antibody targeting of liposomes: cell specificity obtained by conjugation of F (ab') 2 to vesicle surface. *Science*, 210(4469), 539-541.
[109] Schroit, A. J., Madsen, J., and Nayar, R. (1986). Liposome-cell interactions: *in vitro* discrimination of uptake mechanism and *in vivo* targeting strategies to mononuclear phagocytes. *Chemistry and physics of lipids*, 40(2-4), 373-393.
[110] Bakshi, M. S., Thakur, P., Sachar, S., and Banipal, T. S. (2007). Synthesis of nanocomposite gold-semiconductor materials by seed-growth method. *Materials Letters*, 61(17), 3762-3767.
[111] Sharma, P., Sachar, S., Kaur, G., Thakur, P., Bakshi, M. S., and Banipal, T. S. (2007). Mixed Micelle Behavior of Dodecyl-β-D-glucopyranoside with Cationic Gemini Surfactants: A Fluorescence Study. *Journal of Surface Science and Technology*, 23(3-4), 131-147.

[112] Jamwal, D., Kaur, G., Raizada, P., Singh, P., Pathak, D., and Thakur, P. (2015). Twin-Tail Surfactant Peculiarity in Superficial Fabrication of Semiconductor Quantum Dots: Toward Structural, Optical, and Electrical Features. *The Journal of Physical Chemistry C*, 119(9), 5062-5073.

[113] Kawashita, M., Tanaka, M., Kokubo, T., Inoue, Y., Yao, T., Hamada, S., and Shinjo, T. (2005). Preparation of ferrimagnetic magnetite microspheres for *in situ* hyperthermic treatment of cancer. *Biomaterials*, 26(15), 2231-2238.

[114] Jordan, A., Wust, P., Scholz, R., Tesche, B., Fähling, H., Mitrovics, T., and Felix, R. (1996). Cellular uptake of magnetic fluid particles and their effects on human adenocarcinoma cells exposed to AC magnetic fields *in vitro*. *International journal of hyperthermia*, 12(6), 705-722.

[115] Drbohlavova, J., Chomoucka, J., Adam, V., Ryvolova, M., Eckschlager, T., Hubalek, J., and Kizek, R. (2013). Nanocarriers for anticancer drugs-new trends in nanomedicine. *Current drug metabolism*, 14(5), 547-564.

[116] Qiao, R., Yang, C., and Gao, M. (2009). Superparamagnetic iron oxide nanoparticles: from preparations to *in vivo* MRI applications. *Journal of Materials Chemistry*, 19(35), 6274-6293.

[117] Gu, H., Xu, K., Xu, C., and Xu, B. (2006). Biofunctional magnetic nanoparticles for protein separation and pathogen detection. *Chemical Communications*, (9), 941-949.

[118] Sounderya, N., and Zhang, Y. (2008). Use of core/shell structured nanoparticles for biomedical applications. *Recent Patents on Biomedical Engineering*, 1(1), 34-42.

[119] Tanaka, K., Ito, A., Kobayashi, T., Kawamura, T., Shimada, S., Matsumoto, K., and Honda, H. (2005). Heat immunotherapy using magnetic nanoparticles and dendritic cells for T-lymphoma. *Journal of bioscience and bioengineering*, 100(1), 112-115.

[120] Kapp, D. S., Hahn, G. M., and Carlson, R. W. (2000). Principles of hyperthermia.

[121] Faraji, M., Yamini, Y., and Rezaee, M. (2010). Magnetic nanoparticles: synthesis, stabilization, functionalization, characterization, and applications. *Journal of the Iranian Chemical Society*, 7(1), 1-37.

[122] Whitesides, G. M. (2003). The 'right' size in nanobiotechnology. *Nature biotechnology*, 21(10), 1161-1165.

[123] Bae, S., Lee, S. W., Takemura, Y., Yamashita, E., Kunisaki, J., Zurn, S., and Kim, C. S. (2006). Dependence of frequency and magnetic field on self-heating characteristics of nickel ferrite nanoparticles for hyperthermia. *IEEE transactions on magnetics*, 42(10), 3566-3568.

[124] Kim, D. H., Lee, S. H., Kim, K. N., Kim, K. M., Shim, I. B., and Lee, Y. K. (2005). Temperature change of various ferrite particles with alternating magnetic field for hyperthermic application. *Journal of Magnetism and Magnetic Materials*, 293(1), 320-327.

[125] Saptarshi, S. R., Duschl, A., and Lopata, A. L. (2013). Interaction of nanoparticles with proteins: relation to bio-reactivity of the nanoparticle. *Journal of nanobiotechnology*, 11(1), 26.

[126] Jun, Y. W., Seo, J. W., and Cheon, J. (2008). Nano scaling laws of magnetic nanoparticles and their applicabilities in biomedical sciences. *Accounts of chemical research*, 41(2), 179-189.

[127] Yang, H., Zhang, C., Shi, X., Hu, H., Du, X., Fang, Y., and Yang, S. (2010). Water-soluble Superparamagnetic manganese ferrite nanoparticles for magnetic resonance imaging. *Biomaterials*, 31(13), 3667-3673.

[128] Wu, H., Liu, G., Wang, X., Zhang, J., Chen, Y., Shi, J., and Yang, S. (2011). Solvothermal synthesis of cobalt ferrite nanoparticles loaded on multiwalled carbon nanotubes for magnetic resonance imaging and drug delivery. *Acta biomaterialia*, 7(9), 3496-3504.

[129] Su, L., Qin, W., Zhang, H., Rahman, Z. U., Ren, C., Ma, S., and Chen, X. (2015). The peroxidase/catalase-like activities of MFe 2 O 4 (M = Mg, Ni, Cu) MNPs and their application in colorimetric biosensing of glucose. *Biosensors and Bioelectronics*, 63, 384-391.

In: Modified Biopolymers: Challenges and Opportunities ISBN: 978-1-53612-116-2
Editors: D. Pathania, G. Sharma and A. Kumar

Chapter 12

Biopolymer Modifications Using Ionic Liquids for Industrial and Environmental Applications

Sunil Kumar[1,3,*], Neha Rana[1], Manish Kumar[1], Archana Sharma[1], Pragya Katoch[1], Ashvinder Rana[1] and Samjeet Singh Thakur[2]

[1]Department of Chemistry, Sri Sai University, Palampur, India
[2]Department of Chemistry, Govt. Degree College, Jogindernagar, India
[3]Department of Chemistry, Govt. Degree College, Khundian, Kangra, India

Abstract

Biopolymers are produced by a range of microorganisms and plants. Biopolymers are produced by microorganisms using specific nutrients and controlled environmental conditions. Applications of biopolymers in various fields, especially in the field related to nanoscience and nanotechnology with focus on industry and environment is discussed in this chapter. Biopolymers have many applications in various fields including water treatment, biomedical application, energy sector, and food industry. This chapter gives an overview of the important nanotechnology-based applications of biopolymers.

It covers the recent developments in the field of bio-based packaging materials with emphasis on the barrier properties, which are important in terms of food packaging. The state-of-the-art of several biopolymers including pectin, starch, chitosan, xylan, galactoglucomannan, lignin and cellulose nanofibrils is also discussed.

* Corresponding Author Email: sunil678kumar@rediffmail.com.

1. INTRODUCTION

Biopolymers are the polymers which are formed under natural conditions during the growth cycles of all organisms [1-2]. Therefore, they are also named natural polymers. They are formed within cells by complex metabolic processes. For materials applications, cellulose and starch are most important [3]. However there is an increasing attention in more complex hydrocarbon polymer produced by bacteria and fungi, particulary polysaccharides such as xynthan, curdlan pullulan, chitin, chitosan and hyaluronic acid. Biodegradable polymers are growing day by day and current research is focused on producing new biodegradable polymers. Natural biodegradable polymers are, in general, called biopolymers and they have wide applications in various fields [4]. Starch, fibers and cellulose are the most common examples of biopolymers and they can be organized into a variety of biological materials. Biopolymers or renewable polymers such as cellulose, lignin, starch, pectin, chitin, and xylan are the abundantly found polymers in nature. Biopolymers can be converted by different ways into various chemicals, fuels, and materials for the benefit of our society [5]. Biopolymers are produced by a range of microorganisms and plants. Microbial biopolymers can be produced either directly via fermentation or by chemical polymerization of monomers, which are in turn produced through fermentation [6]. Biopolymers produced by microorganisms and many biopolymers are biocompatible; they have no adverse effects on biological system. Some petroleum-based polymers and some biopolymers of plant origin are also classified as biopolymers and are biodegradable [7].

2. BIOPOLYMERS

Pectin

Pectin is a hetero polysaccharide found in the primary cell walls of terrestrial plants such as sugar beet [8]. Pectin films plasticized with glycerol and further associated with nanoclay (>98% montmorillonite) can be prepared by casting onto Petri dishes. In order to establish the sufficient delimitation and nanosized structure of the nanoclay platelets, the powder was first immersed in water using either ultrasonic treatment or high pressure fluidization. The model surfaces were prepared by spin coating onto silica surfaces from the aqueous dispersions after both pre-treatments [9]. The high pressure fluidizer treatment produced an even and more homogenous layer of parallelly oriented nanoclay platelets with the narrower size distribution as compared to particles after ultrasonic treatment [10]. Thus, high pressure fluidizer treatment was selected as an efficient and easily up-scalable method to produce bio-hybrid nanocomposite dispersions. The resulting bio-hybrid nanocomposite films were totally greaseproof and also showed improved barrier properties against oxygen and water vapour. Recently, pectin solutions have been roll-to-roll flexo-coated with semi-industrial pilot-line onto argon/nitrogen plasma-activated oriented polypropylene (OPP) film [11]. Plasma-activation was carried out to increase the surface energy of OPP film before the coating with water-based pectin solution. Pectin was easily spread onto activated surface forming an even and smooth dry coating. Also other biopolymers, such as various starch grades and wood-derived xylan can be coated in a similar manner [12]. Starch films can be produced either by extrusion or casting. Plasticizers, such as glycerol, sorbitol or xylitol, are typically used for reducing the brittleness. Effects of glycerol, sorbitol or xylitol on physical and mechanical properties of starch films were largest for glycerol and smallest for sorbitol. High contents of xylitol and sorbitol resulted in changes in physical and mechanical properties of films probably due to phase separation and crystallization. Water sorption and water vapour permeability were lower for the films plasticized with xylitol-sorbitol mixture compared to the films plasticized with glycerol-xylitol and glycerol-sorbitol mixtures at constant plasticizer contents. Oxygen permeability of the films was strongly dependent on the water content [13]. At low humidity conditions the films are typically excellent barriers against oxygen transmission, but when the water content rises above 20% the barrier is mostly lost. Oxygen permeability of starch films can be decreased by using below 21% of sorbitol. This behavior was related to changes in secondary relaxations which are hindered because of the connections established between starch and sorbitol, leading to decreased diffusion of oxygen molecules. For sorbitol contents above 21%, oxygen permeability increased slightly. Compounding starch and glycerol with PCL (polycaprolactone) decreased the water vapour permeability of extruded film [14].

Chitosan

Chitosan, the β-1-4-linked polymer of 2-amino-2-deoxy-β-d-glucose, can be prepared by the N-deacetylation of chi-tin, the second most abundant natural biopolymer after cellulose [15]. Chitosan is soluble in dilute aqueous acid solutions and has been widely studied due to its good film forming properties. Chitosan can be used as such in cast free standing films or it can be applied as a coating onto paper/board or plastic films. Chitosan dissolved in 1% acetic acid was applied onto copy paper by lab-scale bar coater and dried at 105°C. Chitosan coatings clearly improved both the gloss and oxygen barrier properties of paper. The gloss value in the machine direction was increased as a function of added chitosan. In addition, the effects on the mechanical properties were positive, but not significant. The water vapour permeability of the paper increased as a result of the chitosan coating [16]. Chitosan can also be immobilized onto NH_3- or CO_2-plasma activated polypropylene films by exploiting either carbodiimide or glutaraldehyde chemistries.

Chitosan coatings made film surfaces hydrophilic and prevented clearly oxygen, carbon dioxide and ethylene transmission [17]. Barrier properties can be further improved by using nanoclays incorporated into chitosan. Nanoclay was mixed with chitosan and further dispersed using ultrasonic treatment and coatings were applied onto plasma-activated LDPE-coated paper by lab-scale coater. Also free-standing films were cast onto Petri dishes. Bio-hybrid nanocomposite films and multilayer coatings had improved barrier properties against oxygen, water vapour, grease, and UV-light transmission [18].

Xylan

Xylan is alkali which is extracted from Kraft hardwood pulp. It contains predominantly β-D-xylose units linked as in cellulose. Xylan as such is not suitable for film formation due to its high internal cohesion, which results into fragmented films [19]. Thus, xylan should be plasticized externally or internally. It can be further hydroxypropylated resulting in water soluble derivative with improved film forming properties. The inner plasticization with hydroxypropyl groups can be combined with external plasticization by using glycerol or sorbitol [20].

Coatings of aqueous HPX solutions containing glycerol and sorbitol were applied with a lab-scale bar coater onto a pigment coated board. 5% citric acid was used as a biobased crosslinker to increase its barrier properties. A crosslinked HPX coating shows a good grease and mineral oil barrier as well as low oxygen transmission rate of 200 $cm^3/m^2/day$ at 50% relative humidity. Water vapour permeabilities of the xylan coatings were slightly below that measured for the commercial biopolymer, polylactic acid (PLA), coating. Also the free standing HPX films plasticized with 10% and 20% of sorbitol had relatively low oxygen permeabilities at $<75\%$ relative humidity. In addition, fatty acid esters of hemicelluloses and cellulose have been coated successfully on paper/board [21]. The coatings were hydrophobic and provide a good grease and moisture barrier properties. The best water and water vapour barrier was achieved with the cellulose derivatives esterified with the longest fatty acid. Benzyl ether derivatives of birch xylan with varying degree of substitution have recently been prepared. These derivatives are also hydrophobic and thermoplastic capable of forming free standing films from organic solvents. The paper sheets prepared from the esterified cellulose fibers showed high hydrophobicity [22]. The barrier properties of xylan may be further improved through nanoclay reinforcement in aqueous medium. The nanoclay dispersions were mixed with the plasticized (glycerol) xylan, resulting in homogeneous and viscous coating solutions. Xylan-nanoclay dispersion was further coated onto paper by lab-scale bar coater [23]. The coatings enhance the water vapour barrier properties.

Galactoglucomannan

The main hemicellulose in softwood is O-acetyl-galactoglucomannan (AcGGM). AcGGM has a main chain of β-(1→4)-linked D-mannose and D-glucose with α-(1→6)-linked D-galactose moieties in various amounts. AcGGM is soluble in water and organic solvents [24]. The solubility properties can be applied to a high degree of acetylation combined with a low molecular weight. Cast films have been produced with plasticizers such as glycerol, sorbitol and xylitol, and the biobased polymers such as alginate and carboxymethylcellulose (CMC). The addition of plasticizers results in increased film flexibility and moisture sensitivity. With other biobased polymers, the mechanical strength and resistance towards humidity increased [25].

Lignin

Lignin is one of the most abundant natural polymers, associated with cellulose and hemicellulose. Softwood lignin has been esterified with tall oil fatty acid (TOFA) and applied with a lab-scale bar coater onto paperboard. The WVTR of the paperboard was reduced by 70% with a TOFA lignin ester double coating of 3.9 g/m^2. Also a decrease in oxygen transmission was observed for the TOFA lignin ester as well as the TOFA coated paperboard samples [26]. The coating material did not affect the tensile strength of the paperboard. Esterification with palmitic and lauric acid chloride resulted in coatings with improved moisture barrier properties [27].

3. Modification of Biopolymers

Biopolymers have potentials well as advantages, but they have some drawbacks as well [28]. In spite of increasing production capacity, they are still quite expensive compared to commodity polymers and their properties are also often inferior, or at least do not correspond to the expectation of converters or users [29]. Although natural polymers are available in large quantities and are also cheap.

Modification of these materials is in the focus of scientific research. In order to adjust the properties of biopolymers to the intended application, wide varieties of approaches are used for their modification, like plasticization, association of infillers and reinforcements, blending and impact modifications. In contrast to the development of novel polymeric materials and new polymerization routes, the preparation of heterogeneous polymeric systems is a relatively cheap and fast method to accommodate the properties of plastics.

As a result, this approach may play important role in enhancing the competitiveness of biopolymers [30].

Biopolymers can be modified by the incorporation of fillers and reinforcements which provides various biocomposites [31]. The introduction of fillers or reinforcements into a polymer matrix results in a heterogeneous system. Under the effect of external load heterogeneities Heterogeneous stress distribution and local stress maximums initiate local micromechanical deformations, which determine the deformation and failure behavior, as well as the overall performance of the composites [32].

Although conventional, mineral fillers can also be applied for the modification of biopolymers, most studies focus on the potential use of natural lignocellulosic fibers, i.e., wood flour, sisal, flax, etc. into both conventional and biopolymers [33]. Lignocellulosic reinforcements might present a viable alternative to glass fibers applied traditionally. These biocomposite materials have some applications in the field of building and automotive industry. One of the most important biopolymers, poly (lactic acid) is also often used for the production of particulate filled or fiber reinforced composites [34].

Biopolymer systems containing particles with one or more dimensions in the nanometer scale, i.e., bio-nanocomposites, form a special class of materials possessing a range of unique characteristics with respect to gas and water vapor permeability, thermal stability, fire resistance, mechanical and optical properties, etc [35]. The characteristics of biopolymers can be modified using either natural nanofibers (cellulose nanofibers and nanowhiskers) or inorganic nanofillers (silica, layered silicates), providing different combinations of properties. In order to achieve nanoscale dispersion, i.e., intercalated or exfoliated structure resulting in superior properties, nanoclays must be modified with suitable organic compounds in order to promote the separation of the silicate layers [36].

Plasticization

Plasticization is often used for the modification of biopolymers to improve their processability and/or other properties [37]. Plasticizers exchange the intermolecular bonds among polymer chains to bonds between the macromolecules and the small molecular weight compound thus promoting conformational changes resulting in increased deformability. Both the glass transition and the processing temperature of the material decrease, thus enabling the melt processing of heat-sensitive polymers, like poly (3-hydroxybutyrate) at lower temperatures [38].

The melt processing of starch is impossible without gelatinization, i.e., destroying the crystalline structure of native starch by the use of a plasticizer [39]. The material obtained in this way is referred to as thermoplastic starch (TPS). The nature and concentration of the plasticizer strongly influence the rheological and mechanical properties of TPS [40]. Neat starch has a high glass transition temperature, and its relative large modulus and strength is accompanied by poor deformability and impact resistance due to the limited conformational mobility of its stiff chains. Elastomeric behavior would be advantageous for several applications, which can be achieved, among other methods, by decreasing T_g below ambient temperature via plasticization [41].

Numerous studies and patents describe the plasticization of starch using either low molecular weight or polymeric compounds. The plasticizers applied are mainly low molecular weight polar compounds such as water, glycerol, urea and formamide. In plasticized starch the effect of water depends on the nature and concentration of the plasticizer and on relative humidity, and water content affects the macroscopic properties of TPS to a great extent [42]. Recently, using state of the art visualization techniques Paes *et al.* have shown that moisture content and inter-action with water determines the fracture mechanism in cassava starch films as well as in gelatin and hydroxypropyl cellulose [43].

Although water is a very efficient plasticizer for starch too, usually compounds with higher boiling point are preferred because they decrease water sensitivity and lead to more stable properties [44]. A similar approach can be applied for the modification of poly (lactic acid). Usually the application of PLA is not limited by its processability, but by its mechanical properties. Plasticization is one approach to overcome the stiffness and low impact resistance of the polymer [45]. Several compounds have been identified as possible plasticizers for PLA, e.g., different esters glycerol poly (ethylene glyco l) (PEG) and oligomeric lactic acid [46].

Since the properties of poly (hydroxy alkanoate) can be efficiently and conveniently tailored by controlling the comonomer ratio during polymerization via fermentation, plasticization is not a widespread approach for their modification [47]. However, plasticization should not be neglected as a nonexpensive option to improve the toughness of the inherently stiff poly (3-hydroxylbutyrate). Combined with conventional co-polymerization, the deformability of the material can be increased considerably by this route one of the main reasons hindering the widespread use of cellulose in polymer industry is that without modification it is a highly crystalline material, with high melting and glass transition temperatures, unable to be processed in melt state [48]. Either plasticization or almost complete derivatization of cellulose is needed in order to make it thermo-plastic, although the latter results in decreased biodegradability. In the case of plasticization, miscibility is an important issue, since plasticizers tend to migrate to the surface of the products raising environmental and health issues, as well as altering the macro-scopic properties of the material as an effect of changing plasticizer content. Migration can be hindered by increasing the molecular weight of the plasticizer, which leads to lower diffusion indeed, but it results in a decrease of miscibility and efficiency as well. Another innovative solution to this problem might be the grafting of plasticizer molecules to the polymer chain by reactive processing i.e., internal plasticization [49].

Numerous studies and patents describe the plasticization of starch using either low molecular weight or polymeric compounds. The plasticizers applied are mainly low molecular weight polar compounds such as water, glycerol, urea and formamide [50]. However, as in their elaborate work Lourdin et al. point out, TPS is always a ternary system, since the water content of plasticized starch should always be taken into consideration. In plasticized starch the effect of water depends on the nature and concentration of the plasticizer and on relative humidity, and water content affects the macroscopic properties of TPS to a great extent. Recently, using state of the art visualization techniques Paes *et al.* have shown that moisture content and interaction with water determines the fracture mechanism in cassava starch films as well as in gelatin and hydroxypropyl cellulose [51]. Although water is a very efficient plasticizer for starch too, usually compounds with higher boiling point are preferred because they decrease water sensitivity and lead to more stable properties [52].

Physical blending

Physical blending is the simple mixing of polymeric materials in the melt state with no chemical reactions taking place. It is a convenient way to create new materials with the desired combination of properties. Blending can be carried out by using conventional machinery, i.e., no expensiveness, which is an important aspect for industry [53]. A very wide range of properties can be achieved by this approach to meet the requirements of the targeted application in short time and for low cost compared to the development of new monomer and polymerization techniques.

4. Need for Modification of Biopolymers

In addition to incorporation of nanoclays, the barrier properties of biopolymers may be improved by chemical and physical crosslinking, or with surface treatments, such as grafting and coating [54]. For example, the plasma or heat induced oxidation/degradation can promote chain scission and crosslinking on surfaces. By heating the coatings over melting point and cooling them slowly back to room temperature, the coating density, morphology, crystallinity and spherulite size can be modified towards better barrier properties [55]. Heat treatment at 200°C can be induced irreversible hydrogen bonding to cellulose which improved both barrier properties and mechanical strength of CNF films (Chitosan-nanoclay bio-hybrid films have been successfully crosslinked with glutaraldehyde, genipin and glyoxal. Moisture sensitivity of films decreased as a result of crosslinking which led to improved barrier properties against water vapour and oxygen transmission at high humidity [56]. Recently, the aqueous crosslinked PLA dispersions were coated onto board at pilot-scale. For decreasing the moisture sensitivity, the hydrophobic softwood galactoglucomannan films have been prepared by benzylation. Also styrene grafting of films based on unmodified AcGGM by plasma and vapour-phase treatments have been performed [57]. Finally, lamination of an unmodified AcGGM film with benzylated AcGGM was also investigated in terms of water tolerance and oxygen permeability. Both surface grafting and especially the lamination method decreased the moisture sensitivity and improved the barrier properties. Recently, the cellulose treated enzymatically and has been dissolved in a NaOH/ZnO solvent system and mixed together with poly(ethylene-co-acrylic acid) or poly(acrylamide-co-acrylic acid) polymers, in order to improve the properties and prepare homogeneous cellulose-based blends for films and coatings [58]. An easy method to improve hydrophobicity and oxygen barrier properties of CNF films at very high humidity is to use simple paraffin wax coating. The films were dipped into the melted wax and the excess wax was removed from the surface resulting in improved water vapour and oxygen barrier as well as the high surface hydrophobicity. Multilayer coatings of CNF and shellac have been deposited on the fibre based substrates using a lab-scale bar coater or a spray coating technique. Multi-layer system with CNF as a first layer and shellac as the top layer decreased both oxygen and water vapour transmission rates [59].

5. Ionic Liquids Modified Biopolymers

Ionic liquids (ILs) are regarded as green organic solvents having excellent physical-chemical properties, such as low vapor pressure, satisfactory solvability, designability and biodegradability [60]. The utilization of ILs in dissolution, regeneration and modification of biopolymers has drawn growing attention. In this paper, the dissolution, regeneration and modification (or degradation) of cellulose, chitosan, starch, β-cyclodextrin, lignin and protein in various ILs are reviewed [61]. Specifically, the influencing factors on the dissolution of biopolymers in ILs, such as the kind and volume of cation or anion of IL, temperature, heating method and the source of biopolymers are summarized. The mechanisms that ILs dissolves biopolymers are mentioned. The reactions of biopolymers in ILs as well as the physicochemical properties of biopolymer derivatives are introduced [62].

Modification Types of Ionic Liquids

Modification of Cellulose in Ionic Liquids

The formation of homogeneous cellulose/IL solution provides a convenient method for the preparation of cellulose-based materials by homogeneous reaction [63]. e.g., Cellulose acetate butyrate, cellulose lauric acid ester and methyl phenyl amino acids [64]. The cellulose acetate butyrate (CAB) prepared in AmimCl could dissolve in 2-methyl ethyl ketone, 1,2 dichloroethane, ethyl acetate and butyl acetate and the solubility increases with the increase of butyryl content in CAB [65]. The thermal decomposition temperature of CAB (334-348°C) is lower than that of cellulose (376°C). The larger the content of butyryl in CAB is, the lower the thermal decomposition temperature is. Water soluble cellulose derivatives could be prepared through reactions between cellulose and urea, phthalic anhydride, maleic anhydride or butyloxirane propyl ether in BmimCl [66]. The reactions are recognized by the data of FTIR, 1H NMR and TGA. The solubility of these derivatives in water could be up to25 wt%.

In addition, cellulose derivatives could be prepared in ILs through homogenous reaction under the support of organic catalyst (including ILs themselves). Sun and coworkers found that both 4-dimethylaminopyridine (DMAP) [67] and N-bromosuccinimide (NBS) could catalyze the modification of succinic anhydride on cellulose in BmimCl under homogeneous reaction system, and obtained succinylated cellulose with degree of substitution (DS) in the range of 0.24-2.31. The DS is higher than that without catalyst under the same reaction conditions. The 13C NMR chemical shifts at 67.8 (C2 and C3), 61.1 and 59.8 (C6) ppm disappeared, indicating that the modifying reaction takes place at C2, C3 and C6 sites. An IL can also be used as a catalyst for their special advantages, such as less corrosion, special capability of swelling cellulose and breaking the inter- and intermolecular hydrogen bonds. Zhang et al. synthesized cellulose acetate by environment friendly [68]. Solvent-free method using small amount of 1-vinyl-3-(3-sulfopropyl) imidazolium hydrogensulfate (SO_3H-functionalized acidic IL, abbreviated as $VSimHSO_4$), N-methyl-imidazolium chloride (HmimCl) or N-methyl-imidazolium bisulfate ($HmimHSO_4$) as catalysts. The DS values of resulting cellulose acetate are controllable, ranging from 1.13 to 2.95, by changing the reaction time and the amount of ILs. It is considered that the dual-function of swelling and catalyzing of acidic ILs play an important role in producing "quasi-homogeneous" acetylation system as well as controlling the DS of cellulose acetate [69]. The structure and acidity of ILs play key roles on their catalytic activity. The catalytic activities are in the order of CF_3SO_3- > HSO_4- > OAc-, which is in good agreement with their acidity order.

Modification of Chitosan in Ionic Liquids

Along with the research of the ionic liquid dissolving chitosan, the modification of chitosan in an IL through homogenous reaction has attracted much attention [70]. Wei and coworkers prepared chitosan-IL grafted polymer (CS-IL) in 1-carboxymethyl-butyl-3-methylimidazolium chloride (CmimCl) through homogenous reaction between carboxyl and $-NH_2$ at 100°C. The CS-IL has excellent adsorption capacity for Cr_2O_7 - (0.422 mmol/g) and PF_6- (0.840 mmol/g) anions. The adsorption capacity increases with the grafted degree of CmimCl, as well as the decreases of aqueous pH. It ascribes to the protonation of $-NH_2$ at low pH, which is inclined to interact with anions through electrostatic force. The CS-IL adsorbed anion aggregates in aqueous solution and precipitates from bulk solution, in favor of being separated. Wang and coworkers prepared polycaprolactone grafted chitosan (CS-g-PCL) in EmimCl through the reaction between polycaprolactone and $-NH_2$ in chitosan with stannous octate as catalyst at 100 °C. The structure and grafting degree of CS-g-PCL could be modulated by changing the reaction conditions and reactant materials ratio. The glass transition temperature (Tg) of CS-g-PCL is lower than that of CS, due to the change of polymer crystallinity and polymer composition. The T_g of PCL is about -60 °C. It provides a recent method for the preparation of CS-g-PCL that could be used in tissue engineering, drug and gene transfer. Researchers prepared two kinds of monomethyl fumaric acid (MFA) modified chitosan MF-chitosan) with different chemical structures in ILs [71]. One is an MF-chitosan salt that composes chitosan cation and MFA anion, synthesized in a mixture of 1-ethyl-3-(3-dimethylaminopropyl) carbodiimide (EDC) and N-hydroxysuccinimide. In MF-chitosan molecules, MFA and chitosan are covalently attached.

6. Synthetic Approaches of Modified Biopolymers

Synthetic Approaches for Polymer–Protein Hybrid Structures

The modification of proteins with polymers is one of the most widespread approaches to create hybrid structures and offers tremendous possibilities for functional system [72]. Both polymerization and coupling reactions are used to achieve polymer structures on protein surfaces.-First, three major strategies have to be distinguished. The grafting-to approach involves the synthesis of polymer chains with suitable end-groups that are then linked to proteins. Protein-reactive groups can be introduced to polymers in a post-polymerization step or protein-reactive initiators are used directly for the polymerization reaction [73]. The biomolecules are only involved in the last synthetic step as the polymers are synthesized separately. An excess of polymer is needed for the conjugation reaction, which hampers purification of the protein–polymer hybrid. Unreacted polymer and maybe even protein have to be removed from the conjugate. This disadvantage can be avoided with the grafting-from approach. Here, the polymer chains grow directly from a protein. First, small initiator molecules are linked to the protein surface able to initiate the polymerization from these so-called macroinitiators. Hence the purification of the product is easier because only small molecules, unreacted monomer and the catalyst, have to be removed furthermore, the location of the polymers is defined by the initiator positions and steric hindrance is negligible. Particularly in cases where it is intended to link a high density of polymer chains to a protein, grafting-from is superior to grafting-to because the steric hindrance of the growing, initially short, polymer chains is much smaller. Additionally, the synthesis of hydrophobic polymer–protein conjugates and the preparation of conjugates with high molecular weight homopolymers or block copolymers are easier *via* grafting-from. The drawback associated with the grafting-from approach is that polymerization conditions have to be chosen that do not affect the activity and folded structure of the protein [74]. The third possibility is the grafting-through strategy. Here, monomers that contain proteins/peptides are polymerized so that every repeat unit of the comb-shaped polymer structure has a protein/peptide moiety. Grafting-through potentially allows the incorporation of a high amount of proteins/peptides into the conjugates [75].

Controlled radical polymerization (CRP) techniques allow the synthesis of a wide range of polymers with a narrow molecular weight distribution. The two most widely employed methods for synthesizing polymer–protein hybrid structures are atom transfer radical polymerization (ATRP) and reversible addition fragmentation chain transfer (RAFT) polymerization [76].

7. Applications of Biopolymers

Biodegradable polymers are used especially in three major areas namely, agricultural, medical and consumer goods packaging, of which medical device application have developed faster than the other two.

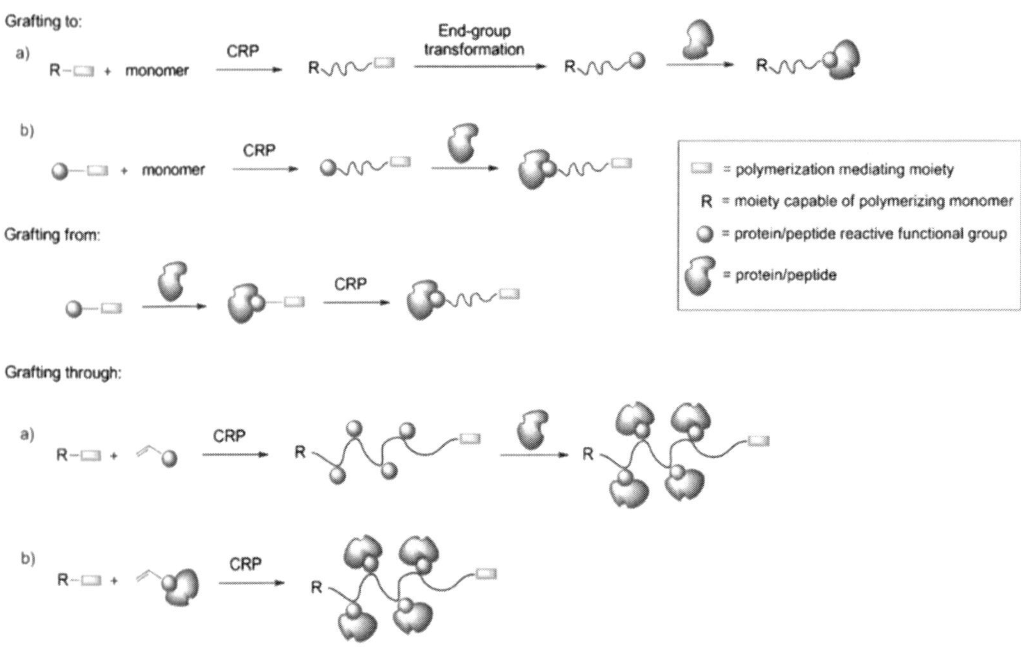

Medical Applications

Biodegradable plastics have broad applications as surgical implants in vascular and orthopedic surgery as implantable matrices for the controlled long-term release of drugs inside the body, as absorbable surgical sutures, and for use in the eye [77]. Recently, the term biomaterial was defined as a nonviable material used in medical device applications that is intended to interact with a biological system. Biocompatibility is the ability of a material to perform with an appropriate host response in a specific application Poly (lactide), poly (glycolide), and other materials such as PDO, PTMC, PCL, and PLGA have been widely used for medical devices.

Surgical Sutures: Loss of structural integrity results from tissue damage and the insertion of some material or device to hold the tissue together may facilitate the healing process [78]. The classic examples are the use of sutures to hold both deep and superficial wounds together. Once the healing is complete, the suture becomes unnecessary and can impose undesirable constraints on the healing tissues. It is preferable to remove the material from the site, either by degradation or by physical ways. Synthetic absorbable sutures were developed in 1960s. They are now extensively used in tracheobronchial surgery as well as general surgery, because of their good biocompatibility in tissues. They are multifilament-typesutures, which have good handle ability. PGA, PLLA and their copolymers, and polyglactin are the most popular and are now commercially available. PGA sutures have reliable knot stability and excellent smoothness.

Drug Delivery Systems: There are numerous degradable polymers that are potentially valuable for drug deliveryapplications [79]. These include a collection of synthetic and natural substances. The use of intentionally degradable polymers in medicine has been brought into importance with new innovations in drug delivery systems, for example, the use of controlled drug delivery systems from which the drug is released at a constant, predetermined rate, and possibly targeted to a particular site.

One of the most approaches is that in which the drug is contained within a polymer membrane or is otherwise encapsulated in a polymer matrix and where the drug diffuses out into the tissues following implantation. Polymer erosion or dissolution gives a way to the release mechanism in some cases. Degradable polymers such as PLA and poly (orthoesters), are used for drug delivery systems.

Agricultural Applications

Plastics, coatings, elastomers, fibers, and water-soluble polymers are utilized in applications that comprise the CR of pesticides and nutrients soil conditioning, seed coatings, gel plantings and plant protection [80]. However degradable plasticsare also of significance as agricultural mulches and agricultural planting containers. Biodegradability is also of some interest in composting as it would permit degradable plastics to be combined with other biodegradable materials and converted into useful soil improving materials.

Controlled Release of Agricultural Chemicals CR is a method by which biologically active chemicals are made available to a target species at a specified rate and for a predetermined time. The polymer serves primarily to control the rate of delivery, mobility, and period of efficacy of the chemical component [81]. The principal benefit of CR formulations is that less chemicals are used for a given time period, thus lowering the impact on nontarget species and limiting leaching volatilization and degradation. The macromolecular nature of polymers is the key to limiting chemical losses by these processes. The natural polymers used in CR systems are starch, cellolose, chitin (476-478), alginic and lignin.

Packaging

The physical characteristics of packaging polymers are influenced by the chemical structure, molecular weight, crystallinity, and processing conditions of the polymers [82]. The physical characteristics required in packaging depend on what item will be packaged as well as the environment in which the package will be stored. The challenge in the development of biodegradable packaging will be to combine polymers that are truly biodegradable into a laminate film or a film blend that has properties as good as those found in synthetic laminates [83]. An equally important function of packaging is to protect the product from physical, chemical, or biological damages.

Cellulose-Based Packaging Materials

Cellulose-based packaging materials incorporated with wheat glutenare biodegradable, gas selective, and permeable, and improve the shelf life of cultivated mushrooms [84].

Food Industry

Replacing the oil-based packaging materials with biobased films and containers might give not only a competitive advantage due to more sustainable and greener image but also some improved technical properties. Biopolymers are currently used in food coatings, food packaging materials, and encapsulation matrices for functional foods [85].

Based food-packaging applications, the most important parameter to be considered are its barrier properties. Hydrophilic polymers usually have poor moisture resistance which cause water vapor transmission through packaging and thus affect the quality of foods. This results in shorter shelf lives, increased costs, and eventually more waste [86]. Another technique to improve the barrier properties of biopolymers is to add various nanofillers like nanoclays, metal oxide Nanoparticles, and so on. Among the bioplastics, polyglycolic acid (PGA) has excellent barrier properties thus it's one of the most promising new commercially available barrier polymers. Its precursor, glycolic acid, can now be produced via a natural metabolic route, the glyoxylate cycle.

Water purification: Safe drinking water is a significant, but simple indicator of development.

Nanotechnology has shown promising developments in providing safe drinking water through effective purifying mechanisms. Several nanomaterials have already proved to have antibacterial and antifungal properties [87].

Environmental Applications

A large number of conventional techniques have been used for the removal of a single class of contaminants (either heavy metals or dyes) [88]. The use of natural polymers as biosorbents for heavy metal removal has several advantages. Nowadays, few studies have reported simultaneous removal of metals and dyes from multicomponent solutions using single functional groups However, both metals and dyes load on the same active site, resulting in inevitable competition for free adsorption sites. Thus, the development of a facile and environmentally friendly approach to simultaneously remove both of them from wastewater is urgent and crucial. A suggested approach is to modify an existing successful method for a single class of contaminants, so it can be applied to simultaneously remove both classes of contaminants. A variety of chemosensors have been reported for the detection of metal ions. Recently, fluorescence chemosensor and functional magnetic Nanoparticles were developed to detect and remove metal ions from aqueous solution simultaneously. Shi and coworkers produced a novel probe of adamantyl (AD)-maleic anhydride (MAH)– rhodamine B (RhB). The probe showed high selectivity and sensitivity to metal ions in an aqueous solution, and the results suggested that AD–MAH–RhB is able to conjugate with metal ions and form the binding complexes with different amounts of probe and metal ions. Moreover, β-CD-modified magnetic nanoparticles (CD-MNP) were developed and used as host materials to form an inclusion complex of CD-MNP and AD–MAH–RhB-metal ions. Then, the metal ions could be removed by an outer magnet. In other research [89], a novel biodegradable β-cyclodextrin-based hydrogel (CAM) was prepared and applied to the removal of Cd^{2+}, Pb^{2+}, and Cu^{2+} ions from aqueous solutions. CAM hydrogel showed outstanding capability for the removal of heavy metal ions. The adsorption capacity was in the order $Pb^{2+} > Cu^{2+} > Cd^{2+}$ under the same experimental conditions. The high adsorption capacity indicates that the eco-friendly hydrogel can be used as an alternative adsorbent to remove heavy metals from an aqueous solution.

Biopolymers can also be used for the separation of toxic metal cations in wastewater, some research involving biopolymers can be cited: a cellulose acetate–tin (IV) phosphate composite material [90], a chitosan–ethyl acrylate (CEA) biopolymer , and geotextiles from polyethylene terephthalate (PET) non-woven material by chitosan functionalization using a coating technique with sodium tripolyphosphate (TPP) as the cross-linking agent . All these biomaterials have shown to be economical and have exhibited potential benefits in the recovery of metal ions from wastewater. There are many adsorbents that can be used to remove heavy metals from aqueous solutions but natural biopolymer-based adsorbents have stood out due to their low-cost, availability, and biodegradability. Recently it was reported that an ethylenediaminetetraacetic acid (EDTA)-cross-linked-β-cyclodextrin (EDTA-β-CD) bifunctional adsorbent for metals and dyes from aqueous solution was synthesized using an easy and green approach. While the cavities of CDs form inclusion complexes with the dye molecules, the EDTA groups act not only as crosslinkers but also as chelating sites for metal ions. The authors have chosen Cu^{2+} and Cd^{2+} metals ions, methylene blue (MB), safranin O (SO), and crystal violet (CV) as dyes for the study. The results indicated that that EDTA-β-CD has the potential for practical applications in the integrative and efficient treatment of coexistenting toxic pollutants. X. Luo *et al.* thinking about the most threatening problems in the water treatment field developed a new kind of effective and recyclable adsorbent for the removal of heavy metal ions from water. They synthesized magnetic (γ-Fe_2O_3) chitosan/cellulose hybrid microspheres, combining renewability and the biocompatibility of chitosan and cellulose as well as the magnetic properties of γ-Fe_2O_3 to create a hybrid system for heavy metal ions removal [91]. Three types of chitosan with high, medium, and low molecular weights as ligands to synthesize chitosan magnesium, calcium, iron (III), and zinc coordination complexes were studied. The aim of the study was the degradation of four organophosphorus pesticides (dichlorvos, omethoate, dimethoate, and chlorpyrifos) by the complexes formed in a heterogeneous system. The study revealed that the complexes derived from chitosan with low molecular weight and transition metals achieved the highest degradation. In addition, the order of degradation ratio was dichlorvos > omethoate = dimethoate > chlorpyrifos. The authors suggested that the type of organophosphorus pesticides impacted the degradation effect and that all degradation processes were more intense with higher pH and temperature. Recent studies report the microbial breakdown of chitosan and its derivative cross-linked chitosans, in both non-contaminated and contaminated conditions in laboratory incubation. Biodegradation of chitosan and its derivatives was affected by the presence of heavy metals [92]. According to authors, among the metals studied, zinc was superior at inhibiting microbial activity than copper and lead. It is estimated that about 100 years is necessary to complete the breakdown of the cross-linked chitosans and up to approximately 10 years for unmodified chitosan [93]. The authors also reported that the binding behavior of chitosan for heavy metals was not affected by the biodegradation process [94].

8. Environmental Benefits of Biopolymers

- These polymers are carbon neutral and can always be renewed. These are sustainable as they are composed of living materials.
- These polymers can reduce carbon dioxide levels in the atmosphere and also decrease carbon emissions. This happens because bio-degradation of these chemical compounds can release carbon dioxide that can be reabsorbed by crops grown as a substitute in their place.
- It is also compostable which means there is less chance of environmental pollution from this compound. This is one of the primary advantages of this chemical compound. However, the materials composed from this compound are not compostable.
- These chemical compounds reduce dependency on non-renewable fossil fuels. These are easily biodegradable and can decrease air pollution. It greatly reduces the harmful effect of plastic use on the environment. Long term use of biopolymer use will limit the use of fossil fuel.

Conclusion

The biodegradable polymers are used in various fields and it has resulted in the development of many commercial products. Because of their high specialization and greater unit values, particularly medical applications have developed more rapidly than the others. An important group of biopolymer is represented by the homopolymers of β-glucans because of its very low-to-negligible toxicity. They have tremendous potential use in a variety of diseases. Modern medical research has reached a phase where the basic mechanisms of glucan effects are known and the relationship between structure and activity has been outlined clearly. For many years, the major focus of drug research has been on the synthesis or discovery of new drugs. The use of biopolymer materials for drug delivery can minimize tissue reaction and allow drugs to be administered in nonconventional ways [95]. The use of biopolymers in these formulations has thus far been restricted to a narrow set of applications. Drug formulations incorporating polymer drug delivery systems are currently undergoing registration and are likely to make an impact on the market in near future. The major area of application for these novel sustained release systems is in the treatment of cancers and diseases of the elderly [96]. In the treatment of hormone disorders and geriatric diseases, biopolymer delivery systems allow drugs to be administered occasionally.

Increasing awareness towards the sustainable development has caused the researchers to think about natural and biodegradable polymers for replacing synthetic polymers for various applications. Nanotechnology has already emerged as the future technology which can make tremendous changes to the current existing technology. Various biopolymers are produced by the bacteria under limiting conditions to store carbon and energy resources. Incorporation of various nanofillers to these biopolymers will improve its mechanical and barrier properties which will improve its various applications. The biocompatibility together with the biodegradability of these materials encourages the use of these materials in day-to-day applications.

REFERENCES

[1] Rao. Gopal., Bharathi. P., and Akila. RM. 2014. "A Comprehensive Review On Biopolymers." India:College of Pharmacy.

[2] Francis. Raju., Sasikumar. Soumya., and Gopalan. Geethy. 2013. "Synthesis, Structure And Properties Of Biopolymers." Wiley- VCH Verlag GmbH and Co. KGaA.

[3] Mohan. Sneha., S. Oluwatobi., Kalarikkal. Nandakumar., Thomas. Sabu., and Songca. Sandile. 2016." Biopolymers-Application in Nanoscience and Nanotechnology." Intech.

[4] Vartiainen. Jari., Vaha-Nissi. Mika., and Harlin. Ali. 2014. "Biopolymer Films and Coating in Packaging Applications-A Review of Recent Developments." Finland:VTT Technical Research Centre of Finland.

[5] Vartiainen. J., Kunnari. V., and Kokko. A. 2012."Bio-hybrid Nanocomposite Roll-to-Roll Coating for Based Materials and Plastic." Finland:VTT Technical Research Centre of Finland.

[6] Matianilainen. K., Harlin. A., Kunnari. V., Tammmelin. T., Vartiainen. J., Pere. J., Vaha-Nissi. M., Mustonen. T., Hartman. J., Peltonen. S., and Harkonen. M. 2011. "Up-Scaling Biohybrid Varrier Coating to Piolet Coater." Naples: International Work-Shop Safe Nano structured Polymer Materials Characterization.

[7] Imre. Balazs. 2013. "Modification of Biopolymers-Chemistry, Interactions, Structure and Properties."Budapest University of Technology and Economics.

[8] Yan. Xiaodeng., Qiao. Congde., Li. Yan., Li. Tianduo. 2016. "Dissolution and Resourcefulization of Biopolymers in Ionic Liquid." China: Jinan.

[9] Swatloski. R. Spear, K. Scott., Holbery. John., and Rogers. Robin. 2002. "Dissolution of Cellulose with Ionic Liquids."America: The university of Alabama.

[10] Krishna. M. Gupta. Jiang. Jianwen. 2015. "Cellulose Dissolution and Regeneration in Ionic Liquid." Sigapore: National University of Singapore.

[11] Yaich. Anas. Edlunal. Ulrica., Albertsson. Ann. 2015. "Enhanced Formability and Mechanical Performance of Wood Hydrolysate Film Through Reductive Amination Chain Extension." Swedan: Royal Institute of Technology, Teknikringen-56.

[12] Ibrahim. Farouk, Maniruzzaman. Muhammad., Yusup. Suzana., Uemura. Yoshimitu. 2015. "Dissolution of Cellulose with Ionic Liquid in Pressurized Cell." Malasiya: Centre of Research In Ionic Liquid.

[13] Mine. Shozaburo. Izawa. Hironori et al. 2009. Acetylation of α-Chitin in Ionic Liquids. Japan: Graduate school of Science and Engineering, Kagashima University.

[14] Y. Wei., W. Huanq., Y. Zohu., S. Zhanq., D. Hua., and X. Zhu. 2013. "Modification of Chitosan with Carboxyl Functionalized Ionic Liquid for Anion Adsorption." China: Medical College of Soochow University.

[15] Li. Xinda., Zhang. Yue., Tang. Jingwen., Lan. Ai., Yang. Yanping., Gibril. Magdi., and Yu. Muhuo. 2016 "Efficient Preparation of High Concentration Cellulose Solution with Complex DMSO/ILs Solvent." Netherland: Springer Netherland.

[16] Li. W., Gin. A., Liu. C., Sun. R., Zhang. A., and Kennedy. J. 2009. "Homogeneous Modification of Cellulose with Succinic Anhydride in Ionic Liquid using 4-Dimethyle Amino Pyridine As A Catalyst." China: South China University of Technology.

[17] Yang. Xiaodeng., Qiao. Congde., Li. Yan.,and Li. Tianduo. 2016. "Dissolution of Resourcfulization of Biopolymers in Ionic Liquid." China: Jian.

[18] Xu. Guilong., Deng. Lili., Wen. Xiufang., Pi. Hui., Zheng. Dafeng., Cheng. Jiang., and Yang. Zhuoru. 2010. "Synthesis and Characterization of Flurine-Containing Poly-Styrene-Acrylate Latax with Core-shell Structure Using Or Reactive Surfactant. "Cina" South China University of Technology.

[19] Glebe. Ulrich., Miranda. Barbara., Rijn. Patrick., and., Boker. Alexander. 2015. "Synthetic Modification of Proteins." Netherland:U niversity of Groningen.

[20] Ahn, Y.; Hu, D. H.; Hong, J. H.; Lee, S. H.; Kim, H. J.; Kim, H. Effect of co-solvent on the spinnability and properties of electrospun cellulose nanofiber. *Carbohydr. Polym.* 2012, *89*, 340–345.

[21] Andanson, J-M.; Bordes, E.; Devemy, J.; Leroux, F.; Padua, A. A. H.; Gomes, M. F. C. Understanding the role of co-solvents in the dissolution of cellulose in ionic liquids. *Green Chem.* 2014,

[22] Barthell, S.; Heinze, T. Acylation and carbanilation of cellulose in ionic liquids. *Green Chem.* 2005, *8*, 301–306.

[23] Borbely, E. Lyocell, the new generation of regenerated cellulose. *Acta Polytech. Hung.* 2008, *5*, 11–18.

[24] Cao, Y.; Meng, T.; Zhang, J.; He, J.; Li, H.; Zhang, Y. Acetone-soluble cellulose acetate prepared by one-step homogeneous acetylation of cornhusk cellulose in an ionic liquid 1-allyl-3-methylimidazolium chloride (AMIMCl). *Carbohydr. Polym.* 2007, *69*, 665–672.

[25] Chanzy, H. Aspects of cellulose structure. In *Cellulose Sources and Exploitation: İndustrial Utilisation Biotechnology and Physico-Chemical Properties*; Kennedy, J. F., Phillips, G. O., Williams, P. A., Eds.; Ellis Horwood: Chichester, UK, 1990; pp. 3–12.

[26] Chausan, G. S.; Mahajan, S.; Guleria, L. K. Polymers from renewable sources: Sorption of Cu2+ ions by cellulose graft copolymers. *Desalination* 2000, *130*, 85–88.

[27] Cheminet, N.; Jarrosson, T.; Lere-Porte, J.-P.; Serein-Spirau, F.; Cury, L.; Moreau, J.; Viau, L.; Vioux, A. One pot synthesis of fluorescent π-conjugated materials: Immobilization of phenylene-ethynylene polyelectrolytes *in silica* confined ionogels. *J. Mater. Chem.* 2011, *21*, 13588–13593.

[28] Cross, C. F.; Bevan, B. T.; Beadle, C. Die chemie der pflanzenfasern: Cellulosen, oxycellulosen, lignocellulosen (in German). *Ber. Dtsch. Chem. Ges.* 1893, *26*, 2520–2533.

[29] Cross, C. F.; Bevan, B. T.; Beadle, C. Thiokohlensäureester der Cellulose (in German). *Ber. Dtsch. Chem. Ges.* 1893, *26*, 1090–1097.

[30] Edgar, K. J.; Arnold, K. M.; Blount, W. W.; Lawniczak, J. E.; Lowman, D. W. Synthesis and properties of cellulose acetoacetates. *Macromolecules* 1995, *28*, 4122–4128.

[31] El-Salmawi, K. E.; Zaid, M. M. A.; Ibraheim, S. M.; El-Naggar, A. M.; Zahran, A. H. Sorption of dye wastes by poly (vinylalcohol)/poly (carboxymethyl cellulose) blend grafted through a radiation method. *J. Appl. Polym. Sci.* 2001, *82*, 136–142.

[32] Erdmenger, T.; Haensch, C.; Hoogenboom, R.; Schubert, U. Homogeneous trytilation of cellulose in 1-butyl-3-methylimidazolium chloride. *Macromol. Biosci.* 2007, *7*, 440–445.

[33] Fischer, S.; Thummler, K.; Pfeiffer, K.; Liebert, T.; Heinze, T. Evaluation of molten inorganic salt hydrates as reaction medium for the derivatization of cellulose. *Cellulose* 2002, *9*, 293–300.

[34] Fischer, S.; Voigt, W.; Fischer, K. The behaviour of cellulose in hydrated melts of the composition LiXċn H2O (X = I−, NO3−, CH3COO−, ClO4−). *Cellulose* 1999, *6*, 213–219.

[35] Fort, D. A.; Swatloski, R. P.; Moyna, P.; Rogers, R. D.; Moyna, G. Use of ionic liquids in the study of fruit ripening by high-resolution 13C NMR spectroscopy: "Green" solvents meet green bananas. *Chem. Commun.* 2006, 714–716. doi:10.1039/B515177P.

[36] Fukaya, Y.; Hayashi, K.; Wada, M.; Ohno, H. Cellulose disooslution with polar ionic liquids under mild conditions: Required factors for anions. *Green Chem.* 2008, *10*, 44–46.

[37] Fukaya, Y.; Sugimoto, A.; Ohno, H. Superior solubility of polysaccharides in low viscosity, polar, and Halogen-free 1,3-dialkylimidazolium formates. *Biomacromolecules* 2006, *7*, 3295–3297.

[38] Gericke, M.; Fardim, P.; Heinze, T. Ionic liquids—Promising but challenging solvents for homogeneous derivatization of cellulose. *Molecules* 2012, *17*, 7458–7502.

[39] Grabovac, V.; Guggi, D.; Bernkop-Schnurch, A. Comparison of the mucoadhesive properties of various polymers. *Adv. Drug Deliv. Rev.* 2005, *57*, 1713–1723.

[40] Guo, Y.; Wang, X.; Shen, Z.; Shu, X.; Sun, R. Preparation of cellulose-graft-poly (ε-caprolactone) nanomicelles by homogeneous ROP in ionic liquid. *Carbohydr. Polym.* 2013, *92*, 77–83.

[41] Gupta, K. C.; Khandekar, K. Temperature-responsive cellulose by ceric (IV) ion-initiated graft copolymerization of *N*-isopropylacrylamide. *Biomacromolecules* 2003, *4*, 758–765.

[42] Hameed, N.; Guo, Q. Blend films of natural wool and cellulose prepared from an ionic liquid. *Cellulose* 2010, *17*, 803–813.

[43] Hebeish, A.; El-Hilw, Z. H. Preparation of DEAE cotton-g-poly (methacrylic acid) for use as ion exchanger. *J. Appl. Polym. Sci.* 1998, *67*, 739–745.

[44] Heinämäki, J. T.; Iraizoz Colarte, A.; Nordström, A. J.; Yliruusi, J. K. Comparative evaluation of ammoniated aqueous and organic-solvent-based celluloseester enteric coating systems: A study on free films. *Int. J. Pharm.* 1994, *109*, 9–16.

[45] Heinze, T.; Liebert, T. Unconventional methods in cellulose functionalization. *Prog. Polym. Sci.* 2001, *26*, 1689–1762.

[46] Heinze, T.; Schaller, J. New water soluble cellulose esters synthesized by an effective acylation procedure. *Macromol. Chem. Phys.* 2000, *201*, 1214–1218.

[47] Heinze, T.; Schwikal, K.; Barthel, S. Ionic liquids as reaction medium in cellulose functionalization. *Macromol. Biosci.* 2005, *5*, 520–525.

[48] Hough, W. L.; Smiglak, M.; Rodriguez, H.; Swatloski, R. P.; Spear, S. K.; Daly, D. T.; Pernak, J.; Grisel, J. E.; Carliss, R. D.; Soutullo, M. D.; et al. The third evolution of ionic liquids: active pharmaceutical ingredients. *New J. Chem.* 2007, *31*, 1429–1436.

[49] Hufendiek, A.; Trouillet, V.; Meier, M. A. R.; Barner-Kowollik, C. Temperature responsive cellulose-*graft*-copolymers via cellulose functionalization in an ionic liquid and RAFT-polymerization. *Biomacromolecules* 2014, doi:10.1021/bm500416m.

[50] Hutchins, F. E. *Toxicity of Pulp and Paper Mill Effluent: A Literature Review*; United States Environmental Protection Agency: Corvallis, OR, USA, 1979.

[51] Hyatt, J. W. Manufacture of Celluloid. U. S. Patent No. 232037, 1880.

[52] IONCELL-F-Cellulosic Fibers from Ionic Liquid Solution. Available online: https://www.youtube. com/watch?v=5bhCbGmNfTQ (accessed on 27 May 2014).

[53] Isik, M.; Gracia, R.; Kollnus, L. C.; Tome, L. C.; Marrucho, I. M.; Mecerreyes, D. Cholinium-based poly(ionic liquid)s: Synthesis, characterization and application as biocompatible ion gels and cellulose coatings. *ACS Macro Lett.* 2013, *2*, 975–979.

[54] Kadokawa, J.; Kato, T.; Setoyama, M.; Yamamoto, K. Preparation of galactomannan-based materials compatibilized with ionic liquids. *J. Polym. Environ.* 2013, *21*, 512–519.

[55] Kadokawa, J.; Murakami, M.; Kaneko, Y. A facile method for preparation of composites composed of cellulose and a polystyrene-type polymeric ionic liquid using a polymerizable ionic liquid. *Compos. Sci. Technol.* 2008, *68*, 493–498.

[56] Kadokawa, J.; Murakami, M.; Takegawa, A.; Kaneko, Y. Preparation of cellulose–starch composite gel and fibrous material from a mixture of the polysaccharides in ionic liquid. *Carbohydr. Polym.* 2009, *75*, 180–183.

[57] Kadokowa, J.-I.; Murakami, M.; Kaneko, Y. A facile preparation of gel materials from a solution of cellulose in ionic liquid. *Carbohydr. Res.* 2008, *343*, 769–772.

[58] Kawasaki, M.; Iwasa, Y. "Cut and stick" ion gels. *Nature* 2012, *489*, 510–511.

[59] Klemm, D.; Heublein, B.; Fink, H.-P.; Bohn, A. Cellulose: Fascinating biopolymer and sustainable raw material. *Angew. Chem. Int. Ed.* 2005, *44*, 3358–3393.

[60] Kohler, S.; Liebert, T.; Heinze, T. Interactions of ionic liquids with polysaccharides. VI. Pure cellulose nanoparticles from trimethylsilyl cellulose synthesized in ionic liquids. *J. Polym. Sci. Part A Polym. Chem.* 2008, *46*, 4070–4080.

[61] Kosan, B.; Michels, C.; Meister, F. Dissolution and forming of cellulose with ionic liquids. *Cellulose* 2008, *15*, 59–66.

[62] Lacroix, C.; Sultan, E.; Fleury, E.; Charlot, A. Functional galactomannan platform from convenient esterification in imidazolium-based ionic liquids. *Polym. Chem.* 2012, *3*, 538–546.

[63] Lecomte, F.; Siepmann, J.; Walther, M.; MacRae, R. J.; Bodmeier, R. Blends of enteric and GIT-insoluble polymers used for film coating: physicochemical characterization and drug release patterns. *J. Control. Release* 2003, *89*, 457–471.

[64] Lehr, C. M. Lectin-mediated drug delivery: The second generation of bioadhesives. *J. Control. Release* 2000, *65*, 19–29.

[65] Liebert, T.; Hansch, C.; Heinze, T. Click chemistry with polysaccharides. *Macromol. Rapid Commun.* 2006, *27*, 208–213.

[66] Liebert, T. F.; Heinze, T. J. Exploitation of reactivity and selectivity in cellulose functionalization using unconventional media for the design of products showing new superstructures. *Biomacromolecules* 2001, *2*, 1124–1132.

[67] Liu, C. F.; Sun, R. C.; Zhang, A. P.; Ren, J. L. Preparation of sugarcane bagasse cellulosic phthalate using an ionic liquid as reaction medium. *Carbohydr. Polym.* 2007, *68*, 17–25.

[68] Liu, J.; Williams, R. O. Long-term stability of heat-humidity cured cellulose acetate phthalate coatedbeads. *Eur. J. Pharm. Biopharm.* 2002, *53*, 167–173.

[69] Liu, W.; Budtova, T. Dissolution of unmodified waxy starch in ionic liquid and solution rheological properties. *Carbohydr. Polym.* 2013, *93*, 199–206.

[70] Lu, B.; Xu, A.; Wang, J. Cation does matter: How cationic structure affects the dissolution of cellulose in ionic liquids. *Green Chem.* 2014, *16*, 1326–1335.

[71] Lunstroot, K.; Driesen, K.; Nockemann, P.; van Hecke, K.; van Meervelt, L.; Görller-Walrand, C.; Binnemans, K.; Bellayer, S.; Viau, L.; Bideau, J. L.; et al. Lanthanide-doped luminescent ionogels. *Dalton Trans.* 2009, 298–306.

[72] Matsumi, N.; Nakamura, Y.; Aoi, K.; Watanabe, T.; Mizumo, T.; Ohno, H. Enhanced ionic conduction in organoboron ion gels facilely designed via condensation of cellulose with boric acids in ionic liquids. *Polym. J.* 2009, *41*, 437–441.

[73] Movassaghian, S.; Barzegar-Jalali, M.; Alaeddini, M.; Hamedyazdan, S.; Afzalifar, R.; Zakeri-Milani, P.; Mohammadi, G.; Adibkia K. Development of amitriptyline buccoadhesive tablets in management of pain in dental procedures. *Drug Dev. Ind. Pharm.* 2011, *37*, 1–12.

[74] Murakami, M.; Kaneko, Y.; Kadokawa, J. Preparation of cellulose-polymerized ionic liquid composite by *in situ* polymerization of polymerizable ionic liquid in cellulose-dissolving solution. *Carbohydr. Polym.* 2007, *69*, 378–381.

[75] Neouze, M.-A.; Bideau, J. L.; Gaveau, P.; Bellayer, S.; Vioux, A. Ionogels, new materials arrising from the confinement of ionic liquids within silica derived networks. *Chem. Mater.* 2006, *18*, 3931–3936.

[76] O'Sullivan, A. C. Cellulose: Structure slowly unravels. *Cellulose* 1997, *4*, 173–207.

[77] Okamura, K. Structure of cellulose. In *Wood and Cellulosic Chemistry*; Hon, D. N.-S., Shiraishi, N., Eds.; Marcel Dekker: New York, NY, USA, 1991; pp. 89–111.

[78] O-Rak, K.; Ummartyotin, S.; Sain, M.; Manuspiya, H. Covalently grafted carbon nanotubes on bacterial cellulose composite for flexible touch screen application. *Mater. Lett.* 2013, *107*, 247–250.

[79] Pabby, A. K.; Rizvi, S. S. H.; Sastre, A. M. *Handbook of Membrane Seperations: Chemical, Pharmaceutical, Food and Biotechnological Applications*; CRC Press, Taylor & Francis Group: New York, NY, USA, 2002.

[80] Petersen, R. C. The chemical composition of wood. In *The Chemistry of Solid Wood*; American Chemical Society: New York, NY, US, 1984; pp. 57–126.

[81] Prasad, K.; Mine, S.; Kaneko, Y.; Kadokawa, J. Preparation of cellulose-based ionic porous material compatibilized with polymeric ionic liquid. *Polym. Bull.* 2010, *64*, 341–349.

[82] Preston, R. D. X-ray analysis and the structure of the components of plant cell walls. *Phys. Rep.* 1975, *21*, 183–226.

[83] Quan, S.-L.; Kang, S.-G.; Chin, I.-J. Characterization of cellulose fiber electrospun using ionic liquid. *Cellulose* 2010, *17*, 223–230.

[84] Rahatekar, S. S.; Rasheed, A.; Jain, R.; Zammarano, M.; Koziol, K. K.; Windle, A. H.; Gilman, J. W.; Kumar, S. Solution spinning of cellulose carbon nanotube composites using room temperature ionic liquids. *Polymer* 2009, *50*, 4577–4583.

[85] Regiani, A. M.; Frollini, E.; Marson, G. A.; Arantes, G. M.; El Seoud, O. M. Some aspects of acylation of cellulose under homogeneous solution conditions. *J. Polym. Sci. Part A Polym. Chem.* 1998, *37*, 1357–1363.

[86] Ries, M. E.; Radhi, A.; Keating, A. S.; Parker, O.; Budtova, T. Diffusion of 1-ethyl-3-ethyl-imidazolium acetate in glucose, cellobiose and cellulose solutions. *Biomacromolecules* 2014, *15*, 609–617.

[87] Rosenau, T.; Potthast, A.; Sixta, H.; Kosma, P. The chemistry of side reactions and by product formation in the system NMMO/cellulose (Lyocell process). *Prog. Polym. Sci.* 2001, *26*, 1763–1837.

[88] Sarko, A. Cellulose: How much do we know about its structure. In *Wood and Cellulosics: Industrial Utilization, Biotechnology, Structure and Properties*; Kennedy, J. F., Ed.; Ellis Horwood: Chichester, UK, 1987; pp. 55–70.

[89] Schenzel, A.; Hufendiek, A.; Barner-Kowollik, C.; Meier, M. A. R. Catalytic transesterification of cellulose in ionic liquids: sustainable access to cellulose esters. *Green Chem.* 2014, *16*, 3266–3271.

[90] Schönbein, C. F. Notiz über eine Veränderung der Pflanzenfaser und einiger andern organischen Substanzen (in German). *Ber. Naturforsch. Ges. Basel* 1847, *7*, 27.

[91] Semsarilar, M.; Ladmiral, V.; Perrier, S. Synthesis of a cellulose supported chain transfer agent and its application to RAFT polymerization. *J. Polym. Sci. Part A Polym. Chem.* 2010, *48*, 4361–4365.

[92] Song, Y.; Sun, Y.; Zhang, X.; Zhou, J.; Zhang, L. Homogeneous quaternization of cellulose in NaOH/urea aqueous solutions as gene carriers. *Biomacromolecules* 2008, *9*, 2259–2264.

[93] Sun, N.; Rahman, M.; Qin, Y.; Maxim, M. L.; Rodriguez, H.; Rogers, R. D. Complete dissolution and partial delignification of wood in the ionic liquid 1-ethyl-3-methylimidazolium acetate. *Green Chem.* 2009, *11*, 646–655.

[94] Swatloski, R. P.; Spear, S. K.; Holbrey, J. D.; Rogers, R. D. Dissolution Cellulose with Ionic Liquids. *J. Am. Chem. Soc.* 2002, *124*, 4974–4975.

[95] Takegawa, A.; Murakami, M.; Kaneko, Y.; Kadokawa, J. A facile preparation of composites composed of cellulose and polymeric ionic liquids by *in situ* polymerization of ionic liquids having acrylate groups. *Polym. Compos.* 2009, 1837–1841. doi:10.1002/pc.20756.

[96] Tome, L. C.; Freire, M. G.; Rebelo, L. P. N.; Silvestre, A. J. D.; Neto, J. P.; Marrucho, I. M.; Freire C. S. R. Surface hydrophobization of bacterial and vegetable cellulose fibers using ionic liquids as solvent media and catalysts. *Green Chem.* 2011, *13*, 2464–2470.

EDITOR CONTACT INFORMATION

Dr. Deepak Pathania
Professor and Head, School of Chemistry
Shoolini University
Solan, India
dpathania74@gmail.com

Dr. Gaurav Sharma
Assistant Professor
School of Chemistry
Shoolini University
Solan, India
gaurav8777@gmail.com

Dr. Amit Kumar
Assistant Professor
School of Chemistry
Shoolini University
Solan, India
mittuchem83@gmail.com

INDEX

A

B

C

D

E

F

G

H

I

T

X